Random Patterns and Structures in Spatial Data

This book presents a general mathematical framework able to detect and to characterise, from a morphological and statistical perspective, patterns hidden in spatial data. The mathematical tools employed are Gibbs Markov processes, mainly marked point processes with interaction, which permits us to reduce the complexity of the pattern. It presents the framework, step by step, in three major parts: modeling, simulation, and inference. Each of these parts contains a theoretical development followed by applications and examples.

Features

- Presents mathematical foundations for tackling pattern detection and characterisation in spatial data using marked Gibbs point processes with interactions
- Includes application examples from cosmology, environmental sciences, geology, and social networks
- Presents theoretical and practical details for the presented algorithms in order to be correctly and efficiently used
- Provides access to C++ and R code to encourage the reader to experiment and to develop new ideas
- Includes references and pointers to mathematical and applied literature to encourage further study

Random Patterns and Structures in Spatial Data is primarily aimed at researchers in mathematics, statistics, and the above-mentioned application domains. It is accessible for advanced undergraduate and graduate students and thus could be used to teach a course. It will be of interest to any scientific researcher interested in formulating a mathematical answer to the always challenging question: what is the pattern hidden in the data?

Radu S. Stoica is a full professor in mathematics at the University of Lorraine, France. His research activity connects stochastic geometry, spatial statistics, and Bayesian inference for probabilistic modeling and statistical description of random structures and patterns. The results of his work are tailored to the data methodologies based on Gibbs Markov models, Monte Carlo algorithms, and inference procedures, which can characterise and detect structures and patterns either hidden or directly observed in the data. The tackled application domains are astronomy, geosciences, image analysis, and network sciences. Prior to his current position, Dr. Stoica was an associate professor at the University of Lille, France. He also worked as a researcher for INRAe Avignon, France, University Jaume I, Spain, and CWI Amsterdam, The Netherlands.

Random Patterns and Structures in Spatial Data

Radu S. Stoica

CRC Press
Taylor & Francis Group
Boca Raton London New York

CRC Press is an imprint of the
Taylor & Francis Group, an **informa** business
A CHAPMAN & HALL BOOK

First edition published 2025
by CRC Press
2385 Executive Center Drive, Suite 320, Boca Raton, FL 33431, U.S.A.

and by CRC Press
4 Park Square, Milton Park, Abingdon, Oxon, OX14 4RN

CRC Press is an imprint of Taylor & Francis Group, LLC

ISBN: 978-1-032-45936-3 (hbk)
ISBN: 978-1-032-46175-5 (pbk)
ISBN: 978-1-003-38038-2 (ebk)

DOI: 10.1201/9781003380382

Typeset in Latin Modern font
by KnowledgeWorks Global Ltd.

Publisher's note: This book has been prepared from camera-ready copy provided by the authors.

To my family,
to my friends,
to my professors.

Contents

Preface

I am an applied mathematician. I have had a passion for mathematics since childhood. And yet, for various reasons, I started my university studies to become an engineer rather than a mathematician. Not without some frustration.

Making computers talk and see is something that many people, whether they have a passion for science or not, can only dream about. In the mid-90s, thanks to the development of mathematics and computing power, these dreams were transformed into structured research projects supported by universities and research institutions around the world. For me, this dynamic came at just the right time in relation to my studies, as it was an opportunity to reconnect with mathematics. During the first five years of my studies, I specialised in signal and image processing. This allowed me to continue with a PhD under the supervision of Xavier Descombes and Josiane Zerubia (Inria Sophia—Antipolis, France). During my thesis, I developed a new mathematical model, the Candy model, a marked point process, for the unsupervised detection of linear networks in satellite and aerial images.

I decided to delve deeper into the mathematics that was the basis of my thesis. This was the start of a long journey across Europe. As a post-doc, I was welcomed into their respective teams by exceptional scientists with wonderful human qualities: Marie-Colette van Lieshout (CWI Amsterdam, Netherlands), Jorge Mateu (Jaume I Castellon University, Spain) and Rachid Senoussi, Joel Chadoeuf, Denis Allard (INRAE Avignon, France). In 2006, I obtained a permanent position as an associate professor at the University of Lille. In 2016, I became a full professor at the University of Lorraine in Nancy.

With patience and delicacy, Marie-Colette carefully guided me through the mysteries and wonders of stochastic geometry and point processes. Thanks to the open-mindedness of Jorge, Rachid, Joel and Denis, I was also able to make new scientific encounters. These encounters opened the way to fields of application that were new to me. In Spain, I met Vicent J. Martinez and Enn Saar, astronomers who study the structures created by the distribution of galaxies in our universe. In Avignon, I met André Kretzschmar, a biologist who studies the structures formed by living organisms, such as the galleries of earthworms, the distribution of cells occupied by bee larvae in beehives, or aggregates of all kinds. It was thanks to André and his love of philosophy that I heard about René Thom and morphogenesis. Then, thanks to Vicent and

Enn, I was able to understand what it's like to build a scientific project that really takes place over a long period of time. They encouraged and helped me to interact, collaborate or just exchange ideas with wonderful cosmologists like Jaan Einasto, Maret Einasto, Alberto Fernandez Soto, Pekka Heinämäki, Bernard T., Jones, Jaan Pelt, Elmo Tempel, Peeter Tenjes, Rien van de Weijgaert and many, many others.

A year after my arrival in Lille, Youri Davydov created the "Stochastic Geometry" group, which I joined with enthusiasm as soon as it was set up. In the heart of this group, thanks to exchanges with participants, I was able to mature and develop my research project on the detection and characterisation of patterns and structures in spatial data. As a professor in Nancy, my project is also progressing thanks to the doctoral projects I am launching on my own or with my new colleagues. For example, we have co-directed several doctoral theses with Guillaume Caumon, an exceptional geologist who, among many other things, uses data to study the shape and geometry of various geological structures.

This book is a travel journal, offering a synthesis of the main points of my scientific itinerary, based on previously published articles.

A common thread can be identified between structures that appear to be very different, such as cosmic filaments, road networks, aggregates formed by galaxies or epidemics, seismic faults or social networks. All these structures are intricate entities with complex geometries and morphologies. These structures can be seen as made of a set of "simple" objects that interact. To illustrate, a road network can be approximated by a set of segments that are locally connected and aligned. In this case, the spatial distribution and interactions of these segments can be governed by a probability law, such as a Gibbs distribution.

This book presents marked Gibbs point processes with interaction and some related models as modelling tools for detecting and characterising patterns and structures in spatial data. The proposed models are accompanied by simulation algorithms and methods for performing real data inference.

It should be noted that Gibbs processes are not the only mathematical framework capable of dealing with the type of structures mentioned above. Other approaches include fractals, wavelets, topological data analysis or neural networks. Gibbs processes have some important advantages. One of them is the capacity to integrate the local properties and interactions of the objects forming the pattern under consideration in the form of a probability distribution. In this context, invariant features of the structure of interest can be computed. These features are moments of the Gibbs probability distribution. The knowledge of these moments allows the control and description of the pattern in question as well as reproducibility of patterns with desired properties.

There is a price to pay, a challenge to overcome, when using Gibbs point processes. Their probability densities are not always fully known, as their normalisation constant is not always available in explicit analytical form. Provided the conditions are not too restrictive, it is possible to construct simulation procedures for these distributions, and in this way to replicate the structures of interest. As a result, statistical inference can be carried out, which makes it possible to answer questions related to the structures under consideration.

The need to write this book became obvious when I began to train students in research and to collaborate with scholars from other disciplines. Interdisciplinary work is not only rewarding, but also time-consuming and challenging. First, from a theoretical perspective, I have to provide concepts and details so that the patterns under consideration can be studied in a mathematical way. Next, from a practical perspective, I have to present examples and applications to make clear that a mathematical description of a given pattern can be developed and used for a variety of data types. Finally, and perhaps most importantly, we all need to be aware that a small change in the problem formulation with respect to a given example can lead to major transformations at all levels of the framework adopted to solve the problem at hand. These transformations can affect everything from the theoretical mathematical framework to the numerical and programming implementation. And this is how research goes on ...

The primary audience for this book is advanced undergraduate and graduate students in mathematics and statistics. At the same time, the book aims to reach engineers and researchers in other fields where the application of modern mathematical techniques is of great importance. For the more applied reader, it may be more convenient to reverse the local order of the chapters. That is, start by reading the applied chapters first and then move on to the more theoretical ones. This book is based on the lecture notes of a master's course that was created by using my research activity. The course has been adapted and transformed to be taught in different contexts and institutions. I have had the opportunity to teach this course at the Universities of Tartu, Lille, Montevideo and Nancy.

I am a researcher who develops his work through scientific collaborations and teaching. I would like to take this opportunity to thank all my co-authors and students. It is thanks to them that I continue to learn, to share and exchange knowledge, and to improve my understanding of our world. I would also like to thank my wife, my son, my entire family and my friends. Their constant and unfailing support are perhaps my most precious source of energy. Christian Lantuéjoul meticulously, patiently and with great kindness read one of my early manuscripts, which then became the project for this book. With the same care and generosity, Aila Särkkä reread parts of this manuscript. In April 2022 she invited me to spend a week at Chalmers University and to give a seminar. After my seminar, Rob Calver of Chapman & Hall contacted me and asked me to write a book bringing together some of the key moments of my research. I would like to thank everyone who has contributed, directly or indirectly, to the realisation of this book.

1

Introduction: what this book is about

1.1 Patterns and structures in spatial data

Observations related to a particular physical phenomenon are called *spatial data* if they can be represented as a set in which each element represents *a location with or without associated features.* A digital image is an example of spatial data. The image is represented by a set of coloured picture elements (pixels) arranged on a regular grid. Each pixel is characterised by an index in a matrix, its position, and some numerical values corresponding to a colour code, its associated features. Catalogues of galaxies in astronomy and databases of dairy farms are also examples of spatial data. A catalogue of galaxies contains the position in our Universe and the characteristics of each galaxy (mass, luminosity, redshift, shape, etc.). A database of dairy farms contains the location in the territory and the characteristics of each farm (type and number of animals, parentage and place of birth of each animal, etc.). In these last two examples, the positions of the features are no longer arranged on a regular grid, and the corresponding features are represented in the form of very rich and complex data structures.

The spatial nature of the observed data indicates that the possible answers to the questions posed exhibit a strong *morphological component.* To illustrate, consider the following examples of questions: extracting the road networks present in a satellite image, characterising the large-scale distribution of galaxies in our Universe, determining the tendency of farms with a high predisposition to a certain disease to cluster together. The information sought is a *pattern*, a *complex structure with specific geometric characteristics.* Indeed, in numerous instances, the ability to analyse spatial data necessitates the formulation of an answer to the question: what is the pattern "hidden" in the data?

This book addresses the detection and characterisation of patterns and structures in observed spatial data. The scope of this topic is excessively broad. In order to provide a focused approach, it is essential to narrow the perspective

DOI: 10.1201/9781003380382-1

by assuming the following characteristics for the patterns:

A1: *the pattern of interest is a complex entity made up of "simple" geometric objects that interact,*

A2: *locally finiteness: the number of objects forming a pattern observed in a bounded domain is finite,*

A3: *the pattern is the realisation of a spatial random process*

In the context of the above assumptions, two different scenarios can be identified. First, the pattern in question is "hidden" in the data. For example, the data is represented by a catalogue of galaxies, and you want to find the filamentary pattern outlined by the positions of the galaxies. In this case, the pattern is not directly observed but has to be inferred from the available data. Second, a pattern is directly observed. Some covariates may or may not be associated with the data. To illustrate this second scenario, suppose that a catalogue of galaxies is observed together with covariates given by the dark matter distribution field. The aim is to characterise the spatial distribution of the observed galaxy positions conditional on the known dark matter distribution.

1.2 Pattern detection problem

Let $\boldsymbol{d}$ be a spatial data set observed in a compact window $W \subset \mathbb{R}^d$. The objective is to detect the hidden pattern $\boldsymbol{x}$ induced by these observations.

The first two assumptions, $A1$ and $A2$, allow to consider the pattern of interest as a finite collection of objects given by

$$\boldsymbol{x} = \{x_i = (w_i, m_i) \,:\, 1 \leq i \leq n\},$$

with $n \in \mathbb{N}$. An object $x = (w, m)$ is defined by its position $w \in W$ and its characteristics $m \in M$ where M is some characteristics space. The assumption $A3$ and the use of the Gibbs modelling framework enable the probability density controlling the pattern to be written

$$p(\boldsymbol{x}|\theta, \boldsymbol{d}) = \frac{\exp[-U(\boldsymbol{x}|\theta, \boldsymbol{d})]}{Z(\theta)} = \frac{\exp[-(U_{\boldsymbol{d}}(\boldsymbol{x}|\theta) + U_i(\boldsymbol{x}|\theta))]}{Z(\theta)}, \tag{1.1}$$

where $\theta \in \Theta$ is the parameter vector, $Z(\theta)$ is the partition function or the normalising constant of the model and Θ is the parameter space. The total energy of the system, $U(\boldsymbol{x}|\theta, \boldsymbol{d})$ is given by the sum of two terms. The first term, $U_{\boldsymbol{d}}(\boldsymbol{x}|\theta)$ represents the *data energy* and concerns the position of objects within the data field $\boldsymbol{d}$ in W. The second term, $U_i(\boldsymbol{x}|\theta)$ represents the *interaction energy* and concerns the interaction between the objects composing the pattern. This term can also be interpreted from a Bayesian perspective as prior knowledge about the distribution of the objects making up the pattern. In the first scenario example mentioned in Section 1.1, the data $\boldsymbol{d}$ is the field of galaxy positions. The configuration of objects $\boldsymbol{x}$ that approximate the unknown filamentary pattern is represented by a set of thin cylinders or segments that connect and align to form filaments in those regions of the data field that are suitable.

The choice of the Gibbs modelling framework allows the pattern of interest to be naturally described by a probability distribution given by (1.1). This probability distribution simultaneously controls the positions and the interactions of the objects forming the pattern. On this basis moments can be calculated. The moments are fixed quantities that describe the distribution, i.e. the pattern, partially or completely. Furthermore, the variational principle for Gibbs processes guarantees that among all probability distributions with the same expected total energy, the Gibbs distribution is the one with the maximum entropy. This means that Gibbs processes are the most disordered models that can describe random configurations of interacting objects that form patterns with desired properties [77, 106, 318].

Assuming the model parameters θ are known, the pattern of interest is estimated by the object configuration that maximises the probability density representing the model (1.1):

$$\hat{\boldsymbol{x}} = \arg\max_{\boldsymbol{x}\in\Omega} p(\boldsymbol{x}|\theta, \boldsymbol{d}) = \arg\min_{\boldsymbol{x}\in\Omega} U(\boldsymbol{x}|\theta, \boldsymbol{d}). \tag{1.2}$$

where Ω is the configuration space containing all the patterns.

The main difficulties in solving this problem are that the normalising constant $Z(\theta)$ is not available in analytically closed form and that the energy function $U(\boldsymbol{x}|\theta, \boldsymbol{d})$ is not guaranteed to be convex, except in some very special cases.

The solution to this problem is constructed in the following way. Firstly, a *well-defined model* (1.1) is constructed. This can be done without knowing the normalising constant. Secondly, a *global optimisation* technique such as simulated annealing can be used to solve the problem (1.2). This algorithm works by sampling the distribution $p(\boldsymbol{x}|\theta)^{1/T}$ as the temperature parameter T is decreased to 0. For sufficiently high initial T values and sufficiently slow cooling schedules, as T approaches 0 the algorithm converges to the uniform distribution over the subspace of pattern configurations that maximises the probability density (1.1). This approach requires two ingredients: *a sampling algorithm and a cooling schedule.* For the sampling procedure, *Markov chain Monte Carlo (MCMC) algorithms* can be constructed that avoid the computation of the normalising constant. Theoretically, mathematical proofs should be derived to show the existence of a suitable cooling schedule that guarantees the convergence of the algorithms. However, such cooling schedules are rather slow. Therefore, in practice, ad hoc faster cooling schedules are derived based on theoretical arguments and numerical simulations.

1.3 Pattern characterisation problem

Let $\boldsymbol{x}$ be an observed pattern within the compact window $W \subset \mathbb{R}^d$ and $\boldsymbol{d}$ a set of covariates. The covariate data set acts as an external field that influences the location within the window W of the observed pattern [12, 54]. Let us recall the example illustrating the second scenario in Section 1.1. In a given region W of the Universe, the observation $\boldsymbol{x}$ represents the pattern given by the positions of the galaxies while the covariate $\boldsymbol{d}$ is given by the dark matter distribution field. The observed pattern $\boldsymbol{x}$ is assumed to be the realisation of a spatial process in W that depends on $\boldsymbol{d}$. Furthermore, the process is assumed to belong to a parametric family of models defined by the probability density functions $\{p(\boldsymbol{x}|\theta, \boldsymbol{d}) : \theta \in \Theta\}$. The aim is to identify the parameters of the model within this family that best characterise the observed pattern.

Within this context, the likelihood function is

$$L(\theta; \boldsymbol{x}, \boldsymbol{d}) = p(\boldsymbol{x}|\theta, \boldsymbol{d}) = \frac{\exp[-U(\boldsymbol{x}|\theta, \boldsymbol{d})]}{Z(\theta)}. \tag{1.3}$$

The Bayesian approach allows to use knowledge on the model parameters via the prior distribution $p(\theta)$ on Θ. The obtained posterior writes as

$$p(\theta|\boldsymbol{x}, \boldsymbol{d}) \propto L(\theta; \boldsymbol{x}, \boldsymbol{d})p(\theta) = \frac{\exp[-U(\boldsymbol{x}|\theta, \boldsymbol{d})]}{Z(\theta)}p(\theta).$$

Hence, a possible answer to the formulated question is given by the *Maximum A Posteriori (MAP)*

$$\hat{\theta} = \arg\max_{\theta \in \Theta} p(\theta|\boldsymbol{x}, \boldsymbol{d}). \tag{1.4}$$

Similarly to the pattern detection problem, the difficulties in tackling the pattern characterisation problem are related to the computation of normalising constants and the optimisation procedure needed to solve (1.4). In this case, much more sophisticated strategies are needed to approximate or bypass the calculation of the normalising constants. On the other hand, by choosing appropriate models and priors, it can be guaranteed that the posterior distribution will be convex. This enables the use of optimisation procedures that allow a more precise control of the obtained solutions of (1.4).

Pattern detection and characterisation problems are dual. In the first problem, some data field $\boldsymbol{d}$ is observed, a model for the unknown pattern to be detected $\boldsymbol{x}$ is assumed and the parameters θ are considered known. The pattern is estimated by maximising $p(\boldsymbol{x}|\theta, \boldsymbol{d})$. In the second problem, a pattern $\boldsymbol{x}$ is observed, covariates $\boldsymbol{d}$ may be available, the parameters θ are unknown and knowledge of them is available through the prior $p(\theta)$. Assuming a parametric model for the pattern conditional on the covariates, the parameters that give the best model characterising the observation are obtained by maximising the posterior $p(\theta|\boldsymbol{x}, \boldsymbol{d})$.

The use of the Gibbs modelling framework and the Bayesian inference allows natural and elegant constructions of these conditional distributions. Furthermore, this combined approach allows the development of both pattern detection and characterisation frameworks. The pattern detection framework can be further developed by using knowledge of the model parameters via a prior distribution. The pattern characterisation framework can be further extended by developing algorithms for sampling the posterior distribution to provide more insight into the fitted model than just pointwise estimation. These are key ideas developed in this book. I like to think that Bayesian inference and Gibbs modelling give hope that modelling the joint distribution of the pattern of interest and its parameters is not such an unrealistic research perspective.

1.4 Structure of the book

The book is divided into three parts, each consisting of two chapters. The first chapter provides the necessary mathematical skills to implement a particular idea related to patterns. The second chapter presents concrete examples and applications of the mathematical tools introduced earlier. The first part defines and establishes, through probabilistic modelling, what the patterns of interest are. The second part constructs the patterns using stochastic simulation algorithms. Finally, the third part describes and characterises the patterns through statistical inference procedures.

Part I

Define the pattern: probabilistic modelling

2

Marked point processes

Point processes are probabilistic models designed to study the random distribution of points in a domain. The points can have features attached to them. And these features or marks can be described by a probability distribution. In this case we use the term *marked point processes.* If the marks represent the properties of geometric objects, we speak of *object point processes.*

A sequence of events occurring over time can be seen as the realisation of a point process. It is a one-dimensional point process. The points in the process are naturally ordered on the time line. It is relatively easy to imagine extensions of this example to multiple dimensions. To illustrate, the positions of trees in a forest in a given region are a point process in two dimensions. If we also consider the characteristics of each tree (e.g. age, size, species, etc.), then we have a two-dimensional marked point process.

The mathematical treatment of multidimensional point processes does not immediately follow from a straightforward generalisation of the theory for stochastic processes in one dimension to several dimensions. This is due to the fact that the natural order between events is not respected in a multidimensional space [68, 69, 181]. Point processes are a fundamental object of study in the field of stochastic geometry [10, 24, 54, 88, 104, 207, 270, 274]. In their most general form, point processes are a special case of random sets [207, 217, 220].

The aim of this chapter is to give the reader the minimum necessary elements to build her own point process with interaction models described by a probability density.

The chapter begins with the construction and definition of unmarked and marked point processes. Characterisation results and moment measures for point processes are then presented. The direct application of these results is the construction of summary statistics for stationary point processes. A general development of summary statistics for marked point processes is not really straightforward due to the complexity that can be introduced by the structure of the mark distribution. Nevertheless, the examples that illustrate the use of summary statistics are applications to real random structures such as road networks and galactic structures, which can also be seen as complex

DOI: 10.1201/9781003380382-2

object point processes. The chapter continues with a presentation of the Poisson point process. This is perhaps the best-known point process. It allows the modelling of completely random point patterns consisting of independent random points to which independent characteristics can be attached. The Poisson process is also a key element in building models of more structured random patterns. The final section of the chapter is devoted to marked point processes with interactions. The construction of probability densities for such processes is presented along with properties and examples. These probability densities use Poisson processes as reference measures. To deal with the complexity of these probability densities, Markov point processes are introduced. The probability density of a Markov point process can easily be written as a Gibbs distribution.

In writing this chapter I have relied primarily on the following sources [24, 54, 101, 181, 186, 229, 300]. The works [68, 69, 168, 254] treat point processes from a general theoretical perspective. For a more applied perspective the following monographs are recommended [11, 12, 104, 142]. The following references focus on point processes applied to specific scientific domains. These include image analysis [89], telecommunications [7, 8], forestry and epidemiology [84], geostatistical modelling and simulation [165] and cosmology [93, 146, 201, 202, 247].

2.1 Construction and definition

Let $(\mathbb{R}^d, \mathcal{B}, \nu)$ be the measured space, where ν is the Lebesgue measure, $\mathcal{B}$ is the Borel $\sigma-$algebra induced by d the Euclidean distance[1]. Let W be the subset of $\mathbb{R}^d$ such that $\nu(W) > 0$ and let $(W, \mathcal{B}_W, \nu)$ be restriction of $(\mathbb{R}^d, \mathcal{B}, \nu)$ to W. The pair (W, d) forms a complete and separable metric space.

For each $n \in \mathbb{N}$, let W_n be the set of all configurations $\boldsymbol{w} = \{w_1, w_2, \ldots w_n\}$ formed by n points $w_i \in W$. The cardinality of the configuration $\boldsymbol{w}$ is denoted by $n(\boldsymbol{w})$. The configurations are not ordered and the points are not necessarily distinct. The empty configuration W_0 contains no points.

Let us consider the space of configurations given by $\Omega_W = \bigcup_{n=0}^{\infty} W_n$, with the additional property that for any $\boldsymbol{w} \in \Omega_W$, the number $n(\boldsymbol{w}_B) = n(\boldsymbol{w} \cap B)$ is always finite when $B \subseteq W$ is a bounded Borel set.

[1]This choice does not affect the generality of what follows.

This space is associated with the σ−algebra $\mathcal{F}_W$ generated by the counting applications

$$\mathcal{F}_W = \sigma(\{\boldsymbol{w} = \{w_1, \ldots, w_n\} \in \Omega_W : n(\boldsymbol{w}_B) = n(\boldsymbol{w} \cap B) = m\})$$

where $B \in \mathcal{B}_W$ bounded and $m \in \mathbb{N}$.

Definition 2.1. A *point process* X on W is a measurable application from a probability space $(\mathcal{S}, \mathcal{A})$ into $(\Omega_W, \mathcal{F}_W)$. Its distribution is given by

$$\mathbb{P}(X \in F) = \mathbb{P}\{\omega \in \mathcal{S} : X(\omega) \in F\},$$

with $F \in \mathcal{F}_W$. The realisation of a point process is a random configuration of points in W.

The point process previously defined is said to be *locally finite.* In a similar manner, the configurations of such a process are locally finite. Furthermore, if W is such that $0 < \nu(W) < \infty$, then the process is *finite.* That is, all its configurations are finite almost surely.

It is possible to add characteristics or marks to each point. In this case, we are dealing with a marked point process. Let us consider the point process on $W \times M$ whose realisations are given by the sets $\{(w_i, m_i)\}_{i=1}^n$. Here the $w_i \in W$ are the positions of each point, while the $m_i \in M$ represent the corresponding marks. The space of marks M is equipped with a σ−algebra $\mathcal{M}$ and a probability measure ν_M.

The construction of a marked point process is similar to that of a point process. For each $n \in \mathbb{N}$, the set Ξ_n is the collection of all locally finite non-ordered configurations denoted by $\boldsymbol{x} = \{x_1, \ldots, x_n\}$. The elements that form these configurations are not necessarily distinct marked points $x_i = (w_i, m_i) \in W \times M$. The empty configuration is designated by Ξ_0. The space of configurations is defined as $\Omega = \bigcup_{n=0}^{\infty} \Xi_n$. This space is equipped with the σ−algebra $\mathcal{F}$ which is generated by the counting of points application in the bounded Borel sets $A \subseteq W \times M$.

Definition 2.2. A point process on $W \times M$ is a *marked point process* if the marginal law of the positions only is a point process on W. A marked point process is a measurable mapping from a probability space to $(\Omega, \mathcal{F})$.

The marks may be, for example, objects of simple geometric shape, such as circles, ellipses, or line segments. In this instance, the mark space represents the domain of definition for the parameters of the objects. For example, one may consider mark spaces such as $M_{\text{circle}} = (0, \infty)$ or $M_{\text{segment}} = (0, \infty) \times [0, \pi]$.

An additional illustration of a marked point process is the point process containing multiple categories of points. In this instance, the set $M = \{1, 2, \ldots, k\}$

represents the various types of points or objects. This is equivalent to a multivariate point process of dimension k. Such a process can be regarded as a k-tuple of point processes, namely $(X_1, X_2, \ldots, X_k)$, where each process $X_1, X_2, \ldots, X_k$ corresponds to a specific object type.

Example 2.1 (The galaxy distribution). The observable Universe contains approximately 200 billion galaxies. A cosmological catalogue is a database that stores information related to the observed galaxies, including their position in the observed region of the Universe, their mass, luminosity, and shape. In this context, it is natural to study the galaxy distribution by assuming it to be the realisation of a marked point process [93, 146, 201, 202, 247].

2.2 Characterisation of point processes

Let X be a point process on W and let N be the associated counting variable

$$N(B) = n(X \cap B), \quad B \in \mathcal{B}_W,$$

that represents the number of points falling in the given set B.

Theorem 2.1. *The distribution of a point process X on a complete and separable metric space (W, d) is entirely determined by the probabilities*

$$\mathbb{P}(N(B_1) \leq n_1, N(B_2) \leq n_2, \ldots, N(B_m) \leq n_m) \tag{2.1}$$

where the sets $B_i \in \mathcal{B}_W$ are bounded, $n_i \in \mathbb{N}$ and $m \in \mathbb{N}^\star$ is finite.

A point process is *simple* if the points in any configuration do not share the same location

$$N(\{w\}) \in \{0, 1\}, \text{ for all } w \in W.$$

It was stated earlier, that for locally finite point processes the random variables $N(B)$ are finite whenever the sets B are bounded. The point processes presented in this book are *locally finite* and *simple*. By a slight abuse of notation, it is considered that these processes are defined by the probability space $(\Omega, \mathcal{F}, \mathbb{P})$ constructed previously.

Theorem 2.2. *The distribution of a simple point process on a complete separable metric space (W, d) is characterised by the probabilities of its empty events*

$$v(B) = \mathbb{P}(N(B) = 0), \tag{2.2}$$

for all bounded $B \in \mathcal{B}_W$.

The probability (2.2) can be written

$$\begin{aligned} v(B) &= \mathbb{P}(N(B)=0) = \mathbb{P}(X \cap B = \emptyset) \\ &= 1 - \mathbb{P}(X \cap B \neq \emptyset). \end{aligned}$$

Point processes are part of a larger family of mathematical objects that are encountered in stochastic geometry. These objects are called *random sets* [54, 207, 220]. In this more general setting when X is a random set, the quantity

$$T_B(X) = \mathbb{P}(X \cap B \neq \emptyset)$$

with B a compact set in W is called *capacity functional.* The following result is known in the literature as the *Choquet theorem* [220].

Theorem 2.3. *Let X be a random set in W. The distribution of X is entirely specified by the capacity functionals $T_B(X)$ determined by all B-compacts in W.*

The characterisation of point processes via the *finite dimensional distributions or fidis* does not require the point process to be simple. However, it is a challenging task to use this characterisation to derive practical statistical inference procedures. In contrast, the characterisation via empty events probabilities is suitable for such statistical procedures if particular sets B are used. The characterisation through empty events theorem can be regarded as a special case of the Choquet theorem. Proofs of Theorems 2.1 and 2.2 are presented in [181]. A proof of the Choquet Theorem 2.3 is provided in [220].

2.2.1 Moment measures. Campbell measures. Palm distributions.

The preceding results demonstrate that the distribution of a point process is determined by the behaviour of counting variables within bounded sets. Consequently, in order to describe point processes, it is necessary to employ tools that characterise these counting variables. These tools should also take into account the sets on which the counting variables act. The first category of tools to be mentioned are known in the literature as the *moment measures.*

Definition 2.3. Let X be a point process on W. For any bounded Borel sets $B_1, B_2 \subseteq W$, the *first and second order moment measures* are defined by the following quantities:

$$\begin{aligned} \mu^{(1)}(B_1) = \alpha^{(1)}(B_1) &= \mathbb{E}[N(B_1)] \\ \mu^{(2)}(B_1 \times B_2) &= \mathbb{E}[N(B_1)N(B_2)] \\ \alpha^{(2)}(B_1 \times B_2) &= \mathbb{E}\left[\sum_{\zeta \in X} \sum_{\eta \in X}^{\neq} \mathbf{1}\{\zeta \in B_1, \eta \in B\}\right]. \end{aligned} \tag{2.3}$$

Here the notation $\sum^{\neq}$ indicates that the sum is calculated for all pairs of points $(\zeta, \eta) \in X^2$ such that $\zeta \neq \eta$.

The α measures are also called *factorial moment measures*, while the measure $\mu^{(1)} = \alpha^{(1)}$ is named *intensity measure.* The measures introduced above are not necessarily finite, even for finite point processes. In this case, a necessary condition for $\mu^{(2)}$ to be finite, hence for $\alpha^{(2)}$ and $\mu^{(1)}$ to be finite, is $\mathbb{E}[N_X(W)^2] < \infty$. Classical results in measure theory ensure that these functions can be extended to a unique symmetric $\sigma-$finite Borel measure, provided that they take finite values on bounded Borel sets [54, 142, 181, 186, 229].

The moment measures may be expressed in a more general form [54, 142, 181, 186, 229]. In fact, for functions $h : W^n \to \mathbb{R}$ that $\mu^{(n)}$ and $\alpha^{(n)}$ integrable, *the nth order moment measure* is

$$\mathbb{E}\left[\sum_{\{w_1,w_2,\ldots,w_n\}\in X} h(w_1,\ldots,w_n)\right] = \int_W \cdots \int_W h(w_1,\ldots,w_n)d\mu^{(n)}(w_1,\ldots,w_n)$$

and *the nth order factorial moment measure* is

$$\mathbb{E}\left[\sum_{\{w_1,w_2,\ldots,w_n\}\in X}^{\neq} h(w_1,\ldots,w_n)\right] = \int_W \cdots \int_W h(w_1,\ldots,w_n)d\alpha^{(n)}(w_1,\ldots,w_n).$$

Here $\sum^{\neq}$ is the sum of all the $n-$tuples of distinct points. The formulas given in the definition (2.3) are obtained with h being the indicator function.

Example 2.2 (Product densities). Let us assume local finiteness and absolute continuity with respect to the Lebesgue measure for the factorial measure $\alpha^{(n)}$. Then $\alpha^{(n)}$ has a density $\rho^{(n)}(w_1,\ldots,w_n)$ *the nth order product density* :

$$\alpha^{(n)}(B_1 \times \ldots \times B_n) = \int_{B_1} \cdots \int_{B_n} \rho^{(n)}(w_1,\ldots,w_n)d\nu(w_1)\cdots d\nu(w_n).$$

Furthermore, for integrable functions $h(w_1,\ldots,w_n)$

$$\mathbb{E}\left[\sum_{\{w_1,w_2,\ldots,w_n\}\in X}^{\neq} h(w_1,\ldots,w_n)\right]$$
$$= \int_W \cdots \int_W h(w_1,\ldots,w_n)\rho^{(n)}(w_1,\ldots,w_n)d\nu(w_1)\cdots d\nu(w_n).$$

The nth product densities $\rho^{(n)}(w_1,\ldots,w_n)d\nu(w_1)\cdots d\nu(w_n)$ may be interpreted as the probabilities of having a point in each infinitesimal volume $d\nu(w_i)$.

The transition from moments of counting variables to the definition of an "average pattern" is not always natural and straightforward [134, 220, 181]. However, it is quite possible to focus on object configurations with some particular properties, by considering moments of the quantities $N(B)\mathbf{1}\{X \in F\}$.

Definition 2.4. Let X be a point process on (W, d) a separable, complete, metric space. The *Campbell measure* of X is

$$C(B \times F) = \mathbb{E}\left[N(B)\mathbf{1}\{X \in F\}\right],$$

for every $B \in \mathcal{B}_W$ and $F \in \mathcal{F}$.

The first-order moment measure is expressed in terms of the Campbell measure:

$$\alpha^{(1)}(B) = \mu^{(1)}(B) = \mathbb{E}[N(B)] = C(B \times \Omega).$$

In an analogous way, higher order Campbell measures may be constructed, such as

$$C^{(2)}(B_1 \times B_2 \times F) = \mathbb{E}\left[N(B_1)N(B_2)\mathbf{1}\{X \in F\}\right],$$

which allows to write

$$\mu^{(2)}(B_1 \times B_2) = \mathbb{E}\left[N(B_1)N(B_2)\right] = C^{(2)}(B_1 \times B_2 \times \Omega).$$

Intuitively, the mechanism by which a complex structure is analysed consists first of studying the properties of each object, followed by an operation of synthesis of these properties. The following result, known in the literature as the *Campbell-Mecke formula*, provides the framework for such a procedure.

Theorem 2.4. *Let $h : W \times \Omega \to [0, \infty)$ be a measurable function. Then*

$$\mathbb{E}\left[\sum_{w \in X} h(w, X)\right] = \int_W \int_\Omega h(w, X) dC(w, X). \tag{2.4}$$

If the function h does not depend on the point process X, the Campbell-Mecke formula becomes

$$\mathbb{E}\left[\sum_{w \in X} h(w)\right] = \int_W \int_\Omega h(w) d\alpha^{(1)}(w).$$

Proofs of the previous theorem may be found in [54, 181, 229].

The analysis of structures can also be done in a conditional way. First, a typical object defining the structure is selected. Then the properties of the structure around the selected object are studied by extending the field of observation around the object. For point processes, this type of analysis is carried out using

Palm measures. The study of point processes using Palm measures is called interior conditioning [181]. Thorough, complete, and detailed presentations related to Palm measures can be found in [9, 46, 47, 54, 101, 69, 181, 208, 229].

Definition 2.5. Let X be a simple point process on W that is defined by $(\Omega, \mathcal{F}, \mathbb{P})$. The *Palm measure* of X or the *Palm probability* of X in w is given by

$$\mathbb{P}_w(F) = \mathbb{P}(X \in F | N(\{w\}) > 0). \tag{2.5}$$

This probability is the distribution of X conditioned by the event that there is a point of the process in $w \in W$.

The utilisation of the Palm measure as defined in (2.5) enables the rewriting of the Campbell-Mecke formula [54, 181, 229]. Consequently, equation (2.4) becomes

$$\mathbb{E}\left[\sum_{w \in X} h(w, X)\right] = \int_W \int_\Omega h(w, \boldsymbol{w}) d\mathbb{P}_w(\boldsymbol{w}) d\alpha^{(1)}(w).$$

In a similar manner, the reduced Palm measure is introduced

$$\mathbb{P}^!_w(F) = \mathbb{P}(X \setminus \{w\} \in F | N(\{w\}) > 0),$$

which is the distribution of $X \setminus \{w\}$ given that $w \in W$ is a point of the process. As for the Palm measure, the utilisation of the reduced Palm measure enables the derivation of the following expression:

$$\mathbb{E}\left[\sum_{w \in X} h(w, X \setminus \{w\})\right] = \int_W \int_\Omega h(w, \boldsymbol{w}) dP^!_w(\boldsymbol{w}) d\alpha^{(1)}(w).$$

which is another form of the Campbell-Mecke formula [54, 181, 229].

Definition 2.6. A point process X is said to be *stationary* if its distribution is translation invariant. The process is *isotropic* if its distribution is rotationally invariant.

Stationarity implies that the average number of points of the process observed through the window $B \subseteq W$ is the same for all the translated windows $B + w$ in W. This can be expressed as follows:

$$\alpha^{(1)}(B) = \mathbb{E}[N(B)] = \int_B \rho(w) d\nu(w) = \rho\nu(B) \tag{2.6}$$

with $\rho(w) = \rho > 0$ representing the *intensity function* of the process. If $\rho = 0$ the process X is said to be degenerate and it contains no points. The isotropy of a process implies that the statistics of the process observed through a window remain the same regardless of the rigid motions of the window. It is possible to construct processes that satisfy the property (2.6), but which are not stationary. To illustrate, consider the restriction to the first quadrant of a point process defined on $\mathbb{R}^2$. In this instance, the process is called *homogeneous*.

Similarly to the characterisation results, the Campbell and Palm measures are not readily applicable in a practical context. However, they are very interesting tools to deal with numerous particular situations. For instance, by assuming the point process to be stationary and isotropic, and by considering particular sets B (such as balls or line segments) or particular functions h (such as indicator functions), summary statistics can be constructed. Their purpose is to provide local information about the morphological and statistical characteristics of an observed pattern.

2.2.2 Summary statistics for stationary point processes

The *spherical contact distribution or the empty space function* is defined as:

$$F(r) = \mathbb{P}(d(w, X) \leq r), \quad \forall w \in W \quad \text{et} \quad r > 0 \tag{2.7}$$

with $d(w, \boldsymbol{x}) = \inf\{d(w, \eta), \eta \in \boldsymbol{x}\}$ the distance from the point w to the nearest point η of the configuration $\boldsymbol{x}$. Due to stationarity, the probability (2.7) does not depend on the choice of w, except for possible edge effects. The contact distribution can be rewritten in the following forms:

$$\begin{aligned} F(r) &= \mathbb{P}(d(o, X) \leq r) = \\ &= \mathbb{P}(X \cap b(o, r) \neq \emptyset) = T_{b(o,r)}(X) \\ &= \mathbb{P}(N(b(o, r)) > 0) \\ &= 1 - \mathbb{P}(N(b(o, r)) = 0) \end{aligned}$$

with $o \in W$ is an arbitrary origin in W and $b(o, r)$ is the ball of radius r centred on o.

Applications of the empty space function to the study of real point patterns can be found in [12, 54, 84, 142]. The following example presents an application of the empty space function to study the behaviour of the linear pattern induced by the road networks built by the logging industry in forest landscapes.

Example 2.3 (Effects of logging on roadless space in forest landscapes). Logging is often associated with the development of road networks. This leads to forest fragmentation. When forest fragmentation increases, it has a negative impact on biodiversity conservation. The definition of the empty space function (2.7) applies to more general random sets, not only to point

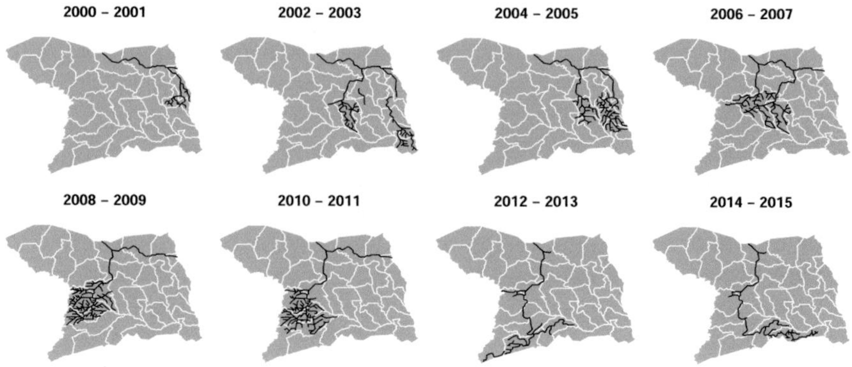

FIGURE 2.1
Annual development of the road network in a logging concession in Cameroon from 2000–2015.

patterns [54, 98, 185]. In [160] the authors used the empty space function to characterise the spatial distribution of road networks in forest landscapes and hence the fragmentation of the forests crossed by these roads. Their main results are presented below.

Figure 2.1 shows the temporal evolution of the road network in a logging concession in Cameroon. The white lines show the division of the concession into annual logging areas, while the black lines show the roads that were open during a two-year interval. This concession is approximately 50 km wide from east to west.

The estimation of the empty space function is affected by edge effects. Here, the border method is chosen to deal with this problem [12, 98]. The expression for the F function estimator is as follows:

$$\widehat{F}_{\mathbb{X}}(r) = \frac{\sum_i \mathbf{1}\{d(w_i, W^c) \geq r\}\mathbf{1}\{d(w_i, \mathbb{X}) \leq r\}}{\sum_i \mathbf{1}\{d(w_i, W^c) \geq r\}}$$

where $\mathbb{X}$ is the road pattern, $d(w, A)$ is the minimum distance from the point w to the set A and $\{w_i, i = 1, 2, \ldots\}$ a finite set of arbitrary points in W. It is common to choose a regular grid for the set of arbitrary points. For more details and references on the construction of summary statistics estimators together with different strategies for preventing edge effects, we recommend [13, 12, 54, 142].

A common indicator for measuring road occupancy in a given domain is the ratio of road length to the area of the domain under consideration. Figure 2.2 shows three toy models representing road networks. For all these models, the area of the considered domain is 4 by 4 units, the road length is 8, so the

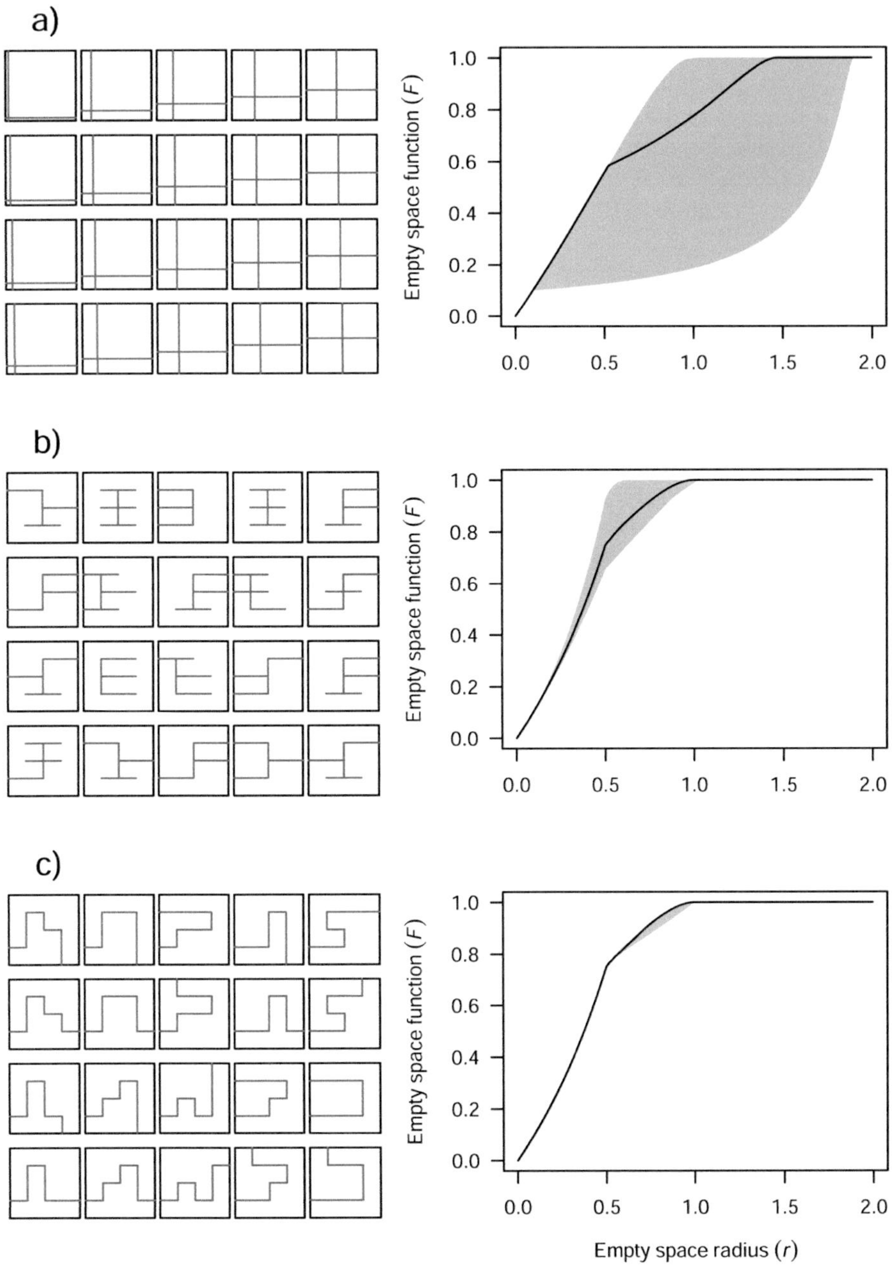

FIGURE 2.2
Left column: 20 realisations for each of the 3 toy models (a) main roads, b) secondary roads, c) – rivers). Right column: median and envelope estimates of the corresponding F function for each model.

road occupancy ratio is 0.5. In these cases, this ratio cannot discriminate between the models presented. A test was carried out to see if the F function was able to discriminate these different situations. This was done by generating a number of 100 simulations for each model and estimating the empty space function for each of these simulations. These simulations were used to estimate the median F function (black lines) and the $5-95\%$ confidence envelopes (grey regions) [12, 142]. It appears that the empty space function is able to discriminate between these models. Clearly, this statistic incorporates important information about the network topology.

The evolution of road networks was recorded and studied in different forest concessions and regions of the Congo Basin. The area under observation was divided into regions characterised by one of the following 5 different management categories: Forest Stewardship Council (FSC) certified, not FSC certified, inside an Intact Forest Landscape (IFL), outside an IFL and national park. The entire observation area was sampled using circular windows with random centres and a fixed diameter of 20 km. For the calculations, only those sampling windows were kept that had more than 90% of their area in one of the above categories. For each window kept, the estimate of the F function related to the observed road network was calculated.

The results are shown in Figure 2.3. The median curves have been calculated for each of the five categories. Panels (b-e) show 2-year steps from 1999 to 2015 (shading from light to dark with change over time). Since the F function represents the probability that a road is located at a distance r from any point in the observed area, it is a natural indicator of roadless space: the smaller the probability, the "less" is the presence of roads in a given area. For example, it can be observed that for national parks it is very likely not to encounter a road within a distance of 15 km around (Fig. 2.3 a). The roadless area decreases over time in both FSC certified and uncertified concessions, but at a faster rate in the uncertified concessions (Fig. 2.3 b,c). The changes in roadless area are less important in the concessions outside the IFL (Fig. 2.3 d,e). There is a slight increase in roadless area in the uncertified concessions outside the IFL (Fig. 2.3 d). For a more detailed analysis of the dynamics of roadless areas in intact forest landscapes in the Congo Basin, the reader is referred to [160].

Replacing the distribution of the process X by the Palm measure in (2.7) gives *the nearest neighbour distance distribution*:

$$G(r) = \mathbb{P}^!_o(d(o, X \setminus \{o\}) \leq r) = \mathbb{P}^!_o(N(b(o,r)) > 0). \tag{2.8}$$

The authors in [189] suggest combining the empty space function and the nearest neighbour distribution to obtain *the $J-$ function*

$$J(r) = \frac{1-G(r)}{1-F(r)}, \tag{2.9}$$

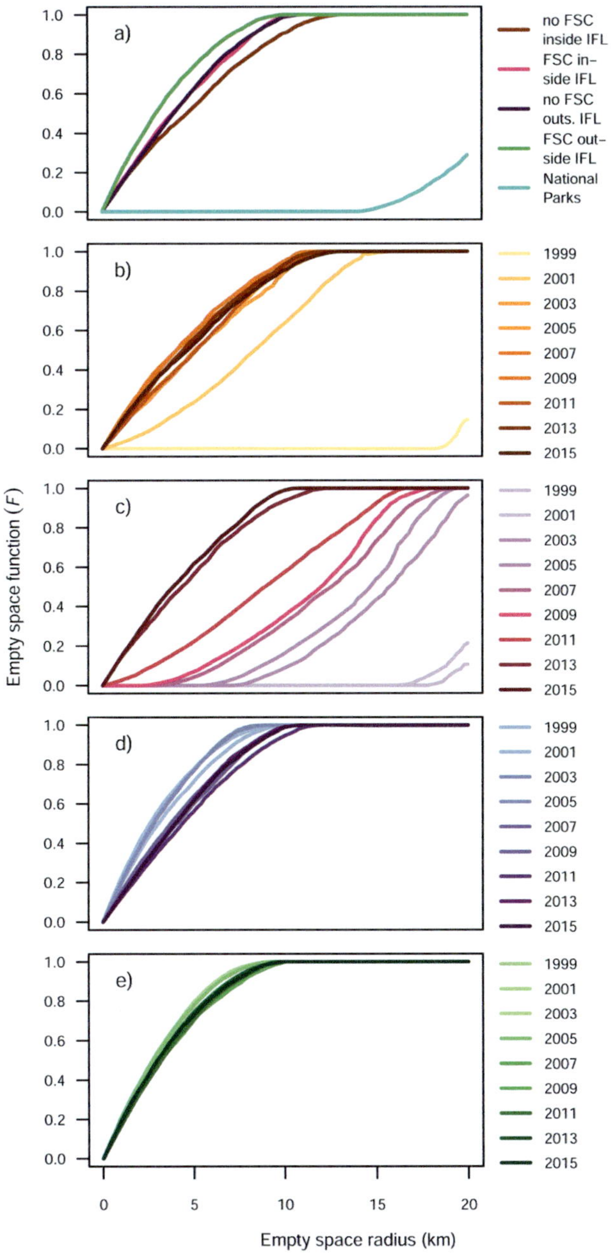

FIGURE 2.3
Median curves estimation of the F function for a) the 5 management categories in 2015, b) uncertified concessions within IFL, c) FSC certified concessions within IFL, d) uncertified concessions outside IFL, and e) FSC certified concessions outside IFL.

which is defined for all $r > 0$ such that $F(r) \neq 1$. This statistic has many interesting and elegant properties. Among them, the next one deserves attention. Suppose X is the point process obtained by superposing two independent stationary point processes X_1 and X_2, that is, $X = X_1 \cup X_2$. Then the $J-$function of X is a convex combination of the $J-$functions of the components X_1 and X_2:

$$J(r) = \frac{\rho_1}{\rho_1 + \rho_2} J_1(r) + \frac{\rho_2}{\rho_1 + \rho_2} J_2(r)$$

where ρ_i and J_i are the intensities and the J functions of the point processes $X_i, i = 1, 2$ respectively.

The $J-$ function is also an indicator of the morphology of an observed point pattern. Let us write

$$J(r) = \frac{\mathbb{P}^!_o(N(b(o,r)) = 0)}{\mathbb{P}(N(b(o,r)) = 0)}.$$

The case $J(r) < 1$ occurs whenever $\mathbb{P}(N(b(o,r)) = 0) > \mathbb{P}^!_o(N(b(o,r)) = 0)$. This indicates that the probabilities of having empty space around a given point of the domain are more important than the probabilities of having a free zone around a point of the process, thus indicating clustering. Conversely, the situation $J(r) > 1$ suggests regularity or repulsion.

Previously, product densities were derived based on the construction of moment measures. *The pair correlation function* is a summary statistic constructed using product densities. For any $w_1, w_2 \in W$, its definition is

$$g(w_1, w_2) = \frac{\rho^{(2)}(w_1, w_2)}{\rho^{(1)}(w_1)\rho^{(1)}(w_2)}.$$

For stationary isotropic point processes the second order product density depends only on $r = \| w_1 - w_2 \|$ and $\rho^{(1)}(w) = \rho = $ ct.. So the pair correlation is

$$g(r) = \frac{\rho^{(2)}(r)}{\rho^2}. \tag{2.10}$$

The interpretation of (2.10) is as follows: given $o \in W$ is a point of the process, the function $\rho^2 g(r)dr$ represents the probability that o has a neighbour inside the ring centred on o with radius r and infinitesimal thickness dr [12, 142].

Further analysis of the moment measures leads to perhaps the best-known summary statistic, the Ripley or $K-$ function [257, 259]. It represents the weighted expectation of the number of neighbours of a point in a point process, and its expression is

$$K(r) = \frac{1}{\rho}\mathbb{E}^!_o\left[N(b(o,r))\right] \tag{2.11}$$

where ρ is the process intensity and $\mathbb{E}^!_o$ is the expectation with respect to the reduced Palm measure. The $K-$function and the pair correlation are directly related. In the general $d-$dimensional case,

$$g(r) = \frac{K'(r)}{db_d r^{d-1}} \tag{2.12}$$

where b_d is the volume of the sphere of unit radius S^{d-1}.

A rigorous presentation of the construction of summary statistics for stationary marked point processes can be found in [142]. The summary statistics presented and discussed include: first order characteristics, mark correlation, mark variogram and mark distribution functions. The presented tools are tailored to the type of mark, qualitative or quantitative, and to the statistical information contained in the pattern of marked points under consideration. A recent survey on the derivation of summary statistics for marked point patterns is given by [86, 87]. Here the authors also consider non-scalar marks such as functions, graphs or manifolds. The authors propose to extend their development to derive summary statistics for marked point patterns on linear networks. Advantages, drawbacks and challenges related to the construction of summary statistics for marked point processes are further analysed in the discussions of this paper [63, 115, 187, 236, 299]. The following example is an application of the ideas and summary statistics tools proposed by [54, 98, 185] to analyse the possible associations between different patterns observed in a cosmological data set [163].

Example 2.4 (Associations between multi-type spatial patterns: cosmic filaments and photometric galaxies). The summary statistics F, G, J and K were originally constructed to characterise point patterns. The previous example showed how the spherical contact distribution F can be extended and applied in order to describe more general random sets, such as the linear patterns. The nearest neighbour distribution (2.8) can be also extended:

$$G_{\mathbb{X},\mathbb{Y}}(r) = \mathbb{P}^!_o(b(o,r) \cap \mathbb{Y} \neq \emptyset)$$

where $\mathbb{P}^!_o$ is the reduced Palm distribution at an arbitrary point of the point process $\mathbb{X}$ and $\mathbb{Y}$ a random set. The border corrected estimator of $G_{\mathbb{X},\mathbb{Y}}(r)$ is

$$\widehat{G}_{\mathbb{X},\mathbb{Y}}(r) = \frac{\sum_{x_i \in \mathbb{X}} \mathbf{1}\{d(x_i, W^c) \geq r\}\mathbf{1}\{d(x_i, \mathbb{Y}) \leq r\}}{\sum_{x_i \in \mathbb{X}} \mathbf{1}\{d(x_i, W^c) \geq r\}}.$$

Applying this development to (2.9) leads to the construction of *the bivariate $J-$function*

$$J_{\mathbb{X},\mathbb{Y}}(r) = \frac{1 - G_{\mathbb{X},\mathbb{Y}}(r)}{1 - F_{\mathbb{Y}}(r)} \tag{2.13}$$

with $F_{\mathbb{Y}}(r) = \mathbb{P}(b(o,r) \cap \mathbb{Y} \neq \emptyset)$. The estimator of the bivariate $J-$function is obtained by introducing in (2.13) the estimators of F and G, respectively. This gives

$$\widehat{J}_{\mathbb{X},\mathbb{Y}}(r) = \frac{1 - \widehat{G}_{\mathbb{X},\mathbb{Y}}(r)}{1 - \widehat{F}_{\mathbb{Y}}(r)}. \tag{2.14}$$

If the patterns $\mathbb{X}$ and $\mathbb{Y}$ are independent, then

$$\mathbb{P}(b(o,r) \cap \mathbb{Y} \neq \emptyset) = \mathbb{P}_o^!(b(o,r) \cap \mathbb{Y} \neq \emptyset)$$

and this gives $J_{\mathbb{X},\mathbb{Y}} = 1$. Thus, the bivariate $J-$function (2.13) can be used as an association indicator between the point pattern $\mathbb{X}$ and the random set $\mathbb{Y}$ as it follows: values of $J_{\mathbb{X},\mathbb{Y}}$ close to 1 may indicate independence, values less than 1 tend to indicate positive association or "clustering", while values greater than 1 tend to indicate negative association or "repulsion". At the same time, it is important to note that the $J_{\mathbb{X},\mathbb{Y}}$ function does not fully characterise the dependence of $\mathbb{X}$ and $\mathbb{Y}$. The authors in [28] present a counterexample in which the bivariate $J-$function is 1 while the patterns are dependent. For more details on the properties of the bivariate $J-$function and related proofs, we recommend [189, 190, 98, 185].

In [163] the authors use the bivariate $J-$function estimate to analyse associations within a multi-type pattern representation of some cosmological data[2]. Some of their results are presented below. The data set used in this paper has three main components: the SDSS spectroscopic galaxies [320], cosmic filaments detected from these spectroscopic galaxies [310], and a catalogue of photometric galaxies [27]. The catalogue of photometric galaxies is a more recent data set containing newly observed galaxies with respect to those presented in the SDSS spectroscopic catalogue. One of the questions raised by these data sets is whether these additional photometric galaxies carry information about the filamentary pattern already detected.

The main difficulty is that the redshift estimate of the photometric galaxies in the catalogue is only roughly known [27]. This means that the position of a galaxy could be anywhere along its line of sight. Figure 2.4 shows the different components of the data sample. The two types of galaxies are represented by their corresponding lines of sight, red for the spectroscopic and green for the photometric. The detected filaments are divided into two categories, perpendicular and non-perpendicular to the line of sight, according to a predefined

[2]Distances to galaxies are estimated by two methods. The first, spectroscopic, uses spectroscopic observations for this estimate. The second, photometric, uses the observed luminosity of a galaxy to estimate the distance. This last estimate is easier to obtain, but the typical errors of these distances are larger than those obtained by spectroscopy. So we call a galaxy spectroscopic if its distance has been obtained by spectroscopy, and photometric if its distance has been obtained by photometry.

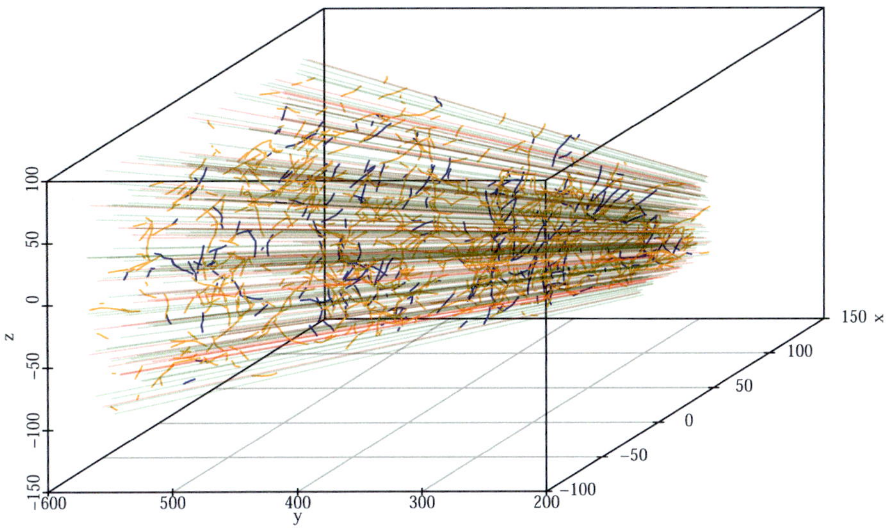

FIGURE 2.4
Data visualisation of a region of our universe, as an observed multi-type pattern : cosmic filaments (blue and orange curves), line of sight of spectroscopic (red) and photometric (green) galaxies. (Credit: Kruuse et al., A & A, 625, A130, 2019, reproduced with permission ©ESO.)

threshold. The perpendicular filaments are coloured purple and the others orange.

Figure 2.5 shows the calculation of the bivariate $J-$function for the data sample presented. The data was first projected onto the unit sphere. Then the bivariate $J-$function was estimated using the spherical distance to compute the empty space function and the nearest neighbour distribution. The analysis of the association between the pattern of observed photometric galaxies and the detected filaments is presented here. This analysis has been carried out by considering the whole set of photometric galaxies and three sets of filaments: all filaments together, the filaments perpendicular to the line of sight, and the filaments not perpendicular to the line of sight. In all these cases, and for all ranges r, it can be observed that the bivariate function has values less than 1, indicating a positive association between these patterns. This, together with the tests presented in [163], suggests that photometric galaxies carry information about the structure of the Universe, and that this unknown information is not independent of what is already known.

To resume, the summary statistics F, G and J are distance based, while g and K are second order statistics. They show relevant features of the pattern under consideration. However, their knowledge is not always sufficient to fully describe the stochastic process under consideration [17, 28].

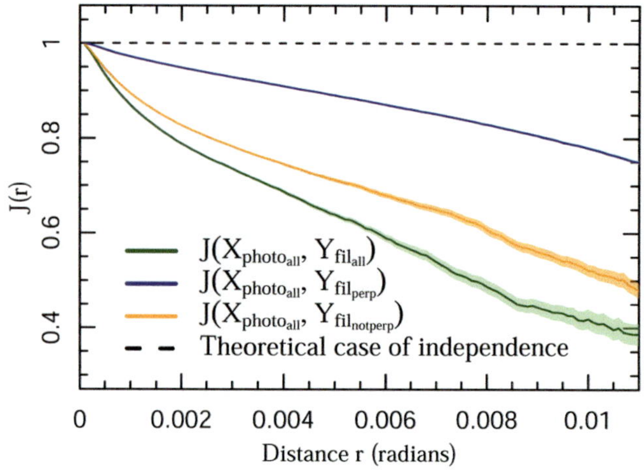

FIGURE 2.5
Results of the bivariate $J-$functions $J(r)_{X_{\text{photo}_{\text{all}}},Y_{\text{fil}_{\text{all}}}}$ (green), $J(r)_{X_{\text{photo}_{\text{all}}},Y_{\text{fil}_{\text{perp}}}}$ (purple) and $J(r)_{X_{\text{photo}_{\text{all}}},Y_{\text{fil}_{\text{notperp}}}}$ (orange). (Credit: Kruuse et al., A & A, 625, A130, 2019, reproduced with permission ©ESO.)

Despite this very natural limitation, summary statistics are a very useful tool. First, there are point process models for which it is possible to derive analytical expressions for these statistics. Second, given a given point pattern, it is perfectly possible to construct estimators of these statistics while controlling for their biases and variances. The computation of these estimators does not require model simulation, which is a very important computational aspect. On this basis, statistical inference can be performed. This is done as follows. Once the data are observed, a model is assumed for which these statistics are available in analytically closed form. Next, these statistics are estimated from the data. Finally, the estimated and analytical quantities are compared to perform statistical tests or parameter estimation. The case of point processes with summary statistics that do not have analytically closed formulas can be treated in a similar way. In this case, the previous analytical quantities are approximated by stochastic simulation. This working paradigm is one of the classic research programmes for constructing statistical inference for point patterns [12, 142, 54, 84, 229, 259].

Important developments and applications of this programme to inhomogeneous, marked, spatio-temporal point processes and more complex spatial patterns should also be mentioned [61, 62, 98, 100, 102, 141, 163, 160, 190, 183, 184, 188]. Research directions for the development of summary statistics also consider mathematical tools such as random measures and stochastic models based on lines, fibres and surfaces [42, 54, 185].

This approach also has a limitation. The asymptotic results that guarantee the quality of the estimators of the summary statistics require the extension of the observed pattern to infinity. However, this is not always possible when dealing with real data. This is simply because the phenomena under consideration do not occur in an infinite domain. Finite point processes described by a probability density are a possible alternative in this context.

2.3 The Poisson point process

This process is of fundamental importance from a theoretical and applied point of view. For a deeper understanding of this subject, in addition to the references cited, we recommend also [155, 165, 168, 217].

Definition 2.7. Let Υ be a locally finite Borel measure on (W, d) a complete separable metric space. A point process X on W is a Poisson process with intensity measure Υ if:

i) $N(B)$ is Poisson distributed with mean $\Upsilon(B)$ for every bounded Borel set $B \subseteq W$;

ii) for any k disjoint bounded Borel sets $B_1, \ldots, B_k$, the random variables $N(B_1), \ldots, N(B_k)$ are independent.

In the following, we assume that the intensity measure is given by

$$\Upsilon(B) = \mathbb{E}[N(B)] = \int_B \rho(w) d\nu(w), \tag{2.15}$$

where $\rho : W \to [0, \infty)$ is the intensity function. It is further assumed, that the intensity measure Υ is locally finite and *atomless (or diffuse)*, i.e. $\Upsilon(\{w\}) = 0$ for $w \in W$.

The Poisson process defined by the intensity measure (2.15) is written as $X \sim \text{Poisson}(W, \rho)$. If the intensity function $\rho(w)$ is constant then the process is *homogeneous* with rate ρ. Otherwise, the process is *inhomogeneous*. The Poisson point process with $\rho = 1$ is called *standard Poisson point process* or *unit rate Poisson point process* on W.

Theorem 2.5. *Let X be a Poisson point process on a complete separable metric space (W, d) with intensity measure $\Upsilon(\cdot)$ and B a bounded Borel set. Then, conditional on $\{N(B) = n\}$, X restricted to B independently distributes its points according to $\frac{\rho(w)}{\Upsilon(B)}$ for any $w \in B$.*

This theorem shows that the points of the Poisson point process do not interact. Therefore, their distribution pattern is considered to be *completely random*. The above-mentioned restricted process is a *binomial point process* [54, 181, 229].

The following result characterises the Poisson point process [229].

Proposition 2.1.
(i) The point process X est Poisson(W, ρ) with intensity measure given by (2.15), if and only if for all $B \subseteq W$ and $F \in \mathcal{F}_W$

$$\mathbb{P}(X_B \in F) = \tag{2.16}$$
$$= \sum_{n=0}^{\infty} \frac{\exp[-\Upsilon(B)]}{n!} \int_B \cdots \int_B \mathbf{1}_F\{w_1, \ldots, w_n\} \left[\prod_{i=1}^{n} \rho(w_i)\right] d\nu(w_1) \ldots d\nu(w_n)$$

where the integral for $n = 0$ is read as $\mathbf{1}_F\{\emptyset\}$.
(ii) If $X \sim$ Poisson(W, ρ), then for all the functions $h : \Omega_W \to [0, \infty)$ and $B \subseteq W$ with $\Upsilon(B) < \infty$

$$\mathbb{E}[h(X_B)] =$$
$$= \sum_{n=0}^{\infty} \frac{\exp[-\Upsilon(B)]}{n!} \int_B \cdots \int_B h(\{w_1, \ldots, w_n\}) \left[\prod_{i=1}^{n} \rho(w_i)\right] d\nu(w_1) \ldots d\nu(w_n).$$

If W is bounded or $\Upsilon(W) < \infty$, then the existence of the associated Poisson process follows from Proposition 2.1. For the case where W is unbounded, the existence and uniqueness of the Poisson process is determined by the void probabilities

$$v(B) = \exp[-\Upsilon(B)],$$

with $B \subseteq W$ bounded [229].

The following result follows from the application of the Campbell-Mecke formulae to Poisson point processes. It is known as the Slyvniak-Mecke theorem and is a valuable tool for calculating average quantities attached to objects resulting from a Poisson process [54, 181, 229].

Theorem 2.6. *If $X \sim$ Poisson(W, ρ), then for $h : W \times \Omega_W \to [0, \infty)$*

$$\mathbb{E} \sum_{w \in X} h(w, X \setminus \{w\}) = \int_W \mathbb{E} h(w, X) \rho(w) d\nu(w), \tag{2.17}$$

(where the left-hand side is finite if and only if the right-hand side is finite).

A consequence of the previous result is that for a stationary Poisson point process its distribution and its reduced Palm measure coincide

$$\mathbb{P}(\cdot) = \mathbb{P}^!_o(\cdot).$$

Example 2.5 (Summary statistics for stationary Poisson point processes). The pattern induced by the distribution of points from Poisson point processes is considered to be the simplest or most random, as there are no point interactions. This is due to the independence property. It also allows the analytical calculation of certain summary statistics.

Let X be a Poisson point process on $W = \mathbb{R}^d$ with intensity $\rho =$ ct.. Then, its empty space function is:

$$\begin{aligned} F(r) &= \mathbb{P}(N(b(o,r)) > 0) = 1 - \mathbb{P}(N(b(o,r)) = 0) \\ &= 1 - \exp[-\rho b_d r^d]. \end{aligned}$$

The Slyvniak-Mecke theorem 2.17 implies that

$$\mathbb{P}(d(o,X) \leq r) = \mathbb{P}^!_o(d(o, X \setminus \{o\}) \leq r),$$

so the expression for the nearest neighbour distribution is

$$G(r) = F(r) = 1 - \exp[-\rho b_d r^d].$$

It follows immediately that

$$J(r) = 1.$$

Similarly, for the K function we get

$$K(r) = b_d r^d,$$

and by taking its derivative as in (2.12), assuming the process is isotropic, the pair-correlation function is

$$g(r) = 1.$$

Due to the independence property, the pair correlation is 1 for any Poisson point process.

It is not possible to calculate exact formulae for these statistics for all stationary point processes. However, it is possible to estimate the statistics F, G, J, K and g from an observed point pattern and then compare them with the corresponding theoretical values of a known process. Very often the process used as a reference is the Poisson process [12, 142, 84, 181, 229, 259].

For example, suppose we observe a configuration of points in $W \subseteq \mathbb{R}^2$. Without giving details for the moment, consider that we can make tests of hypotheses about K from its estimate $\widehat{K}$. If the hypothesis $K(r) > \pi r^2$ is accepted, this means that the average number of points in a ball $b(o, r)$ is greater than in the case of a stationary Poisson process. Therefore, the observed data shows a tendency towards aggregation compared to the stationary Poisson case. Similarly, the $K(r) < \pi r^2$ situation indicates a smaller average number of points in $b(o, r)$ than in the Poisson case. In this case, the data have a repulsive character with respect to the Poisson reference. Finally, if the hypothesis $K(r) = \pi r^2$ is not rejected, then the data tend to behave like a stationary Poisson process with respect to this feature.

These summary statistics are quantitative and morphological descriptors that allow us to decide whether an observed pattern is completely random or not. And in the latter case, these statistics can give an indication of the direction in which the difference from a completely random pattern occurs, i.e. clustering or repulsion.

In the following, two operations on point processes are defined. The results presented show that the class of Poisson processes is closed under these operations [229].

The disjoint union $\cup_{i=1}^{\infty} X_i$ of the point processes $X_1, X_2, \ldots$ is called *superposition*. Consider the function $q : W \to [0, 1]$ and let X be a point process on W. The point process $X_{\text{thin}} \subset X$ is obtained by putting the points $w \in X$ into X_{thin} with the probability $q(w)$, and it is called the *independent thinning* of X. More precisely, we have

$$X_{\text{thin}} = \{w \in X : R(w) \leq q(w)\},$$

where $R(w) \sim \mathcal{U}[0, 1]$, $w \in W$ are mutually independent random variables and independent of X.

Lemma 2.1. *For $i = 1, 2$, let $X_i \sim$ Poisson(W, ρ_i) be independent processes of intensity measures $\Upsilon_i(W) = \int_W \rho_i(w) d\nu(w) < \infty$, respectively. Furthermore, for any A such that $\Upsilon_i(A) \leq \infty$ there exists a measurable $B \subset A$ such that $\Upsilon_i(B) = \frac{1}{2}\Upsilon_i(A)$. Then two processes X_1 and X_2 are almost surely disjoint, that is*

$$\mathbb{P}(X_1 \cap X_2 = \emptyset) = 1.$$

The previous condition imposed on intensity measures is also called the *bisection property*. The Lebesgue measure has the bisection property. The lemma is known in the literature as the Disjointness Lemma [45, 155, 156].

Proposition 2.2. *If the processes $X_i \sim$ Poisson(W, ρ_i) , $i = 1, 2, \ldots$ are mutually independent and if $\rho = \sum \rho_i$ is locally integrable, then $X = \cup_{i=1}^{\infty} X_i$ is almost surely a superposition and $X \sim$ Poisson(W, ρ).*

Proposition 2.3. *Let $X \sim$ Poisson(W, ρ)on which we apply an independent thinning operation according to the probabilities $q(w)$ with $w \in W$ and let $\rho_{thin} = q(w)\rho(w)$. Then, X_{thin} and $X \setminus X_{thin}$ are independent Poisson processes with densities ρ_{thin} and $\rho - \rho_{thin}$, respectively.*

The pattern induced by a Poisson process distribution keeps its Poisson character when the superposition and thinning operations are applied. This property is used to simulate more complex point processes. For example, an inhomogeneous Poisson point process can be obtained by thinning a homogeneous Poisson process.

Corollary 2.1. *Let X be a Poisson(W, ρ) process with ρ bounded above by a positive constant C. Then X is distributed as an independent thinning of Poisson(W, C) with probabilities $p(w) = \rho(w)/C$.*

The definition of a Poisson point process can be extended to the framework of marked point processes. Although this is not a priori difficult, it must be done with some precautions. Indeed, the distribution of the marks can affect the independence property as well as the stationarity and isotropy properties of the process.

Let Y be a Poisson(W, ρ) process and let the marks $\{m_w : w \in Y\}$ be conditionally on Y mutually independent. Then $X = \{(w, m_w) : w \in Y\}$ is a *marked Poisson process*. If the marks have the same distribution ν_M on $(M, \mathcal{M})$, then ν_M is the *mark distribution*.

Example 2.6 (Marked Poisson point processes and Boolean models). If $M = \{1, 2, \ldots, k\}$, we have a multitype Poisson process. It is a superposition of k independent Poisson point processes. If $M = \{K \subseteq W : K \quad \text{compact}\}$ and ν_M is the probability distribution selecting compact objects $\Gamma \in M$, then by the previous definition, X is a marked Poisson process with marks given by independent compact objects.

As noted above, Y is a Poisson(W, ρ) process and let $\Gamma_1, \Gamma_2, \ldots$ be a sequence of independently random compact sets (e.g. random balls or random segments) in W that are independent of Y. Replace each point of $Y(\omega) = \{w_1, w_2, \ldots, \}$ by the corresponding shifted set and take their union.

$$\Gamma = \bigcup_{i:w_i \in Y} (w_i + \Gamma_i).$$

The set Γ is called the *Boolean model*, the points w_i are called the *germs* and the random set Γ_0 is the *"typical" grain*. This model is known as the *Poisson germ grain model*.

The Boolean model is one of the first mathematical models developed for the construction of random patterns. Again, the structure induced by these models is considered to be the simplest, or most random, because there is no interaction between objects. This absence of interaction is guaranteed by the property of independence of the processes. It also allows us to derive precise analytical formulas for different properties of these structures.

As an illustration, we present the solution given by [54], for the calculation of the capacity functional

$$T_\Gamma(K) = \mathbb{P}(\Gamma \cap K \neq \emptyset) = 1 - \exp\left[-\rho \mathbb{E}(\nu(\check{\Gamma}_0 \oplus K))\right],$$

where $\oplus$ is the *Minkowski addition*, i.e. $A \oplus B = \{a + b : a \in A, b \in B\}$ and $\check{A}$ is the symmetric set of A with respect to the origin.

Let K be a fixed compact set. A new thinned process can be created from the original process Y. The thinning is done by deleting w_n from Y if $(w_n + \Gamma_n) \cap K = \emptyset$. So

$$Y_{\text{thin}} = \{w_n \in Y : (w_n + \Gamma_n) \cap K \neq \emptyset\}.$$

A point is deleted or kept by this thinning procedure independently of what happens to the other germs. This implies that Y_{thin} is a Poisson process with intensity measure

$$\Upsilon_{\text{thin}}(B) = \rho \int_B p(w) d\nu(w)$$

with $B \in \mathcal{B}_W$ and $p(w) = \mathbb{P}((w + \Gamma_0) \cap K \neq \emptyset)$ the probability that a germ placed at w will not be deleted.

Clearly, the capacity functional is

$$T_\Gamma(K) = 1 - \mathbb{P}(N(Y_{\text{thin}}) = 0).$$

If $\Upsilon_{\text{thin}}(W)$ is finite then,

$$\mathbb{P}(N(Y_{\text{thin}}) = 0) = \exp\left[\Upsilon_{\text{thin}}(W)\right] = \exp\left[\rho \int_W p(w) d\nu(w)\right].$$

The probability $p(w)$ is given by

$$p(w) = \mathbb{P}(w \in \check{\Gamma}_0 \oplus K)$$

since $w + \Gamma_0$ meets K precisely when $w \in \check{\Gamma}_0 \oplus K$. So

$$\begin{aligned}
\Upsilon_{\text{thin}}(W) &= \rho \int_W \mathbb{P}(w \in \check{\Gamma}_0 \oplus K) d\nu(w) \\
&= \rho \int_W \mathbb{E}\left[\mathbf{1}\{w \in \check{\Gamma}_0 \oplus K\}\right] d\nu(w) \\
&= \rho \mathbb{E}\left[\nu(\check{\Gamma}_0 \oplus K)\right].
\end{aligned}$$

This completes the proof.

The Boolean model is not a marked process because the point positions are not always available. The crystallisation process and the Lilypond models are processes of random geometric objects whose object positions are Poissonian, but with features that depend on the configuration of objects [71, 66, 67]. These processes do not really fit the framework of either marked Poisson processes or Boolean models. They can be considered as more general germ-grain models. For further reading we recommend [54, 145, 165, 207, 217, 220].

Let X_1 and X_2 be two Poisson processes on W. The law of X_1 is *absolutely continuous* with respect to that of X_2 if and only if $\mathbb{P}(X_2 \in F) = 0$ implies $\mathbb{P}(X_1 \in F) = 0$ whenever $F \in \mathcal{F}_W$. According to the Radon-Nikodym theorem, there exists a function $f : \Omega_W \to [0, \infty)$ such that

$$\mathbb{P}(X_1 \in F) = \mathbb{E}[\mathbf{1}\{X_2 \in F\} f(X_2)], \quad F \in \mathcal{F}_W.$$

The function f is called the *density* of the process X_1 with respect to X_2 [37, 229].

Proposition 2.4. *Let X_1 and X_2 be two Poisson processes on W with intensities ρ_1 and ρ_2, respectively.*
(i) If the intensities ρ_1 and ρ_2 are constants and $W = \mathbb{R}^d$, then X_1 is absolutely continuous with respect to X_2 if and only if $\rho_1 = \rho_2$.
(ii) If the intensities ρ_1 and ρ_2 are integrable functions on the bounded domain W, such that $\rho_2(\xi) > 0$ if $\rho_1(\xi) > 0$. Then, X_1 is absolutely continuous with respect to X_2. Furthermore, its Radon-Nikodym derivative is

$$f(\boldsymbol{w}) = \exp[\Upsilon_2(W) - \Upsilon_1(W)] \prod_{w \in \boldsymbol{w}} \rho_1(w)/\rho_2(w),$$

for finite point configurations $\boldsymbol{w} \subset W$ (and by considering $0/0 = 0$).

This proposition shows that two Poisson processes are not always absolutely continuous with each other. However, a Poisson process is always continuous with respect to the standard Poisson process when W is bounded. Two versions of this result, along with proofs and discussions can be found at [208, 229]. This result is important for understanding the construction of point processes with interacting points described by a probability density with respect to the reference measure given by the standard Poisson process [54, 108, 181, 186, 229].

2.4 Point processes with interactions

There are many strategies that allow the construction of random patterns that exhibit some observed or desired features. One approach that has been widely studied in the literature is based on the control of the intensity measure [12, 68, 84, 142, 173, 229, 186]. The very first category of methods that perform this control assumes that the objects forming the observed pattern are distributed in an inhomogeneous space. This inhomogeneity is determined by the presence of covariates. Assuming that the objects forming the pattern are independent, an intensity function can be fitted to the data to determine an inhomogeneous Poisson process capable of characterising the observations. The next category of methods for controlling the intensity function considers the inhomogeneity of the space as a realisation of a random function. For instance, the intensity measure may be given by a random field or a point process. The realisation of the intensity measure can also be convolved with a probability kernel. Then, conditional on the intensity measure realisation, the object positions follow the distribution of an inhomogeneous Poisson point process. For example, the Neymann-Scott or, more generally, the Cox processes are point process models based on these ideas. It is also worth mentioning the log-Gaussian Cox processes, which are a very interesting class of point processes proposed independently by astronomers and statisticians [59, 227]. These processes exhibit a one-to-one correspondence between the mean and covariance of the intensity measure and the first and second order product densities of the associated point process. A more general class of processes are the Lévy-based Cox processes, where the intensity measure is governed by a Lévy basis [135]. Cox processes are an extremely versatile class of processes because they allow the computation of summary statistics and statistical inference can be performed on this basis. Some the domains in which they have been successfully applied include astronomy, biology, earth and environmental sciences [44, 43, 6, 201, 229, 237]. Two limitations should be noted. First, when only one observation is available, it is not possible to distinguish between inhomogeneous Poisson and Cox processes. Second, the independence assumption does not allow to explain possible relationships and interactions between points.

Hawkes point processes are a special class of spatio-temporal point processes introduced by [131, 130, 132]. The points in the processes arrive either independently at a constant rate or due to the presence of another point in the process. They are also called self-exciting processes. Earthquakes are a good example of this type of process, since the events can be viewed as occurring independently or as replicas of an initial earthquake [240, 241, 242]. Neuronal activity is also a phenomenon that has been studied using Hawkes processes [255]. The work of [50] presents new asymptotic results for Hawkes processes exhibiting inhibition, thus opening perspectives for new applications of these processes. However, the embedding of marks and multiple interactions within the modelling framework of Hawkes processes remains an open problem. Exact and approximate simulation methods capable of dealing with boundary effects and computational issues for marked and unmarked Hawkes processes are proposed and discussed by [225, 226]. The pair correlation function of the stationary spatial Hawkes process is derived in [228]. Likelihood-based inference is also available [68, 239]. The review by [253] presents advantages and disadvantages of using maximum likelihood estimation for these procedures. Expectation maximisation (EM) algorithms can sometimes be used. The paper [275] states that this strategy may have its limitations due to the convergence properties of EM algorithms. Therefore, new research directions in inference based on second-order moments, Campbell-Mecke and Georgii-Nguyen-Zessin theorems are recommended. For a deep analysis of the properties of Hawkes processes, and also for their connection with branching processes, we recommend [68, 47].

Permanental and determinantal point processes originated in particle physics as models for boson and fermion fields [199, 276]. In the last two decades, they have appeared in numerous fields such as probability theory, physics, algebra or spatial statistics [22, 137, 209, 170, 169]. From a more applied perspective, the permanental processes are capable of modelling clustered point patterns, while the determinantal processes are suitable for modelling regular or repulsive point patterns. Under smooth conditions, both processes exhibit a probability density with respect to the standard Poisson process. These probability densities have extremely elegant expressions and depend on the permanent or determinant of the matrix of a covariance function. Thus, knowledge of the entire process is determined by its second-order moments. Unlike the determinant, the approximate or exact computation of the permanent of a matrix is an extremely complex task and is an active area of research. Today, the available inference procedures concentrate more on determinantal point processes. For this purpose, by considering parametric families for the allowed covariance functions, a likelihood can be built and moments can be computed [170]. The degree of repulsion of an observed pattern can also be quantified [39]. Contrast functions can also be used [40]. For the general class of non-stationary point processes, the asymptotic normality of parameters obtained from function estimators is derived in [171]. The authors also adapt this result to the particular case of non-stationary determinantal point processes. The likelihood of the determinantal point process is analytically intractable. For stationary determinantal point processes constructed using parametric families of correlation functions with known Fourier transforms, an asymptotic approximation of the likelihood is introduced by [249]. A review of existing methods for the simulation of determinantal point processes, together with improvements, is presented in [172]. The spatio-temporal extension of determinantal point processes is given in [315]. As for the Hawkes processes, the case of marked points and multiple interactions in the framework of determinantal processes is, at best, still an open problem. Nevertheless, it is worth mentioning the work of [169], who obtains point processes with long-range attraction and short-range repulsion via a thinning procedure of determinantal point processes, among other models considered.

The *Gibbs point processes* framework allows the natural embedding of multiple interactions for multiple types of objects. The probability density describing these models is constructed directly from the object interactions [12, 54, 68, 142, 181, 186, 229]. The theory of Gibbs processes comes from statistical physics. The aim is to give a mathematical characterisation of the states of physical systems consisting of a very large number of interacting objects. This is achieved by a probability measure describing the distribution and the interactions of the objects forming the system under consideration [78, 268, 106]. If the interactions of the objects are local in a certain sense, an object can only interact with objects that are within the neighbourhood induced by the considered interaction. This property can lead to a simplified, more synthetic description of the pattern of interest. However, there is a price to pay. The normalising constant of the considered probability measure is often not available in analytically closed form. Therefore, its practical use requires appropriate computational and numerical strategies. The natural way of modelling a real situation, together with the growth of machine computing power and the development of Monte Carlo techniques, encouraged applied mathematicians and engineers to take up this challenge. The result of their work has been the development of robust and reliable spatial models with interactions that have been built to answer questions coming from different fields such as image analysis, pattern recognition, computer science, telecommunications, environmental sciences and astronomy [7, 8, 12, 24, 54, 89, 84, 104, 138, 145, 165, 181, 186, 201, 202, 229, 260, 284, 313, 318].

2.4.1 Point processes described by a probability density

Let W be a compact subset of $\mathbb{R}^d$ and let X be a finite point process on W given by the probability density p with respect to the standard Poisson process on W .

The Proposition 2.1 allows to write for each $F \in \mathcal{F}_W$:

$$\begin{aligned}
&\mathbb{P}(X \in F) = \\
&\quad = \sum_{n=0}^{\infty} \frac{\exp[-\nu(W)]}{n!} \int_W \cdots \int_W \mathbf{1}(\{w_1, \ldots, w_n\} \in F) p(\{w_1, \ldots, w_n\}) \\
&\quad \times d\nu(w_1) \ldots d\nu(w_n).
\end{aligned}$$

For $n = 0$, the corresponding term in the sum is $\exp[-\nu(W)]\mathbf{1}(\emptyset \in F)p(\emptyset)$. If $\nu(W) = 0$ then $P(X = \emptyset) = 1$. For the practical applications considered it is always assumed that $\nu(W) > 0$.

In general, such a probability density is known up to a constant $p \propto h$. This is due to the fact that for most of these models, it is difficult to calculate analytically the *normalising constant*

$$c = \sum_{n=0}^{\infty} \frac{\exp[-\nu(W)]}{n!} \int_W \cdots \int_W h(\{w_1, \ldots, w_n\}) d\nu(w_1) \ldots d\nu(w_n). \quad (2.18)$$

In statistical physics the quantity c is called *partition function.*

Example 2.7 (Normalising constant for the Poisson point process). For a Poisson point process with intensity function ρ, its un-normalised probability density with respect to the standard Poisson process is written as

$$p(\boldsymbol{w}) \propto \prod_{w_i \in \boldsymbol{w}} \rho(w_i).$$

Its intensity measure is $\Upsilon(B) = \int_B \rho(w) d\nu(w)$, and it is further assumed that $0 < \Upsilon(B) < \infty$ for each $B \subseteq W$.

By using (2.18), one gets

$$c = \exp[-\nu(W)] \sum_{n=0}^{\infty} \frac{\Upsilon(W)^n}{n!} = \exp[\Upsilon(W) - \nu(W)],$$

and the complete expression of the probability density

$$p(\boldsymbol{w}) = \exp[\nu(W) - \Upsilon(W)] \prod_{w_i \in \boldsymbol{w}} \rho(w_i)$$

Here we have specified the point process probability density with respect to the standard Poisson process. The Radon-Nykodim theorem and the Proposition 2.4 allow to construct point process probability densities with respect to reference measures given by Poisson processes with finite intensity measure [37, 229]. This may be of interest for some image processing applications [89].

This framework for constructing probability densities can be extended to the case of marked point processes.

Let X be a marked point process on $W \times M$ given by the probability density p with respect to the standard marked Poisson process. According to the previous definitions, the probability density p is concentrated on the space

Ω of finite configurations of marked points. This space is provided with the σ algebra $\mathcal{F}$ constructed from the counting measures on the bounded Borel sets $A \subseteq W \times M$. The standard marked Poisson process chooses the number of points in a configuration according to a Poisson distribution given by the parameter $\nu(W)$, then it distributes the points independently and uniformly in W. Finally, to each point is assigned a mark, chosen independently of the points and the other marks, according to the law ν_M, thus forming the objects and the object pattern.

By applying Proposition 2.1, the distribution of X is written for each $F \in \mathcal{F}$:

$$\begin{aligned} &\mathbb{P}(X \in F) = \\ &= \sum_{n=0}^{\infty} \frac{\exp[-\nu(W)]}{n!} \int_{W\times M} \cdots \int_{W\times M} p(\{(w_1, m_1), \ldots, (w_n, m_n)\}) \\ &\times \mathbf{1}(\{(w_1, m_1), \ldots, (w_n, m_n)\} \in F) \\ &\times d\nu(w_1) \ldots d\nu(w_n) d\nu_M(m_1) \ldots d\nu_M(m_n). \end{aligned} \tag{2.19}$$

The reference measure given by the standard marked Poisson process is noted μ. Its expression is obtained from (2.19) by setting

$$p(\{(w_1, m_1), \ldots, (w_n, m_n)\}) = 1 \text{ for all } n.$$

hence obtaining

$$\begin{aligned} &\mu(F) = \\ &= \sum_{n=0}^{\infty} \frac{\exp[-\nu(W)]}{n!} \int_{W\times M} \cdots \int_{W\times M} \mathbf{1}(\{(w_1, m_1), \ldots, (w_n, m_n)\} \in F) \\ &\times d\nu(w_1) \ldots d\nu(w_n) d\nu_M(m_1) \ldots d\nu_M(m_n). \end{aligned} \tag{2.20}$$

In the following, we consider only the probability densities of marked point processes constructed with respect to the reference measure (2.20).

The probability density of a marked point process with interacting objects is constructed by first specifying the *interaction functions*

$$\phi(x_{i_1}, \ldots, x_{i_k})^{(k)} : \Omega \to \mathbb{R}^+$$

for each $k-$tuple of objects, where $x_i = (w_i, m_i)$ are the objects or the marked points. Then, the probability density is given by the product of all the interaction functions

$$p(\boldsymbol{x}) \propto \prod_{x_i \in \boldsymbol{x}} \phi(x_i)^{(1)} \cdots \prod_{\{x_{i_1}, \ldots, x_{i_k}\} \in \boldsymbol{x}} \phi(x_{i_1} \ldots, x_{i_k})^{(k)} \tag{2.21}$$

The probability densities (2.21) constitute a highly flexible framework for modelling provided that they are integrable on the configuration space

$$\alpha = \int_{\Omega} p(\boldsymbol{x}) d\mu(\boldsymbol{x}) < \infty.$$

The following two results guarantee the integrability of the probability density of a marked point process. These conditions of integrability are known in the literature as the *Ruelle stability conditions* [108, 229, 269]. The presented corresponding proofs are those given by [108].

Definition 2.8. Let X be a marked point process with probability density p with respect to the reference measure μ. If there exists $\Lambda > 0$ such that

$$p(\boldsymbol{x}) \leq \Lambda^{n(\boldsymbol{x})}, \quad \forall \boldsymbol{x} \in \Omega, \tag{2.22}$$

where $n(\boldsymbol{x})$ is the number of marked points in the configuration $\boldsymbol{x}$, the process X is *stable* in the sense of Ruelle. And if

$$p(\boldsymbol{x} \cup \{\eta\}) \leq \Lambda p(\boldsymbol{x}), \quad \forall \boldsymbol{x} \in \Omega, \eta \in W \times M \setminus \boldsymbol{x}, \tag{2.23}$$

the process is *locally stable.*

Proposition 2.5. *The probability density of a stable marked point process is integrable.*

Proof. The integrability of $p(\boldsymbol{x})$ follows directly from the stability condition:

$$\begin{aligned} \int_\Omega p(\boldsymbol{x})\mu(d\boldsymbol{x}) &\leq \int_\Omega \Lambda^{n(\boldsymbol{x})}\mu(d\boldsymbol{x}) \\ &= \sum_{n=0}^{\infty} \frac{\exp[-\nu(W)][\Lambda\nu(W)]^n}{n!} = \exp[\nu(W)(\Lambda - 1)]. \end{aligned}$$

□

Proposition 2.6. *A locally stable marked point process is stable in the sense of Ruelle.*

Proof. It is easy to show by recurrence that

$$p(\boldsymbol{x}) \leq p(\emptyset)\Lambda^{n(\boldsymbol{x})}, \quad \forall \boldsymbol{x} \in \Omega.$$

Hence the local stability of a point process (2.23) implies stability in the sense of Ruelle (2.22) hence its integrability.

□

Let X be a finite marked point process on $W \times M$ with probability density $p \propto h$ with respect to μ. The *Papangelou conditional intensity* of X is given by

$$\lambda(\eta; \boldsymbol{x}) = \frac{p(\boldsymbol{x} \cup \{\eta\})}{p(\boldsymbol{x})} = \frac{h(\boldsymbol{x} \cup \{\eta\})}{h(\boldsymbol{x})}, \quad \boldsymbol{x} \in \Omega, \eta \in W \times M \setminus \boldsymbol{x}, \tag{2.24}$$

provided $p(\boldsymbol{x}) \neq 0$, set $\lambda(\eta; \boldsymbol{x}) = 0$ otherwise [12, 148, 181, 186, 229].

The Papangelou intensity does not depend on the normalising constant of p. If $\lambda(\eta; \boldsymbol{x})$ is bounded, indicating that the undergoing process is locally stable and therefore stable, then its un-normalised probability density h is integrable. The reverse is not true. The Lennard-Jones process is a point process that can be used to model short-range repulsion with clustering at larger scales. This process is stable but not locally stable [54, 181]. Furthermore, convergence proofs can be given for simulation algorithms built to sample probability densities of locally stable point processes [30, 31, 108, 152, 181, 192, 195, 229].

A point process is *hereditary* if

$$p(\boldsymbol{x}) > 0 \Rightarrow p(\boldsymbol{y}) > 0 \quad \text{for} \quad \boldsymbol{y} \subset \boldsymbol{x}.$$

The next result shows that for a hereditary point process there is one-to-one correspondence between its probability density and its conditional intensity [186, 229].

Theorem 2.7. *Let X be a finite hereditary marked point process on $W \times M$ with probability density p with respect to the reference measure μ. Then*

$$p(\{x_1, \ldots, x_n\}) = p(\emptyset) \prod_{i=1}^{n} \lambda(x_i; \{x_1, \ldots, x_{i-1}\}).$$

The product does not depend on the labelling of the points $x_1, \ldots, x_n \in W \times M$, $n \in \mathbb{N}^\star$.

The conditional intensity also provides information about the morphology of the structure induced by the realisation of an object process. A process X is said to be *attractive* if $\boldsymbol{x} \subseteq \boldsymbol{y}$ involves

$$\lambda(\eta; \boldsymbol{x}) \leq \lambda(\eta; \boldsymbol{y}), \tag{2.25}$$

and *repulsive* in the opposite case

$$\lambda(\eta; \boldsymbol{x}) \geq \lambda(\eta; \boldsymbol{y}). \tag{2.26}$$

The morphology of a repulsive process shows some regularity in the sense that the centres of the objects tend to move apart. In an attractive process, the objects tend to form one or more clusters. There are processes that are neither repulsive nor attractive. The properties of attraction and repulsion are to be taken into account for the construction of exact simulation algorithms for hereditary locally stable point processes [30, 31, 152, 229, 181, 195].

2.4.2 Markov point processes

The conditional intensity can be interpreted as the conditional probability

$$\lambda((w, m); \boldsymbol{x}) d\nu(w) d\nu_M(m)$$

that a marked point $\eta = (w, m)$ lies in the infinitesimal volume $d\nu(w)d\nu_M(m)$ given the configuration $\boldsymbol{x}$. After (2.21) the conditional intensity is written as

$$\lambda(\eta; \boldsymbol{x}) = \phi(\eta)^{(1)} \prod_{x_i \in \boldsymbol{x}} \phi(x_i, \eta)^{(2)} \cdots \prod_{\{x_{i_1}, \ldots, x_{i_k}\} \in \boldsymbol{x}} \phi(x_{i_1}, \ldots, x_{i_k}, \eta)^{(k+1)} \tag{2.27}$$

This interpretation is called *external conditioning* [186, 181, 229] because it studies the contribution of a new object that is external to a given pattern. The dual way of studying marked dot patterns is the *interior conditioning.* In this case, conditional on an existing object, one studies the evolution of the pattern containing that point. This other approach leads to the construction of the Palm measures presented earlier [54, 181, 186, 229].

The conditional intensity in its form (2.27) is difficult to handle. To simplify the study of a structure, it is natural to consider a limit for the order of the interaction. For example, one can consider only interactions of pairs of objects. Another natural simplification is to assume that the interactions between objects are *local*, i.e. that an object can only interact with its "neighbours". All these simplifications impose a Markovian character on the studied patterns. This is a key assumption for the modelling framework adopted by the applications presented in this book. In this case, the conditional intensity plays a role similar to the conditional probability at a location in the case of discrete Markov random fields [101, 121, 113, 176, 318].

Let $\sim$ be a symmetric and reflexive interaction relation between objects with positions in W and characteristics in M. It is typically a neighbourhood relation based on a metric (Euclidean, Hausdorff, etc.) or on an intersection of sets.

Definition 2.9. A *clique* is a configuration $\boldsymbol{x} \in \Omega$ such that $\eta \sim \zeta$ for any $\eta, \zeta \in \boldsymbol{x}$. The empty set is a clique.

Definition 2.10. Let X be a marked point process on $W \times M$ with probability density p with respect to the reference measure μ. The process X is Markovian if for any $\boldsymbol{x} \in \Omega$ such that $p(\boldsymbol{x}) > 0$, the following conditions are satisfied simultaneously:

(i) $p(\boldsymbol{z}) > 0$ for all $\boldsymbol{z} \subseteq \boldsymbol{x}$ (hereditary)

(ii) $\frac{p(\boldsymbol{x} \cup \{\zeta\})}{p(\boldsymbol{x})}$ depends only on ζ and its neighbours $\partial(\zeta) \cap \boldsymbol{x} = \{\eta \in \boldsymbol{x} : \eta \sim \zeta\}$.

This type of process is known in the literature as a Markov process in the sense of Ripley-Kelly [181, 261].

Example 2.8 (Markovianity of the Poisson process). With respect to the reference measure μ, the probability density of a marked Poisson process marked on $W \times M$ given by the intensity function $\rho(w,m) > 0$ is

$$p(\boldsymbol{x}) = \left[\prod_{i=1}^{n(\boldsymbol{x})} \rho(w_i, m_i)\right] \exp\left[\nu(W) - \int_{W\times M} \rho(w,m) d\nu(w) d\nu_M(m)\right]. \quad (2.28)$$

It is clear that $p(\boldsymbol{x}) > 0$ for any configuration of objects $\boldsymbol{x}$. Its Papangelou intensity is

$$\lambda((w,m);\boldsymbol{x}) = \rho(w,m)\mathbf{1}\{(w,m) \notin \boldsymbol{x}\}.$$

The Poisson process is therefore Markovian, and it is independent of the interaction functions $\phi^{(k)}$ with $k \geq 2$. This confirms the choice of the Poisson process as a completely random structure [142, 181].

The property (ii) of the Markov point processes definition is generalised by the following result, known as the *spatial Markov property*. It can also be seen as the spatial equivalent of the temporal Markov property [181, 229, 261].

Theorem 2.8. *Let X be a Markov marked point process on $W \times M$ with probability density $p(\cdot)$, and let A be a Borel subset of $W \times M$. Then the conditional distribution of $X \cap A$, knowing $X \cap A^c$ depends only on the restriction of X to the neighbourhood*

$$\partial(A) \cap A^c = \{u \in W \setminus A : u \sim a \quad \textit{for some} \quad a \in A\}.$$

The following result is known in the literature as the Hammersley-Clifford factorisation theorem [15, 181, 186, 229, 261].

Theorem 2.9. *The point process with probability density $p : \Omega \to \mathbb{R}^+$ is Markov with respect to the relation $\sim$ if and only if there exists a measurable function $\phi_c : \Omega \to \mathbb{R}^+$ such that*

$$p(\boldsymbol{x}) = \prod_{\textit{cliques} \quad \boldsymbol{z}\subseteq\boldsymbol{x}} \phi_c(\boldsymbol{z}), \quad \alpha = \phi_c(\emptyset) \quad (2.29)$$

for each $\boldsymbol{x} \in \Omega$.

This theorem simplifies writing the probability density of a point process with interactions (2.21) by considering only the product of interaction functions on cliques. For a Markov point process, the equivalence between (2.29) and (2.21) is obtained by setting $\phi_c(\boldsymbol{z}) = 1$ when $\boldsymbol{z}$ is not a clique.

Markov point processes are known in statistical physics as *Gibbs point processes* [54, 229]. Gibbs processes describe systems with a large number of particles. The probability density of a Gibbs point process is given by

$$p(\boldsymbol{x}) = \frac{1}{Z}\exp\left[-U(\boldsymbol{x})\right] = \frac{1}{Z}\exp\left[-\sum_{\text{cliques} \quad \boldsymbol{z}\subseteq\boldsymbol{x}} U_c(\boldsymbol{z})\right],$$

where Z is the partition function, U is the energy of the system and $U_c = \log\phi_c$ is the potential of a clique. All Markov processes can be seen as Gibbs processes. The reciprocal is not true.

The following examples analyse the Markovian character of some well-known point processes [181, 186, 229].

Example 2.9 (Markovianity of the Poisson process – continued). The probability density of the Poisson point process (2.28) from the previous example can be expressed as the product over cliques of the following interaction functions

$$\begin{aligned} \phi_c(\emptyset) &= \exp\left[\nu(W) - \int_{W\times M}\rho(w,m)d\nu(w)d\nu_M(m)\right], \\ \phi_c(\{(w,m)\}) &= \rho(w,m) \end{aligned}$$

and $\phi_c \equiv 1$ when the cliques have two or more objects. It follows that the potential of cliques consisting of one object is

$$U_c(\{(w,m)\}) = -\log\rho(w,m),$$

with $U_c = 0$ for cliques with more than one object. This confirms the absence of interaction in the Poisson process, as well as the choice of this process to model simple patterns with no particular morphological structure [142, 181].

Example 2.10 (The Widom-Rowlinson model). This model is known in the literature as the penetrable sphere mixing model proposed for the study of liquid-vapour equilibrium [181, 317]. It is described by the space of marks $M = \{1, 2\}$ and the probability density

$$p(\boldsymbol{x}) = \alpha \prod_{(w,m)\in\boldsymbol{x}} \beta_m \prod_{(u,1),(v,2)\in\boldsymbol{x}} \mathbf{1}\{\| u - v \| > r\} \tag{2.30}$$

with respect to the standard marked Poisson point process μ on $W \times M$ with $\nu_M(1) = \nu_M(2)$. Here α is the normalising constant. The spheres of different types are at least at a distance $r > 0$ from each other. This results in clusters of particles of one type in the space left free by the particles of the other type. The parameters β_1 and β_2 control the intensity of the particles of type 1 and 2 respectively.

The conditional intensity for adding the object $(w,1) \notin \boldsymbol{x}$ to the configuration $\boldsymbol{x}$ is

$$\lambda((w,1);\boldsymbol{x}) = \beta_1 \mathbf{1}\{\| u - w \| > r \text{ for all } (u,2) \in \boldsymbol{x}\},$$

while an analogous expression is available for adding an object of type 2. The model Widom-Rowlinson is hereditary and locally stable with

$$\Lambda = \max\{\beta_1, \beta_2\}.$$

Furthermore, $\lambda((w,m);\boldsymbol{x}') \geq \lambda((w,m);\boldsymbol{x})$ for all $\boldsymbol{x}' \subseteq \boldsymbol{x}$ and $(w,m) \in W \times M$.

The interaction functions are

$$\begin{aligned} \phi_c(\emptyset) &= \alpha, \\ \phi_c(\{(w,m)\}) &= \beta_m, \\ \phi_c(\{(u,1),(v,2)\}) &= \mathbf{1}\{\| u - w \| > r\}, \end{aligned}$$

with $\phi_c \equiv 1$ if the cliques contain two objects or more of the same type.

Example 2.11 (Multi-type pairwise interaction processes). Let $M = \{1,\ldots,I\}$ be the space of marks with $I \in \mathbb{N}$ and ν_M the uniform distribution on M. The multitype pairwise interaction processes [15, 181, 261] are defined by their density p with respect to μ and it has the expression

$$p(\boldsymbol{x}) = \alpha \prod_{(w,m)\in\boldsymbol{x}} \beta_m \prod_{(u,i)\neq(v,j)\in\boldsymbol{x}} \gamma_{ij}(\| u - v \|). \tag{2.31}$$

The parameters $\beta_m > 0$, $m \in M$ control the intensity of points of type m. The interaction between each pair of objects is described by the measurable function $\gamma_{ij} : [0,\infty) \to [0,1]$. It is assumed that the interactions between objects are symmetrical, i.e. $\gamma_{ij} \equiv \gamma_{ji}$ for all $i,j \in M$. The model (2.31) is hereditary and locally stable with

$$\Lambda = \max_{m\in M} \beta_m.$$

For $(w,m) \notin \boldsymbol{x}$, the Papangelou conditional intensity

$$\lambda((w,m);\boldsymbol{x}) = \beta_m \prod_{(u,i)\in\boldsymbol{x}} \gamma_{im}(\| u - w \|)$$

is decreasing in its second argument with respect to the order relation induced by set inclusion.

The interaction functions are

$$\begin{aligned}\phi_c(\emptyset) &= \alpha,\\ \phi_c(\{(w,m)\}) &= \beta_m,\\ \phi_c(\{(u,i),(v,j)\}) &= \gamma_{ij}(\| u-v \|),\end{aligned}$$

with $\phi_c \equiv 1$ if the cliques have three or more objects.

Example 2.12 (Quermass interaction processes). These point processes are a particular family of Gibbs processes that generalise germ-grain models by introducing interactions. These interactions are specified by linear combinations of Minkovski functionals. In $\mathbf{R}^2$ these functionals are the area, the perimeter, the Euler-Poincaré characteristic. Markov properties and stability conditions of the model are given by [154]. The existence of infinite volume quermass interaction processes in a general setting of non-locally stable interaction and unbounded convex grains is proved by [76]. Percolation of the model when the grains are assumed to be circular with random radii on a compact support is considered by [60]. Aspects of statistical inference for some particular models can be found in [222, 223, 79].

Quermass interactions can also be used to describe probability densities of interacting marked point processes. The Widom-Rowlinson model or the area interaction process can be considered as particular forms of the general quermass interaction modelling framework [154]. The area interaction process is introduced in the next chapter from the perspective of an application, cluster detection.

3

Applications

This chapter presents models of Gibbs marked point processes with interactions that have been built specifically for practical applications. Two examples of random graph models and one of random tessellation are also given. All the models presented in this chapter belong to the family of Gibbs processes. The practical applications for which all these models were built are pattern detection and pattern characterisation.

3.1 Galactic filaments: Candy and Bisous models

3.1.1 Data presentation and problem formulation

Galaxies are not uniformly distributed throughout the Universe. They form an intricate cosmic web of filaments, clusters and sheets, surrounded by vast regions of void. The filaments are the dominant feature of this network. The identification and mapping of these structures is of great importance to cosmologists as they outline the distribution of dark matter in the Universe [164, 146, 201, 202, 247]. For an overview and comparison of existing methods for tracing the cosmic filamentary network, the reader is referred to the collective work [179].

The data used here are the positions of the galaxies observed in W, a finite volume window. The galactic filaments are given by the linear network induced by the position of the galaxies. Figure 3.1 shows a field of galaxies observed within a solid angle. Due to the distance, only the brightest galaxies are observed. The further away the observer is, the more the density of the observed field of galaxies decreases. We will therefore restrict ourselves to a rectangular volume, so that the field of galaxies within it retains a certain homogeneous character.

The pattern detection context is adopted here. The cosmic filament network is approximated by a configuration of random thin cylinders that connect and align within the regions indicated by the observed data field. This idea is

DOI: 10.1201/9781003380382-3

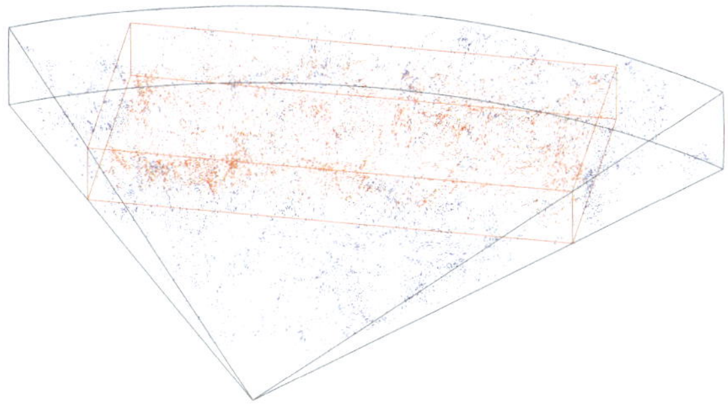

FIGURE 3.1
Rectangular sample in the northern hemisphere of the catalogue 2dF redshift survey.

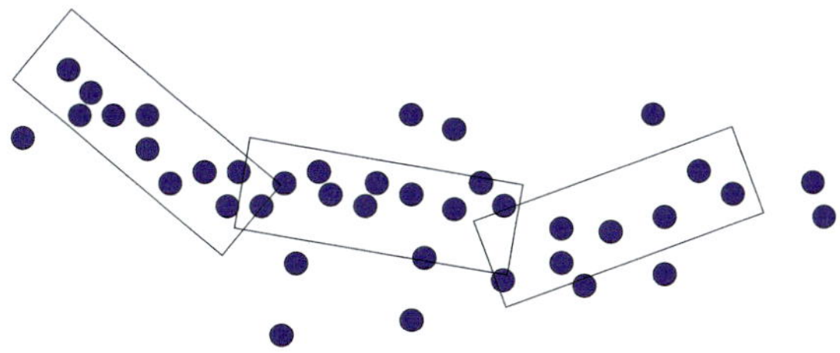

FIGURE 3.2
Filamentary pattern approximation: the purple points represent the observed galaxies field $\boldsymbol{d}$; the thin rectangles are the two-dimensional projection of a configuration of cylinders $\boldsymbol{x}$ that is hidden in the data. This unknown configuration of cylinders is an approximation of the unknown filamentary pattern induced by the positions of galaxies.

illustrated in Figure 3.2. The purple points in the figure represent the observed data set $\boldsymbol{d}$ given by the positions of galaxies. The configuration of cylinders is the hidden pattern of objects $\boldsymbol{x}$ that has to be detected. This hidden pattern approximates the unknown filaments induced by the galaxies positions.

The realisations such patterns of cylinders are controlled by the probability density of a marked Gibbs point processes on $W \times M$. The points of the process are points in W and they represent the centres of the cylinders. The marks correspond to the parameters of the cylinders and they take values in M. The probability density (1.1) is defined with respect to the reference measure μ, the standard unit intensity Poisson process with independent marks on $W \times M$. The process described by this equation can also be seen as the superposition of an inhomogeneous Poisson process and an interaction process

$$p(\boldsymbol{x}|\theta, \boldsymbol{d}) = p_{\boldsymbol{d}}(\boldsymbol{x}|\theta) p_i(\boldsymbol{x}|\theta). \tag{3.1}$$

The inhomogeneous Poisson process is described by

$$p_{\boldsymbol{d}}(\boldsymbol{x}|\theta) = \frac{\exp[-U_{\boldsymbol{d}}(\boldsymbol{x}|\theta)]}{Z_{\boldsymbol{d}}(\theta)}, \tag{3.2}$$

where $U_{\boldsymbol{d}}(\boldsymbol{x}|\theta) = \sum_{x \in \boldsymbol{x}} u(x|\theta)$ and u is a measurable function, while the interaction process is written in the form

$$p_i(\boldsymbol{x}|\theta) = \frac{\exp[-U_i(\boldsymbol{x}|\theta)]}{Z_i(\theta)}, \tag{3.3}$$

where $Z_{\boldsymbol{d}}(\theta)$ and $Z_i(\theta)$ are the normalisation constants associated with to (3.2) and to (3.3) respectively.

Below the data term $U_{\boldsymbol{d}}(\boldsymbol{x}|\theta)$ and the interaction term $U_i(\boldsymbol{x}|\theta)$ are presented.

3.1.2 Modelling the data term

There is no precise method to construct the data term (3.2). In general, it should contain the definition of what we are looking for, it should be easy to compute, and it should ensure that the complete model is well-defined, i.e. integrable. In the following, ideas for the construction of the data attachment term for the detection of galactic filaments are presented [294, 295, 296, 298, 310, 308].

The working hypothesis adopted is that the cosmic filament network is approximated by a configuration of interacting random thin cylinders. The data term should favour the positioning of the cylinders in the observed window W in those regions where $\boldsymbol{d}$ locally indicates the presence of a piece of the filament network.

Let $\boldsymbol{x} = \{x_1, \ldots, x_n\}$ be a configuration of random segments. A segment x is specified by its random centre w, its random orientation vector m and its fixed length. The length of all segments is set to $2r_h$. A cylinder $s(x) = s(w, m)$ is attached to each segment x. The cylinder is centred at w, its height is $2r_h$, the orientation of its symmetry axis is m and its radius is a fixed value r. Each cylinder is associated with an extra cylinder with parameters identical to the previous one except for the radius, which is $2r$. Let $\tilde{s}(x)$ be the shadow of x, which is the set difference between the extra cylinder and the cylinder as shown in Figure 3.3. The cylinder $s(x)$ is divided into three parts of equal volume, which we will note as $s_1(x)$, $s_2(x)$ and $s_3(x)$.

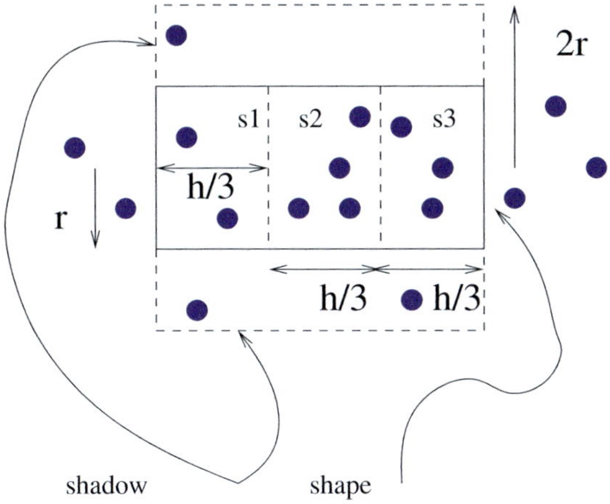

FIGURE 3.3
The two-dimensional view of a cylinder and its shadow within a galaxy field. The cylinder $s(x)$ and its shadow $\tilde{s}(x)$ contain 9 and 3 galaxies respectively.

A cylinder can be part of the network if it fulfils several conditions. First, the density of galaxies inside the cylinder must be greater than the density of galaxies in its shadow. This condition is written as

$$\mathbf{1}\{\text{"density"}\} = \mathbf{1}\left\{\frac{n(\boldsymbol{d}\cap s(x))}{\nu(s(x))} > \frac{n(\boldsymbol{d}\cap \tilde{s}(x))}{\nu(\tilde{s}(x))}\right\},$$

with $n(\boldsymbol{d}\cap s(x))$ and $n(\boldsymbol{d}\cap \tilde{s}(x))$ denoting the number of galaxies covered by the cylinder and its shadow, of the respective volumes $\nu(s(x))$ and $\nu(\tilde{s}(x))$.

Then the cylinder should contain enough galaxies that are rather homogeneously distributed inside it. This is formulated as follows:

$$\mathbf{1}\{\text{"homogeneity"}\} = \prod_{i=1}^{3} \mathbf{1}\{n(\boldsymbol{d}\cap s_i(y)) > 1\},$$

where $n(\boldsymbol{d}\cap s_i(x))$ is the number of galaxies in $s_i(y)$.

In this context, the *potential* of a cylinder is defined as the function

$$\begin{aligned} u(x) &= \mathbf{1}\{\text{"density"}\}\mathbf{1}\{\text{"homogeneity"}\}[n(\boldsymbol{d}\cap s(x)) - \\ &- n(\boldsymbol{d}\cap \tilde{s}(x)) + u_{\max}] - u_{\max}, \end{aligned} \tag{3.4}$$

where $u_{\max}$ is a positive fixed value introduced to give a small chance to a cylinder that does not meet all the criteria to be part of the filamentary network. Depending on the choice of this parameter, more complete networks can be obtained. It also influences the mixing properties of the simulation dynamics of the model [294, 295, 298, 310, 308].

The data energy term is obtained by summing the potentials (3.4) for all cylinders in a configuration

$$U_{\boldsymbol{d}}(\boldsymbol{x}|\theta) = -\sum_{x \in \boldsymbol{x}} u(x). \tag{3.5}$$

Since the domain W contains only a finite number of galaxies, it can be seen that there exists a constant $\Lambda_{\boldsymbol{d}} \geq \exp(-u_{\max})$ such that

$$u(\eta) > -\log \Lambda_{\boldsymbol{d}}$$

for any $\eta \in W \times M$. This implies that the marked point process (3.2) given by the data energy (3.5) is locally stable.

The construction of the data term presented is simple, very intuitive, without seeking any optimality. At the time, this choice was made for pedagogical reasons that remain valid today. We also hoped that cosmologists would respond by using their physical knowledge to give a local definition of the properties of a filament. This strategy was successful. A more elaborate version of the presented data term was considered to study the correlation of the orientation vectors of galaxies with the orientation of the filament of which they are a part [309]. A specification of the local properties of the filaments was then made using statistical tests. An additional condition checking the proximity of the galaxies to the main axis of the cylinder was also introduced. More details can be found in [298, 310, 308]. The significance of the proximity condition has been statistically validated in [140]. All this work opens up the prospect of fitting complex marked point processes to galactic catalogues.

3.1.3 Modelling objects interactions: Candy and Bisous models

You must not forget the suspenders[1], so please remember the key hypothesis that the galactic filaments are approximated by a random configuration of interacting cylinders. Since the cylinders here all have the same fixed radius and height, for simplicity we can speak of segments, keeping in mind the main axis of symmetry of the cylinders. The data term presented earlier gives high probabilities to cylinders located in regions in W where the data field d seems to locally indicate the presence of filaments. However, this term does not align or connect segments to form filaments. This is the role of the interaction energy term (3.3).

The *Candy model* is a marked point process that has as realisations configurations of segments that interact. The interactions considered are: alignment, repulsion, connection. Depending on the weight of each interaction, the model is

[1]as Rudyard Kipling would say.

able to form networks with different topological properties. The model was initially developed as a solution to the problem of extracting thin networks such as roads or rivers from satellite and aerial images [82, 282, 289, 81, 288, 290]. The theoretical properties and parametric inference of the model have also been studied [282, 83, 192].

On this mathematical basis, a generalisation called the *Bisous model* was then proposed [117, 292]. The motivation behind this was to build a general model capable of reproducing, within the same mathematical framework, different patterns such as filaments, sheets and voids, by analogy with the structures exhibited by the distribution of galaxies [146, 247, 201, 202]. The Bisous model is a marked point process capable of producing these different structures. Once constructed, the model has been successfully applied to the detection and to the statistical and morphological characterisation of galactic filaments [294, 295, 296, 283, 309, 310, 298, 308, 163].

In order to construct a random pattern of interacting objects, it is necessary to define the objects and their interaction. The information provided is then embedded in a marked point process probability density. The following is a synthesis of the main ideas presented in [292].

Here the considered objects are of deterministic size and shape, while their position and orientation are random. Depending on the intended application, objects with more complex random characteristics can be considered [192, 288, 290, 305].

The mark represents the orientation of the object. If $W \subset \mathbb{R}^2$, then the mark space is $M = [0, \pi)$. If $W \subset \mathbb{R}^3$, then $M = [0, 2\pi) \times [0, 1]$. In this case the orientation is given by the vector

$$m(m_1, m_2) = (\sqrt{(1 - m_2^2)} \cos(m_1), \sqrt{(1 - m_2^2)} \sin(m_1), m_2).$$

For the distribution of the marks ν_M, the uniform distribution is chosen.

It is convenient to assign a finite number of extremities to each object, as a set of rigid points $\{e_u, 1 \leq u \leq q\}$. These points are used to construct the attraction region of the object.

The *attraction region* $a(w)$ of the object $x = (w, m)$ is the disjoint union

$$a(x) = \bigcup_{u=1}^{q} b(e_u, r_a),$$

where $b(e_u, r_a)$ is the ball of radius r_a centred at e_u.

To form connected structures, objects interact through their attraction regions. The *attraction rule* $\mathcal{C}$ of two objects x_i and x_j is the set of index pairs corresponding to the extremity points allowed to form an attraction interaction.

Consider two given objects $x_1 = (w_1, m_1)$ and $x_2 = (w_2, m_2)$. The objects exhibit an *attraction interaction* $x_1 \sim_a x_2$ if the set $\{(u,v) : 1 \leq u, v \leq q, d(e_u(x_1), e_v(x_2)) \leq r_a\}$ has only one element in $\mathcal{C}$. The *repulsion interaction* $x_1 \sim_h x_2$ occurs whenever $d(w_i, w_j) \leq 2r_h$ for some fixed value $r_h > 0$. The objects are *aligned* $x_1 \sim_{\|} x_2$, if

$$m_1 \cdot m_2 \geq 1 - \tau_{\|}$$

where $\tau_{\|} \in (0,1)$ is a fixed curvature parameter and $\cdot$ is the scalar product of the orientation vectors. [2]. Finally, the objects are said to be *connected* $x_1 \sim_s x_2$ if the following three conditions are satisfied simultaneously

$$\begin{aligned} x_1 &\sim_a x_2, \\ x_1 &\sim_{\|} x_2, \\ x_1 &\not\sim_h x_2. \end{aligned}$$

For objects in $W \times M$, let $\mathcal{R}$ be the set of all pairwise interactions that are local[3], symmetric, and induce a measurable application on $\mathcal{F}$. All previously introduced interactions belong to $\mathcal{R}$, except $\sim_{\|}$, which is not local.

An object is $s-$*connected* if it is connected through exactly s of its extremities. If $s = 0$, the object is said to be *free*. Two objects are said to be of the *same type* if they are defined on the same parameter space, if they have the same number of extremities, and if they follow the same attraction rule $\mathcal{C}$. In the following, only patterns made of objects of the same type are considered.

Example 3.1 (Generating objects for linear, planar and cluster patterns). A *linear pattern* can be generated from interacting segments. Figure 3.4a shows a segment of length $2(r + r_a)$, centred at the origin. It has the following $q = 2$ extremity points

$$e_u = (0, 0, (r + r_a) \sin \frac{(2u-1)\pi}{2}), \quad 1 \leq u \leq q$$

and its orientation vector is $\omega = (0, 0, 1)$. The repulsion radius is $r_h = r$ while the attraction rule is $\mathcal{C} = \{(1,2),(2,1)\}$. So two segments can be connected through extremity points of different indices. Any segment of any centre and orientation in $\mathbb{R}^3$ can be obtained from the segment in this example, by one translation and two rotations [133].

[2]Here it was considered that $W \subset \mathbb{R}^3$.

[3]A local interaction is an interaction with finite range.

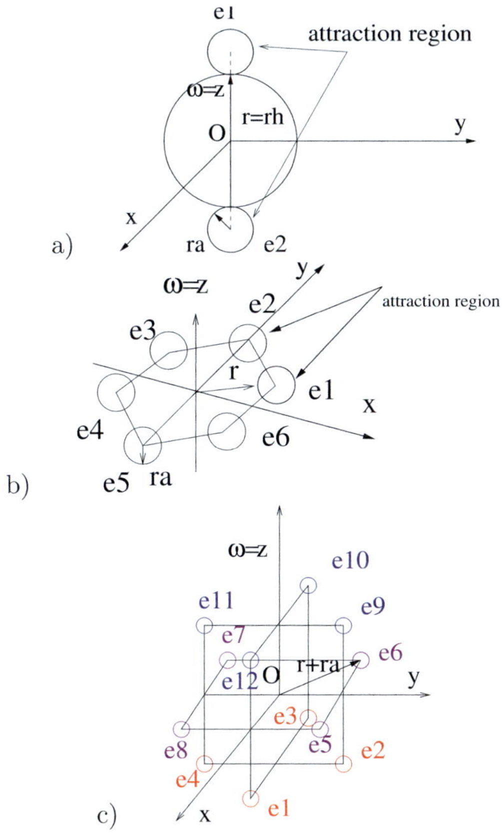

FIGURE 3.4
Generating elements: a) linear structure, b) planar structure, c) cluster (volumic) structure.

Figure 3.4b shows a hexagon which can be the generating element of a *planar pattern*. The hexagon is centred on the origin. Its orientation vector is $\omega = (0, 0, 1)$ and its $q = 6$ extremity points are given by

$$e_u = ((r + r_a)\cos\frac{(2u-1)\pi}{6}, (r + r_a)\sin\frac{(2u-1)\pi}{6}, 0), \quad 1 \leq u \leq q.$$

The repulsion region is identical to the previous case. The attraction rule is defined as follows: the pair of extremities $(e_u(x_i), e_v(x_j))$ satisfies the attraction conditions if $|u - v| = 3$. This gives the index pairs $\mathcal{C} = \{(1,4), (4,1), (2,5), (5,2), (3,6), (6,3)\}$.

To create a *cluster (volumic) pattern* we will start with a cuboctahedron as a generating object. Figure 3.4c shows a centred cuboctahedron, with orientation $\omega = (0, 0, 1)$. It has $q = 12$ extremity points which are written as follow is

$$e_u = ((r+r_a)\cos(\alpha_u+\beta_u), (r+r_a)\sin(\alpha_u+\beta_u), (r+r_a)\sin(\beta_u/2)) \quad 1 \leq q \leq 12,$$

with $\alpha_u = \frac{\pi}{4}(2u-1)$ and $\beta_u = \frac{\pi}{2}\left[\frac{u-5}{4}\right]$. Around the object, the repulsion region is constructed as in the previous two cases. The attraction rule $\mathcal{C}$ between two objects x_i and x_j is established by the pairs of extremity points with indices $(1, 10)$, $(2, 11)$, $(3, 12)$, $(4, 9)$, $(5, 7)$, $(6, 8)$ and their symmetrical counterparts, respectively.

Let $\boldsymbol{x} = \{x_1, \ldots, x_n\}$ be a configuration of objects of the same type. The *Bisous model* is defined by the probability density function

$$p(\boldsymbol{x}) = p(\emptyset)\left[\prod_{s=0}^{q} \gamma_s^{n_s(\boldsymbol{x})}\right] \prod_{\kappa \in \Gamma \subset \mathcal{R}} \gamma_\kappa^{n_\kappa(\boldsymbol{x})} \tag{3.6}$$

with respect to the reference measure given by the standard Poisson process μ. Here $\gamma_s > 0, \gamma_\kappa \in [0, 1]$ represent the model parameters, $p(\emptyset)$ is the normalising constant and Γ is the subset of all the pairwise interactions from $\mathcal{R}$ that were considered to build the pattern. If $\gamma_k = 0$, the interaction $\sim_\kappa$ is said to be hard-core (here we assume $0^0 = 1$). For each s and κ the sufficient statistics $n_s(\boldsymbol{x})$ and $n_\kappa(\boldsymbol{x})$ respectively represent, the number of $s-$connected objects and the number of pairs of objects with interaction κ, in the configuration $\boldsymbol{x}$.

Lemma 3.1. *The Bisous model defined by* (3.6) *is locally stable.*

Proof. Let W be a compact subset of $\mathbb{R}^3$. If $\sim_h \in \Gamma$ and if $\gamma_h = 0$, then there exists a maximum number of objects $n_{\max}$ that can be contained in W. Then the conditional intensity is bounded

$$\lambda(\zeta; \boldsymbol{x}) \leq \prod_{s=0}^{q} \max\{\gamma_s, \gamma_s^{-1}\}^{n_{\max}} = \Lambda.$$

The model is locally stable even if Γ does not contain any hard-core interactions. In fact, the maximum number of unconnected objects that can connect to another object is 12. This number is known as the "kissing number"[4]. The bound for the conditional intensity is thus

$$\Lambda = \prod_{s=0}^{q} \max\{\gamma_s, \gamma_s^{-1}\}^{12}. \tag{3.7}$$

[4]The problem to which it is the answer was originally a mathematical discussion between Isaac Newton and David Gregory (http://en.wikipedia.org/wiki/Kissing_number_problem). The name "kissing number" also inspired the name of the Bisous model.

Since connectivity also assumes aligned objects, it is possible to impose tighter bounds on the conditional intensity (3.7) that depend on the curvature parameter τ.

□

Theorem 3.1. *The unnormalised density* (3.6) *of the Bisous model specifies a well-defined object point process.*

Proof. The sufficient statistics of the Bisous model are measurable with respect to $\mathcal{F}$ [192, 292]. This together with the definition of $\mathcal{R}$ implies the $(\Omega, \mathcal{F})-$measurability of the Bisous model. The local stability implies the Ruelle stability (2.22), hence the integrability of the model. □

The Bisous model is based on pairwise interactions that are symmetric and local, with a finite range. For example, the $s-$ connectivity is symmetric and its range is $4r_h + 2r_a$. Based on this, and assuming that the empty sets and the singletons are cliques, the authors in [192, 292] prove that the Candy and Bisous models are Markovian in the sense of (Ripley-Kelly).

Figure 3.5 shows patterns built using the Bisous model, which controls different types of objects. It can be seen that, depending on the generating objects and the interactions chosen, the Bisous model is able to construct structures that represent linear networks, form sets of two-dimensional surfaces and group to produce volumic clusters.

The interaction component (3.3) of the galactic filament model (3.1) is given by the Bisous model, which is able to reproduce linear structures built from interacting segments. Its probability density is

$$p_i(\boldsymbol{x}|\theta) = \gamma_\kappa^{n_\kappa(\boldsymbol{x})} \prod_{s=0}^{2} \gamma_s^{n_s(\boldsymbol{x})},$$

with the previously defined parameters γ_s and statistics n_s. The interaction $\sim_\kappa$ combines the repulsion $\sim_h$ and the orthogonal interactions $\sim_\perp$. Two objects are said to be *orthogonal* $x_1 \sim_\perp x_2$ if and only if $|m_1 \cdot m_2| \leq \tau_\perp$, with $\tau_\perp \in (0,1)$ a predefined value. Thus, for any two objects x_1 and x_2, we write $x_1 \sim_\kappa x_2$ if we simultaneously have $x_1 \sim_h x_2$ and $x_1 \sim_\perp x_2$. For the model to be well-defined, we need $\gamma_\kappa \in [0,1]$. The statistic n_κ represents the number of pairs of segments exhibiting the interaction $\sim_\kappa$.

The interaction energy of the model is written as:

$$U_i(\boldsymbol{x}|\theta) = -n_\kappa \log \gamma_\kappa - \sum_{s=0}^{2} n_s(\boldsymbol{x}) \log \gamma_s \tag{3.8}$$

Since both components, the data term (3.5) and the interaction term (3.8) are both locally stable, the same holds for the complete model (1.1). Thus, the proposed model is well-defined.

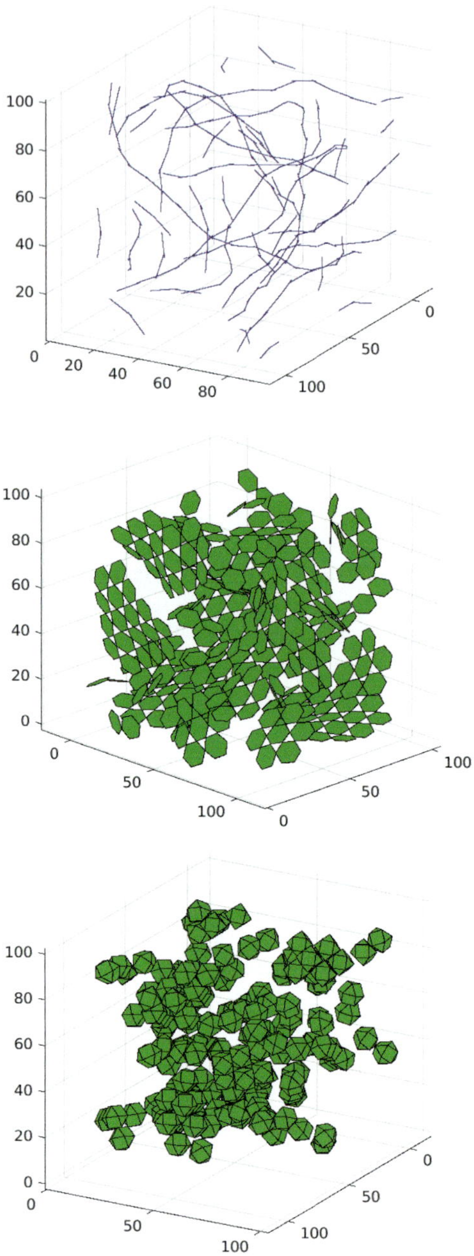

a)

b)

c)

FIGURE 3.5
Patterns generated using the Bisous model: a) linear structure, b) planar structure, c) cluster (volumic) structure.

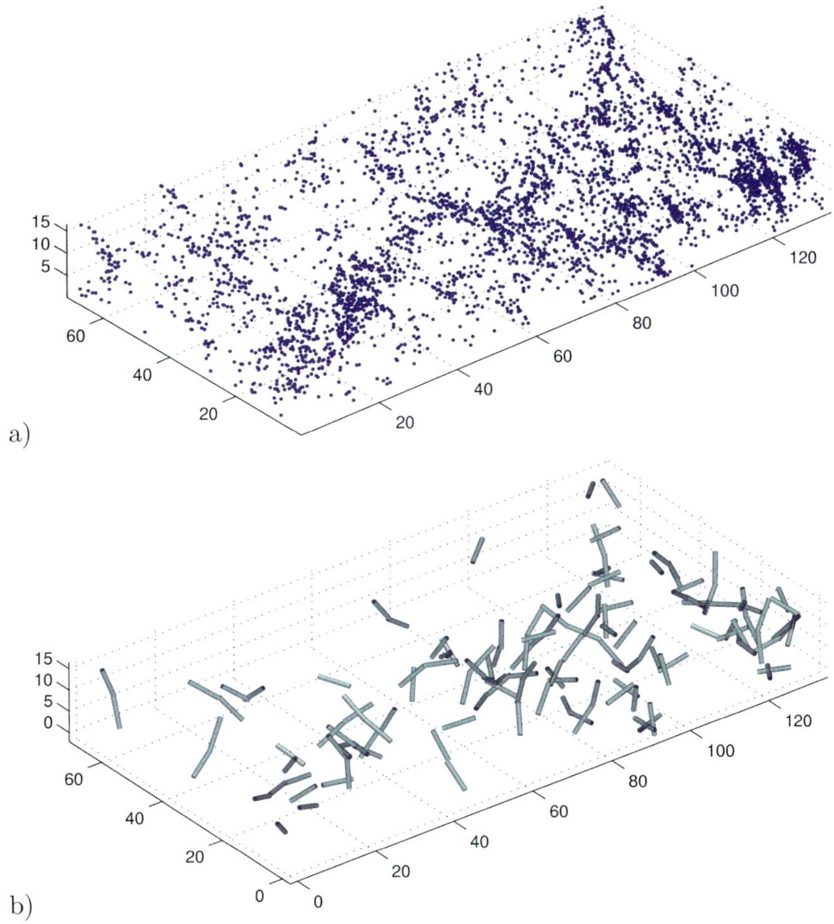

FIGURE 3.6
a) Dataset: a sample extracted from the galaxy catalogue $NGP200$. b) Configuration of cylinders that approximate the filamentary network.

The parameters defining the objects, as well as the parameters defining the data and the interaction energy, have a crucial influence on the quality of the final result. A possible strategy for choosing these parameters is presented in [282, 285, 286]. The role of the choice of the reference measure for the general model is also discussed in [89]. Various methods for estimating parameters of marked point processes with interactions will be presented and discussed later in this book.

A filament detection result obtained using the proposed model is shown in Figure 3.6. It can be seen that the cylinders tend to connect and that they are positioned in regions where visual inspection suggests the presence of filaments.

The following example illustrates the influence of each component, data and interaction, of the model. Finding the perfect balance between these two components is an open classical problem in applied mathematics and machine learning. It is also known as the *hyperparameter optimisation problem.*

Example 3.2 (Pattern detection problem: role of the data and interaction energies). Figure 3.7 illustrates the role of the components of the model (1.1). The top left panel (a) shows a two-dimensional simulated

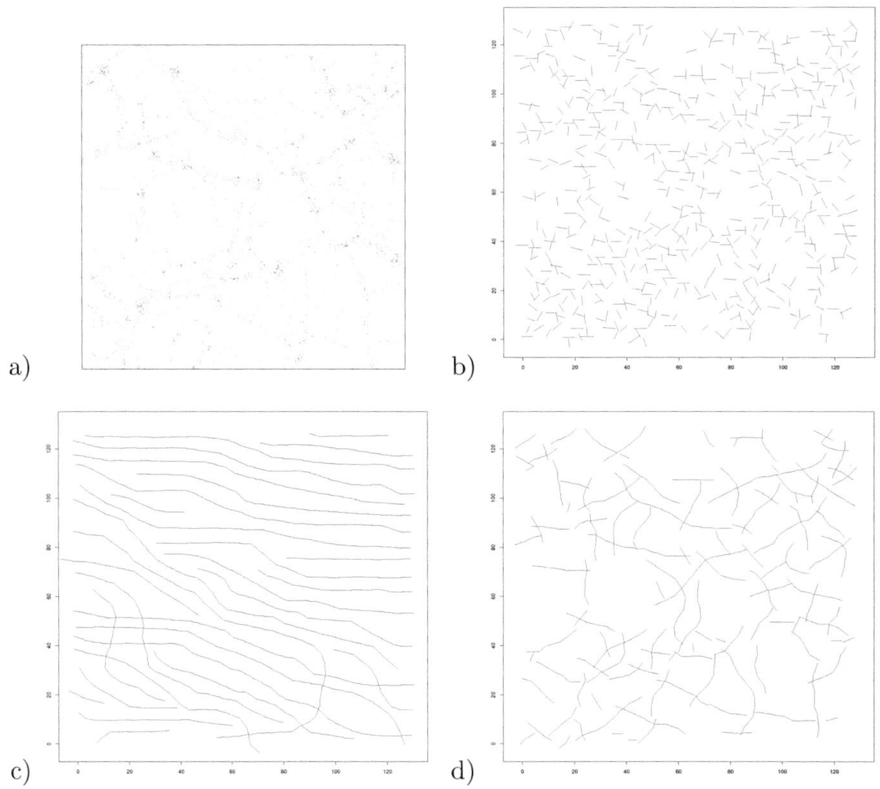

FIGURE 3.7
Influence of the data and interaction components of the Gibbs model (3.1) for filaments detection in cosmological catalogues: a) original cosmological catalogue, b) result obtained using only the data term, c) result obtained using only the interaction term, d) result obtained using the complete model.

cosmological data set exhibiting filaments. It was used as a proof of concept for the filament pattern detection method based on interacting marked point processes [294]. The top right panel (b) shows a realisation of the model build using only the data energy term $U_d(\boldsymbol{x}|\theta)$. The segments forming the pattern tend to be located in regions where the data show filamentary structures. However, their spatial distribution does not form filaments. The lower left panel (c) shows a realisation of the model built with only the interaction term $U_i(\boldsymbol{x}|\theta)$, an adaptation of the Candy model [192, 294]. The segments connect, align and form long filaments but they are independent of the data. Finally, the configuration of segments in the lower right panel (d) is a realisation of a model that includes both data and interaction energies. The segments tend to connect and to be located in the regions where the data field shows filaments.

Example 3.3 (Seismic faults interpretation using the Candy model). This example presents part of the results published in [305]. Faults extracted from seismic data can appear as linear patterns in two-dimensional images. Figure 3.8a illustrates this point. The original data is a two-dimensional image representing a slice in a 3D seismic cube. A likelihood map is computed from this data, as shown in Figure 3.8b. The likelihood is constructed using convolutional neural networks [319]. To each pixel is assigned a probability of belonging to the faults network. Figure 3.8c shows the skeleton of the likelihood. The skeleton is constructed by a thinning procedure. The distance map in Figure 3.8d represents the minimum distance from the centre of a given pixel to the fault skeleton. This image of the distance map can be interpreted as the "territorial influence" of the faults network.

Geologists are not only interested in extracting the faults network. They also want to be able to give a morphological and statistical characterisation of the detected faults. At the same time, it is very important to be able to simulate different possible faults networks that match the observed data. For these purposes, the Candy model gives the possibility to approximate the two-dimensional visualisation of the faults network by a configuration of random segments that interact. The authors in [305] adapt the data term and the interaction term of the Candy model in order to interpret the two-dimensional faults likelihood computed from seismic data.

Figure 3.9 shows the interpretation results from the data described above. Two sets of parameters were used. For each parameter set, 100 simulations of the Candy model were run. For these simulations, the empty space functions were estimated and the corresponding mean values and envelopes were calculated. The first row in Figure 3.9 shows the last simulations for each parameter set. It can be seen that, depending on the choice of parameters, the corresponding configuration of segments almost completely covers the regions with a likelihood greater than 0.75. The distance maps computed for these configurations can be compared visually with the distance map produced by the likelihood

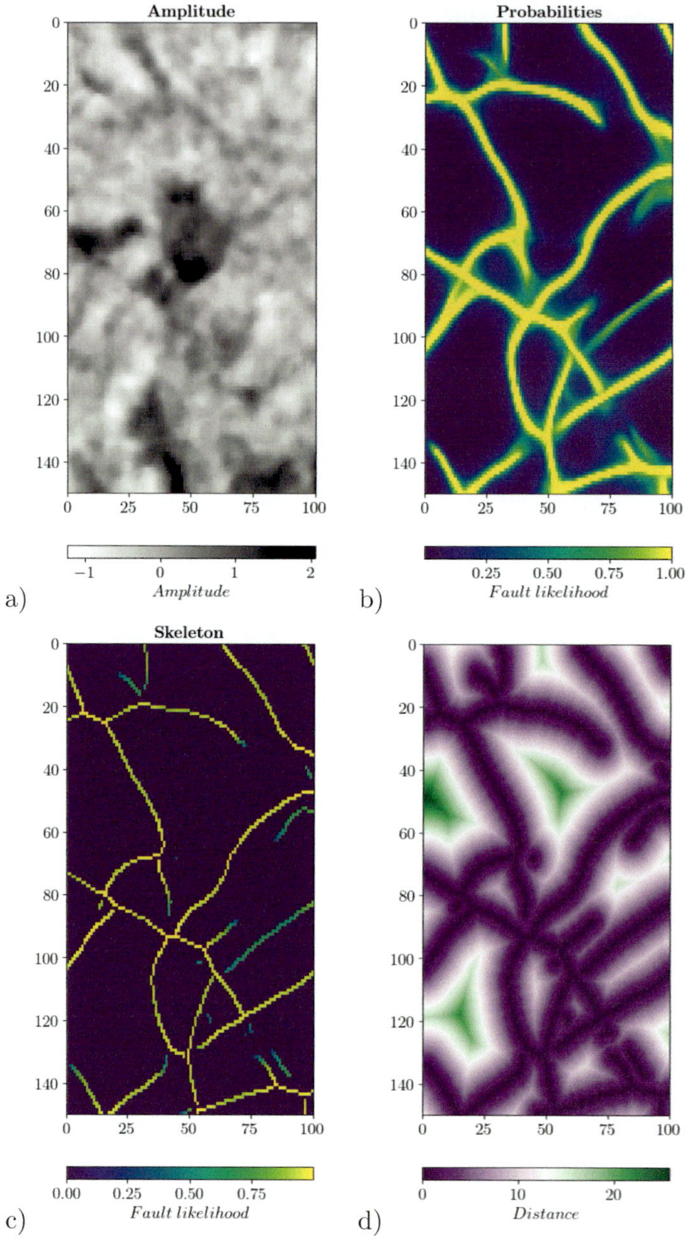

FIGURE 3.8
a) Seismic faults extraction: original data, b) likelihood computed using convolutional neural networks, c) faults skeleton or thinned likelihood of the faults, d) distance map of the faults skeleton.

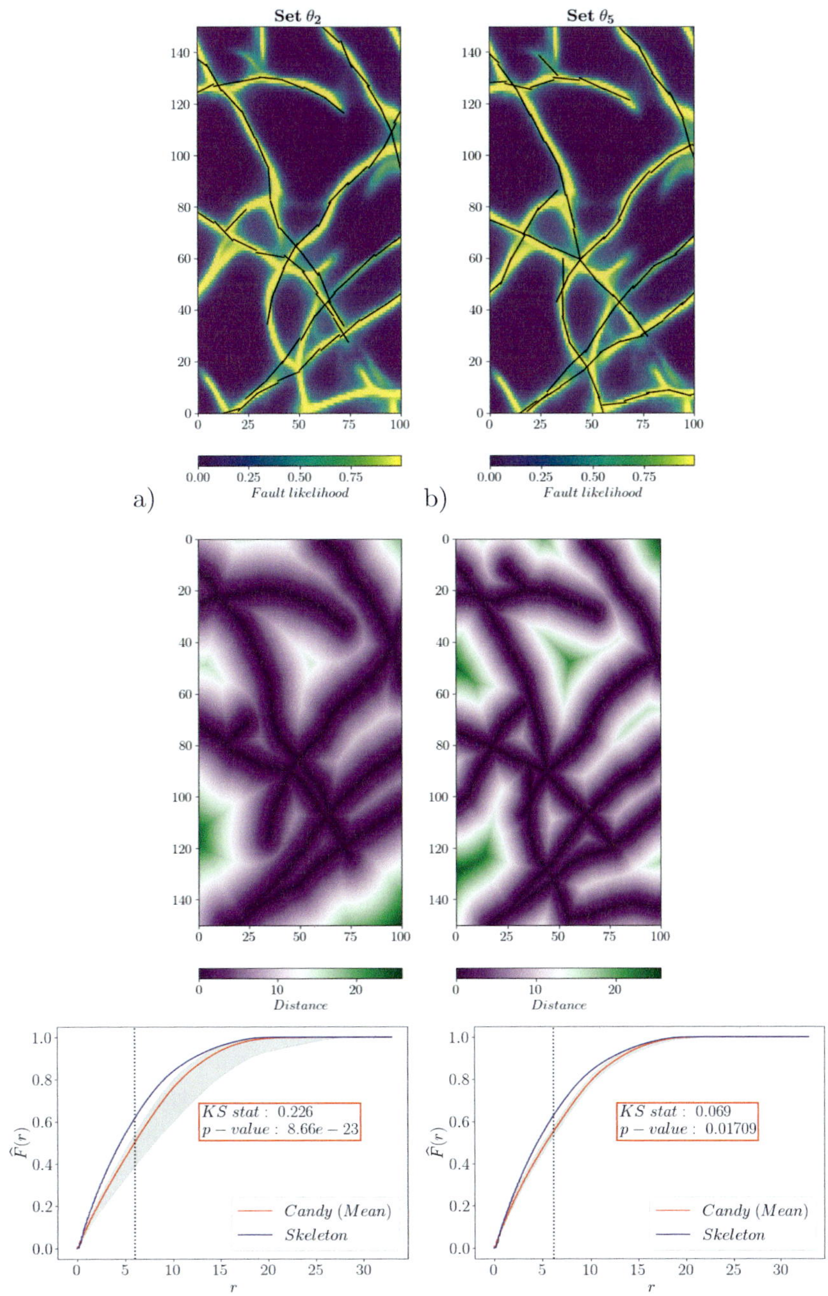

FIGURE 3.9

Results of the interpretation of detected seismic faults using the Candy model. Each column (a and b) presents for a given set of interaction parameters of the Candy model: a configuration of the sampled model, the distance map associated to the configuration and the mean empty space function estimated for each model considered.

skeleton. The networks induced by both models tend to produce more empty space than the likelihood skeleton. Envelope tests confirm this tendency. The empty space function of the skeleton lies outside and above the envelopes produced by the Candy model. The difference in the "territorial influence" of these patterns is significant, as indicated by a two-sided Kolmogorov-Smirnov test. The implemented test compares the mean of the empty space functions of the considered Candy models with the skeleton empty space function. These differences are neither unexpected nor undesirable. Obviously, different parameter choices produce different network scenarios with different statistical properties, while controlling the matching of the obtained patterns with the observed data.

3.2 Clusters in animal epidemiology

Apart from networks, other morphological structures often encountered in applications are aggregates or clusters. The previous paragraph mentioned clusters formed by the position of the centres of galaxies.

The nature of the data, however, imposes a different definition of the pattern that one is trying to detect. In seismology or epidemiology, we also speak of clusters. In these fields, several elements of a dataset can have the same position, because in many cases it can be assumed that earthquakes or epidemics can occur several times in the same place. In addition, compared to the data structure of the previous problem, the position of an element can be associated with certain characteristics, such as the magnitude of an earthquake or the propensity of a disease [174, 173].

Figure 3.10 shows a set of points $\boldsymbol{d}$ observed in a rectangular window W. The set consists of purple and white points that form a structure of three clusters. Each cluster induces a subregion of W. In the regions bounded by a continuous line, the purple points dominate. This situation is reversed in the region bounded by a dashed line. This example illustrates the following assumptions:

- *cluster*: a set of nearby points with "similar" statistical properties
- *cluster pattern*: a set of clusters.

Detecting clusters therefore means either finding the set of points in $\boldsymbol{d}$ that form the cluster pattern, or determining the geometric properties of the regions induced by the clusters (positions, contours, perimeters, areas, etc.).

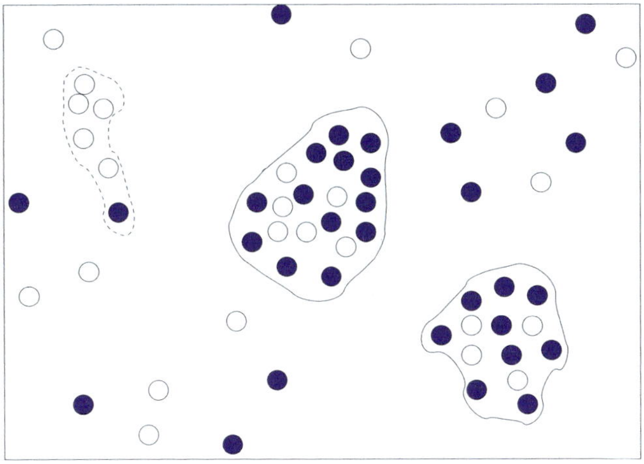

FIGURE 3.10
The observed data point field has three clusters.

3.2.1 Data presentation and problem formulation

The dataset analysed here is a point field with the following properties. Each data point represents the location of a farm in France, where Holstein dairy cows are raised. The position of a farm is given by the centre of the commune to which the farm belongs. To each farm is attached a positive numerical value, which is an annual indicator of sub-clinical mastitis for the dairy herds bred on the farms. The higher the value, the more "sick" the farm is. Typically, one year's data contains approximately 30000 registered farms [291].

Sub-clinical mastitis is a disease that is transmitted from one animal to another within a farm, but not from one farm to another. The "uniform" spread of this disease throughout the observed territory is considered a normal phenomenon by specialists [91]. For the farms that are part of a detected cluster, this may indicate the presence of poor practices and a lack of hygiene, which exhibit a rather regional character.

The question that arises is to what extent the observed point field exhibits a cluster pattern.

In order to address this question, the strategy adopted was to model the regions where points form clusters [291]. In this context, the region created by a cluster is assumed to possess the following properties:

· *locally*, the data points situated in such a region have the "same" statistical properties

· *geometrically*, the region induced by the cluster presence can be approximated by a collection of random discs that overlap.

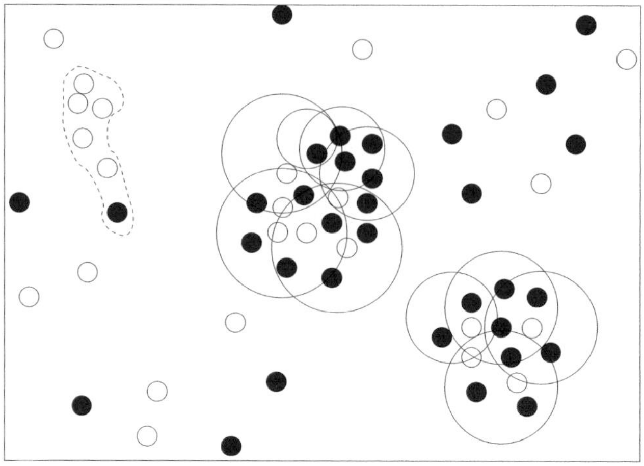

FIGURE 3.11
A configuration of randomly distributed discs that mimic the cluster pattern. The interior of each disc contains a greater proportion of purple points than white ones. The general shape of each connected component is an approximation of its corresponding original cluster shape (see Figure 3.10).

Figure 3.11 provides a visual representation of these ideas. The region induced by the cluster pattern is approximated by a configuration of randomly positioned discs that superpose. The discs are situated only in those regions of W where the number of purple dots is predominant.

The key hypothesis is to assume that the pattern of discs that approximates the cluster pattern is the realisation of a marked Gibbs point process. The realisation of such a process is denoted by $\boldsymbol{x} = \{(w_i, mi), i = 1, \ldots, n\}$, where $w_i \in W$ represents the position of the centre of disc and $m_i \in [r_{\min}, r_{\max}]$ the associated radius which serve as mark. The previously presented modelling framework is equally applicable in this case. The proposed model is a Gibbs point process whose probability density incorporates both a data term and an interaction term. The following section presents the two components of the model.

3.2.2 Modelling the data term

The data energy term, $U_{\boldsymbol{d}}(\boldsymbol{x}|\theta)$, specifies the positions of the discs that form the cluster pattern. In the context of sub-clinical mastitis, the discs are to be placed in those regions where there are a large number of farms with high disease scores.

The number of farms $n_{\boldsymbol{d}}(x)$ covered by the disc $x = (w, m)$ is considered to be of importance if $n_{\boldsymbol{d}}(x) > n_0$, where n_0 is a given value. In order to determine whether the farms covered by a disc have a relatively high disease score, a statistical test is employed. In contrast to a threshold procedure, a statistical test is more effective in accounting for disease behaviour and the heterogeneity of the data. The test compares the mean of the disease scores of the farms covered by a disk to a threshold value, d_0. Assuming that the disease scores follow a Gaussian distribution, let $W_{\boldsymbol{d}}(x, d_0, \alpha)$ be the rejection region obtained using the Student test at level α. For a specific data set, the d_0 value retained was the sum of the empirical mean and the standard deviation.

Within these hypotheses, the potential of a disc is written

$$u(x) = \mathbf{1}\{n_{\boldsymbol{d}}(x) > n_0\}\mathbf{1}\{\bar{\boldsymbol{d}}(x) \notin W_{\boldsymbol{d}}(x, d_0, \alpha)\}[\bar{\boldsymbol{d}}(x) - d_0 + u_{\max}] - u_{\max} \quad (3.9)$$

where $\bar{\boldsymbol{d}}(x)$ is the average of the disease scores of the farms covered by the disc x and $u_{\max}$ is pre-fixed value. The definition of the potential function (3.9), implies that the probability density of the model will give high values to configurations with discs located in regions where a large number of farms have a high disease score. As for the cylinders forming a filament network, the role of the term $u_{\max}$ is to penalise discs that do not fulfil all the criteria, without however prohibiting them completely. This modelling approach represents a robust prevention strategy, as any region encompassed by a disc is susceptible to being designated as an affected area by the disease.

The sum of the potentials of all the discs in a configuration defines the data energy term of the model:

$$U_{\boldsymbol{d}}(\boldsymbol{x}|\theta) = -\sum_{i=1}^{n(\boldsymbol{x})} u(x). \quad (3.10)$$

The exponential of the energy (3.10) yields the first component of the cluster pattern model. This component is an inhomogeneous Poisson point process (3.2). This process is locally stable.

3.2.3 Modelling objects interaction: mixture of the Strauss and the area-interaction processes

Building a model using only the data term can have undesirable effects. First, maximising the probability density of the Poisson process given by (3.10) will place all the discs in the region of maximum data energy, resulting in the detection of a single cluster. Furthermore, this maximisation favours the appearance of configurations with an increasing average number of discs. This effect is cancelled by introducing a repulsive interaction between the discs. Secondly, the data energy term assigns the same probability to discs of different sizes. This phenomenon is illustrated in Figure 3.12. From a mathematical

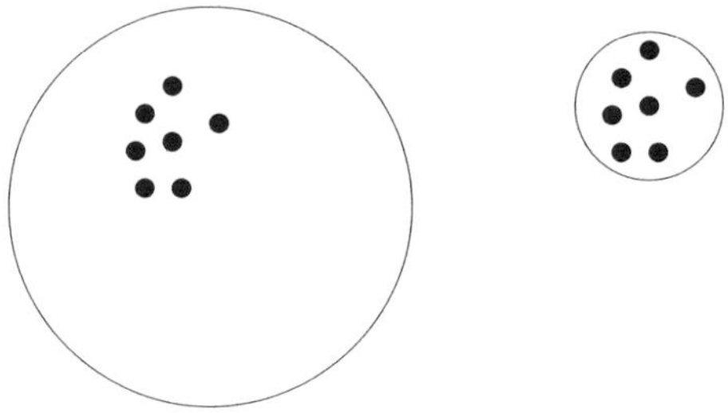

FIGURE 3.12
Two discs with different radii around the same configuration of points.

point of view, it leads to smoothing of the final result. From an application point of view, it unduly enlarges the regions affected by the disease. To remove this effect, the interaction term must aggregate the discs, taking into account the area occupied by the cluster pattern.

In this context, the interaction energy of a configuration of discs is written as

$$U_i(\boldsymbol{x}|\theta) = -n_o(\boldsymbol{x}) \log \gamma_o + \nu[Z(\boldsymbol{x})] \log \gamma_a. \tag{3.11}$$

The first component of (3.11) is a pairwise interaction model known in the literature as the *Strauss process* [149, 229, 181, 54, 302]. Two discs $x_i = (w_i, m_i)$ and $x_j = (w_j, m_j)$ have a *repulsion interaction or overlap* if $d(w_i, w_j) \leq m_i + m_j$. In this case we write $x_i \sim_o x_j$. The term $log\gamma_o$ is the potential associated with such a pair of discs. The statistic $n_o(\boldsymbol{x})$ is the number of pairs of overlapping discs in the configuration $\boldsymbol{x}$. Figure 3.13 shows three realisations of the Strauss model in a square window $W = [0, 1.5]^2$, with the parameter γ_o varied. The discs around the points have a fixed radius of $r = 0.025$. Thus, two intersecting circles indicate a repulsive interaction, since the distance between their respective centres is less than 0.05. For $\gamma_o = 1$ the Strauss component in (3.11) is a Poisson point process, so the discs overlap to form a completely random pattern. When $\gamma_o = 0.5$, the configurations containing many repulsive discs receive a lower probability, so they are penalised. It can be observed that the corresponding statistic $n_o(\boldsymbol{x})$ takes lower values than for the Poisson case. Finally, if $\gamma_o \to 0$, the configurations with overlapping discs get a probability close to zero. In the limit case we are dealing with a hardcore interaction process. Note that for the configuration shown in Figure 3.13c $n_o(\boldsymbol{x}) = 0$. This type of repulsive interaction has already been used for the construction of the Bisous model.

The second component of (3.11) is the *area-interaction process* proposed by [21, 181]. Here $Z(\boldsymbol{x})$ is the set union of the discs of a configuration. Consequently, $\nu[Z(\boldsymbol{x})]$ represents the area of this union. The parameter $\log \gamma_a$ controls the morphology of $Z(\boldsymbol{x})$. If $\log \gamma_a < 0$, then the configurations occupying more space get a higher probability. This also induces a more regular distribution of points. If $\log \gamma_a > 0$, then the discs in a configuration tend

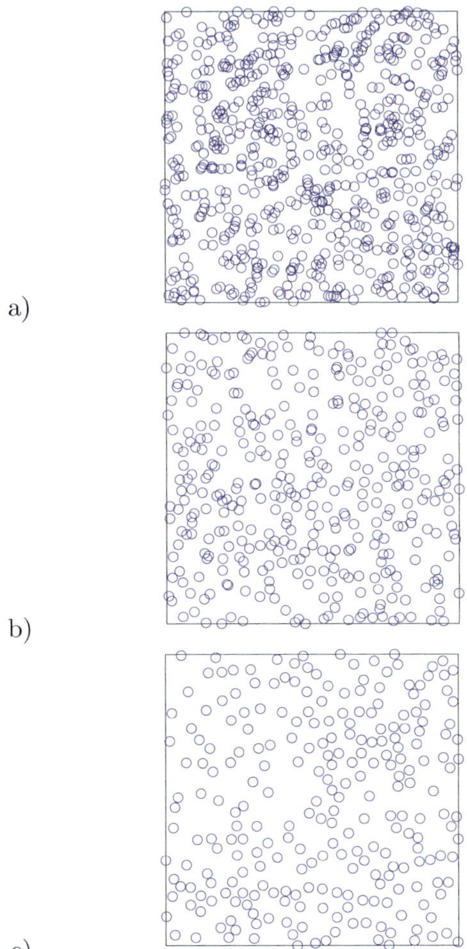

FIGURE 3.13
Realisations of Strauss processes. The γ_o parameter is as follows: a) $\gamma_o = 1$ (Poisson), b) $\gamma_o = 0.5$ (soft repulsion) and c) $\gamma_o \to 0$ (hard-core repulsion).

to reduce the occupied area. This induces a clustering of discs. As before, if $\log \gamma_a = 0$, the model is a Poisson process. The area occupied is produced by a completely random pattern of discs. Figure 3.14 shows realisations of the area-interaction model depending on the parameter γ_a. The process was simulated in the square window $W = [0, 1.5]^2$, the discs in a configuration all have the same radius $r = 0.05$.

It is interesting to note that Strauss proposed his model to generate aggregate structures [302]. Kelly and Ripley determined for which parameters the model is integrable and proved that if the number of points is not fixed, Strauss'

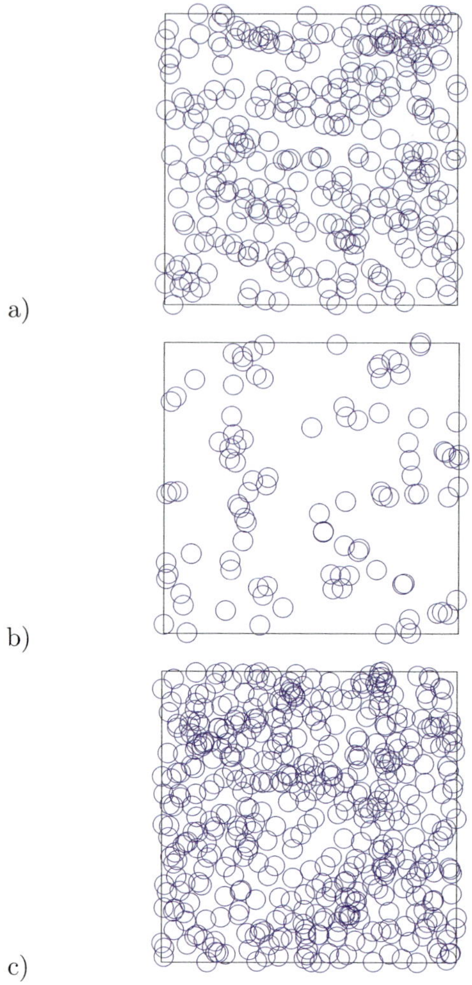

FIGURE 3.14
Realisations of the area-interaction process. The $\log \gamma_a$ parameter is as follows: a) $\log \gamma_a = 0$ (Poisson – completely random area occupation), b) $\log \gamma_a = 2$ (clustered area occupation) and c) $\log \gamma_a = -2$ (regular area occupation).

model can only simulate repulsive structures [149]. This drawback was one of the motivations for Baddeley and van Lieshout to develop the area- interaction model [21]. Four our practical application, the proposed model synthesises both approaches. The $Z(\boldsymbol{x})$ related statistics describe the geometry of the cluster pattern, while the $n_o(\boldsymbol{x})$ can be interpreted as its strength or intensity.

These two components are both locally stable and are Markov with respect to their interactions $\sim_o$ and $\sim_a$ respectively [21, 149, 229, 181, 302]. As a result, the model (3.11) is a well-defined Markov point process. Since the

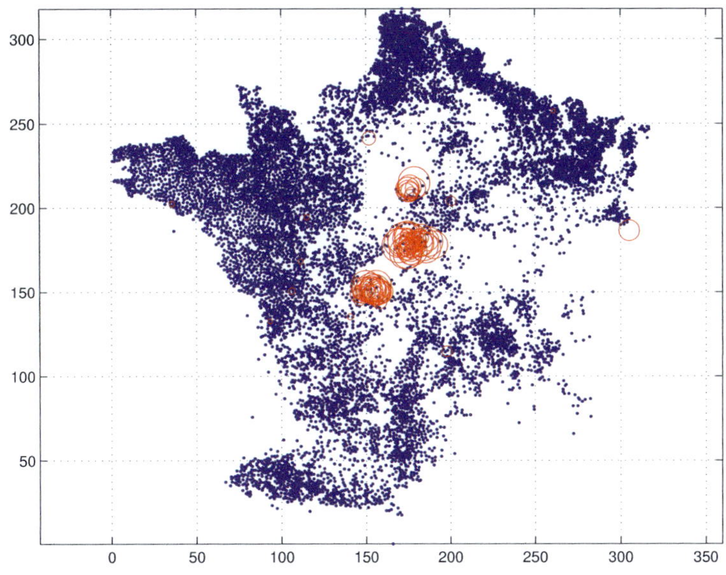

FIGURE 3.15
Cluster pattern made of interacting random discs (red) within a data field (purple).

data term (3.10) is locally stable, it follows that the complete model given by the (1.1), which includes both data and interaction terms, is also a well-defined Markov point process.

Figure 3.15 shows a realisation of the complete model. The data field is represented by purple dots. Only the position of the farms is visible, but the disease scores associated with them are well taken into account in the data term. The formation of large strong clusters in the centre of the French territory can be observed. Some isolated weak clusters also appear in other regions.

3.3 Habitat fragmentation

Pairwise interaction point processes can be regarded as limits of random fields defined on a grid [32, 35]. This enables the mathematical framework to be adapted to an existing practical situation, and to switch from continuous to discrete and vice versa at the appropriate time. For example, the Widom-Rowlinson or the multi-variate pairwise interaction processes can be regarded as the continuous counterparts of discrete random fields analogous to the Ising or to the Potts models [229, 186, 318].

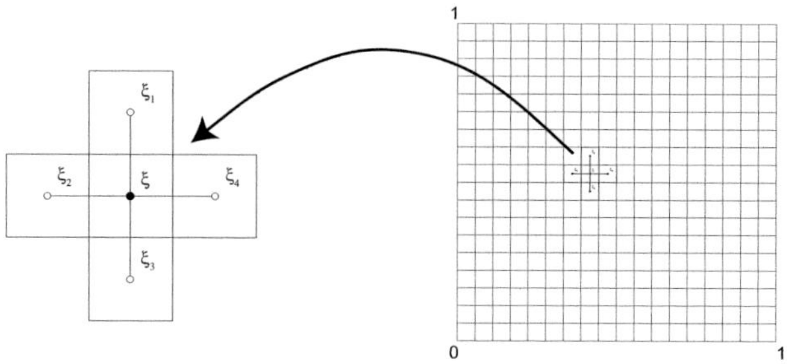

FIGURE 3.16
The 4- neighbourhood system: an element ξ of Ξ and its neighbours ξ_i, $i = 1, \ldots, 4$.

This section presents a conditioned Ising-like model, which is controlled by a positioning parameter. This model was proposed by [266].. The patterns obtained with this model have been employed to investigate the impact of landscape fragmentation on population dynamics in periodic environments. The integration of simulated Ising-like landscape patterns into a population dynamics model demonstrated that the less fragmented the landscape, the more likely the species under study was to survive.

Let $W = [0,1] \times [0,1]$ be a set representing a cell in a periodic environment. Furthermore, let $W = \bigcup_{i=1}^{n_W} W_i$ be a finite partition of W. It is assumed that the cells W_i have the same volume. The grid Ξ is constructed from the centres x_i of the cells W_i, together with a neighbourhood system V_ξ. Let μ^+ and μ^- be two real numbers satisfying the inequality $\mu^+ > \mu^-$. Let us now consider $\boldsymbol{x}$ the realisation of a random field defined on the grid Ξ. The random field assigns a value from the set $\{\mu^+, \mu^-\}$ to each site $\boldsymbol{x}_\xi$. The number n_+ of sites with value μ^+ is a fixed quantity. In this context, let us define the space of all possible states as Ω.

The random field $\boldsymbol{x}$ represents a landscape in which the sites where $\mu(\boldsymbol{x}) = \mu^+$ are interpreted as "habitat"; the others for which $\mu(\boldsymbol{x}) = \mu^-$ correspond to "non-habitat". The landscape pattern is defined by the random field $\boldsymbol{x}$ and the associated function μ. Figure 3.16 presents the four-neighbourhood system considered in this study. In order to account for periodic environments, the domain is wrapped on a torus.

For each $\boldsymbol{x} \in \Omega$, the statistic

$$s(\boldsymbol{x}) = \frac{1}{2} \sum_{\eta, \xi \in \Xi, \xi \sim \eta} \mathbf{1}\{\boldsymbol{x}_\xi = \boldsymbol{x}_\eta\},$$

represents the number of pairs of neighbours $\xi \sim \eta$ such that $\boldsymbol{x}_\xi = \boldsymbol{x}_\eta$. The fragmentation of the habitat is directly linked to $s(\boldsymbol{x})$: a landscape is either aggregated or fragmented depending on whether $s(\boldsymbol{x})$ is large or small.

First landscape model

Let the model be described by the Gibbs measure $\mathbb{P}$ defined on Ω

$$P(X=\boldsymbol{x})=\frac{1}{Z}\exp\left[-U(\boldsymbol{x})\right], \tag{3.12}$$

where $Z=\sum_{\boldsymbol{x}\in\Omega}\exp\left[-U(\boldsymbol{x})\right]$ is the normalisation constant or the partition function, and $U(\boldsymbol{x})$ is the energy function given by

$$U(\boldsymbol{x})=-\frac{\beta}{2}\sum_{\eta,\xi\in\Xi,\ \xi\sim\eta}\mathbf{1}\{\boldsymbol{x}_\xi=\boldsymbol{x}_\eta\}=-\beta s(\boldsymbol{x})$$

with $\beta>0$.

The probability distribution describing the model (3.12) is written as

$$P(X=\boldsymbol{x})=\frac{1}{Z}\exp[\beta s(\boldsymbol{x})]. \tag{3.13}$$

The larger β is, the more the model favours the same value between neighbouring sites. So the β parameter controls the morphology of the landscape. Those familiar with this type of modelling will notice that the model (3.13) is an Ising model conditioned by the number of "spin" cells habitat and non-habitat.

Realisations of the model (3.13) are shown in Figure 3.17. The total of cells is $n_W=50^2$ and the habitat abundance has been set to 20%, which means that $n_+=0.2n_W=500$. As expected, we observe that the parameter β controls the state of aggregation of the landscape.

Figure 3.18 shows two landscapes with two identical components. Under the assumptions of the model, both landscapes have the same probability, whereas the components in Figure 3.18a are closer than those in Figure 3.18b. From a more realistic point of view, it is desirable that a model could differentiate between these two situations [273].

Second landscape model

In light of the preceding remarks, a second model is considered. Let κ be a random point chosen uniformly in W. The model is defined, conditionally on the point κ, by the Gibbs measure

$$P(X=\boldsymbol{x}|\kappa)=\frac{1}{Z_\kappa}\exp\left[-U_\kappa(\boldsymbol{x})\right]. \tag{3.14}$$

The energy function is

$$U_\kappa(\boldsymbol{x})=-\beta s(\boldsymbol{x})+\frac{1}{\sigma^2}\tau_\kappa(\omega),$$

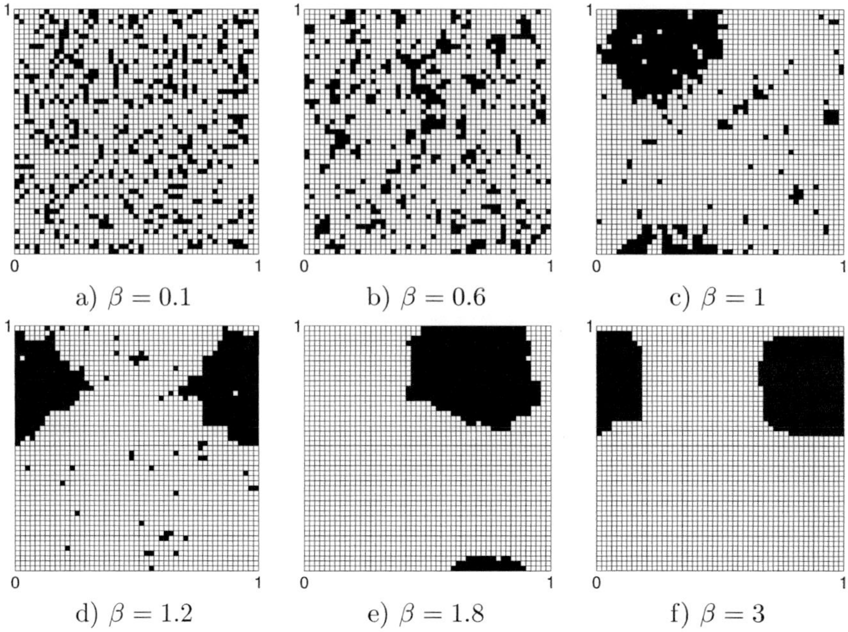

FIGURE 3.17
Patterns obtained via realisations of the landscape model (3.13) with different values for the parameter β.

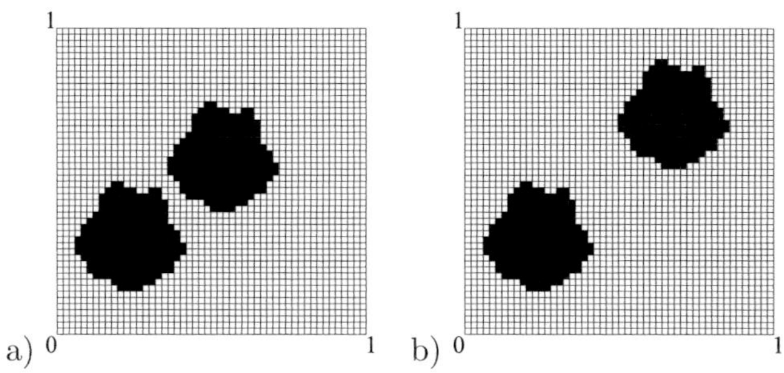

FIGURE 3.18
Two landscape patterns made of two identical components. In a) the components are closer than in b). The model gives the same energy to both configurations.

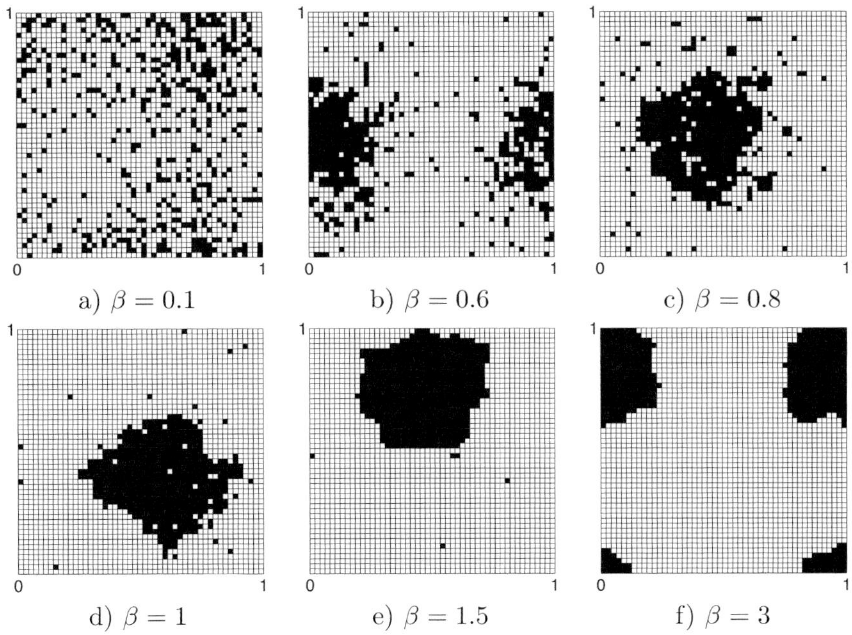

FIGURE 3.19
Patterns obtained via realisations of the conditioned landscape model (3.14) with $\sigma = 0.4$ and different values for the parameter β.

with $\sigma > 0$ and

$$\tau_\kappa(\boldsymbol{x}) = \sum_{\eta \in \Xi,\ \boldsymbol{x}_\eta = \mu^+} d(\eta, \kappa)^2.$$

where $d(\eta, \kappa)$ is the Euclidean distance on W with toroidal symmetry.

The term $\frac{1}{\sigma^2}\tau_\kappa(\boldsymbol{x})$ acts as an external field. This field has the form of a truncated Gaussian centred at κ and of variance σ^2. This choice is justified by the fact that we know that, at least in the case where the difference $|\mu^+ - \mu^-|$ is not too small and the habitat is 20%, the optimal configuration of a landscape tends to become circular [265]. Here, the optimal landscape configuration is that which gives the best chance of survival to the species studied.

Some realisations of the model (3.14) are shown in Figure 3.19. The same values have been employed for n_W and n_+ as were used previously. Furthermore, the value of σ was set to 0.4. For high values of the parameter β, the landscape tends to be structured around a centre.

3.4 Collaborative social networks

From a mathematical perspective, collaborative social networks can be considered complex entities made of simple objects that interact in a manner analogous to the previous examples. Graphs are the usual mathematical tool for representing networks. These graphs form structures and patterns that should be described from both a morphological and statistical perspective.

The work of [166, 167] considered social networks induced by the activity of scientists working for LORIA[5] a well-known French laboratory in computer sciences. Figure 3.20 shows examples of the studied networks. The primary network represents the co-authorship graph for the year 2018. Each node in the graph represents a researcher. An edge is created whenever two researchers co-signed scientific publications during the considered year. The colour attached to each node indicates the affiliation of the researcher to a laboratory. For instance, all LORIA members are coloured in yellow, while the members of the other laboratories have a different colour. It can be observed that the graph pattern evolves over years. The question that arises is how to describe and quantify this evolution.

The key idea of the authors in [166, 167] was to study these networks by using their *dual or line graph*, that is the graph obtained from the initial network graph by transforming nodes into edges and edges into nodes [99]. In the example presented in Figure 3.21a the nodes are coloured with respect to the affiliation: nodes A and B belong to the organisation 1 (yellow), while nodes C and D belong to the organisations 2 (red) and 3 (blue), respectively. The corresponding line graph is shown in Figure 3.21b. The colour of the nodes represent the type of collaboration: intra-organisational (green) and inter-organisational (orange). The line graph can be completed by taking into account the nil edges as shown in Figure 3.21c.

In the following, the observation of an augmented line graph, as previously presented, is modelled as the realisation of a Markov Random Field (MRF) on the complete graph $\mathcal{G} = (\mathcal{W}, \mathcal{E})$, where $\mathcal{W}$ is a finite set of vertices and $\mathcal{E}$ the associated edges set. A MRF on $\mathcal{G}$ is a stochastic process X that controls the labels of each node in $\mathcal{W}$. In this case, the set of labels is given by $\mathcal{M} = \{0, 1, 2\}$, and thus the configuration space for X is Ω, which is the set of all possible labelings $\mathcal{M}^{\mathcal{W}}$. Readers interested in a detailed mathematical description and treatment of MRFs may wish to consult [121, 176, 318].

A node $w \in \mathcal{W}$ represents the potential collaboration of two researchers during a specific year. If the label associated with this node, $m(w)$, is equal to

[5]In French these initials stand for: Laboratoire Lorrain de Recherche en Informatique et ses Applications.

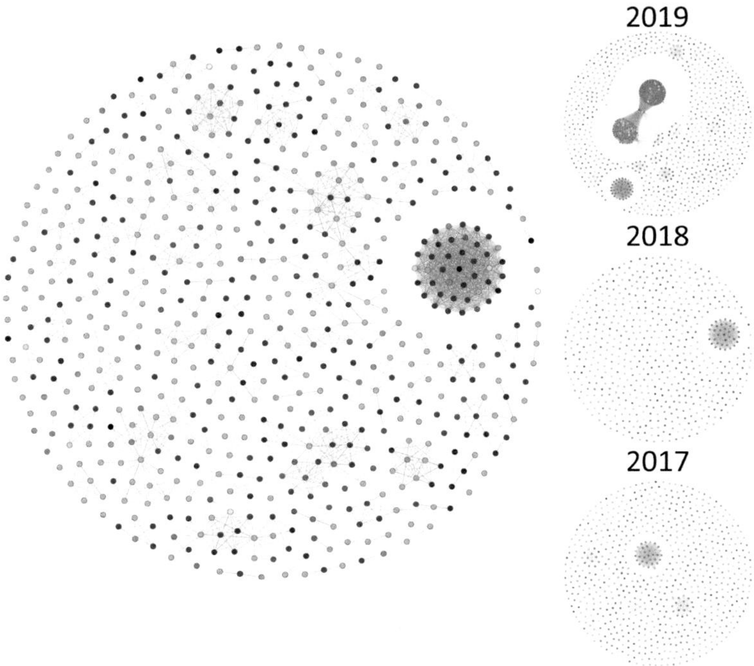

FIGURE 3.20
Collaborations among researchers within the LORIA laboratory during the period 2017-2019.

0, this indicates that the two researchers did not collaborate during the specified period. Conversely, if $m(w) \neq 0$, this indicates that the two researchers co-signed publications during the same period. If the label is 1, this signifies that the two researchers are affiliated with the same organisational entity. Furthermore, if the label is 2 this indicates that the two researchers are employed by different organisations.

The general Gibbsian framework allows to write the probability distribution of X such as

$$\mathbb{P}(X = \boldsymbol{x}|\theta) = \frac{\exp\left[U(\boldsymbol{x}|\theta)\right]}{Z(\theta)}$$

with $\boldsymbol{x} = \{m(w_i) : w_i \in \mathcal{W}\}$, θ the model parameters, $U(\cdot|\cdot)$ the Gibbs energy function and $Z(\theta)$ the normalising constant. The model proposed by [167] is given by the energy function:

$$\begin{aligned} U(\boldsymbol{x}|\theta) &= \theta_{11} \sum_{w_i \sim w_j} \mathbf{1}\{m(w_i) = m(w_j) = 1\} \\ &+ \theta_{22} \sum_{w_i \sim w_j} \mathbf{1}\{m(w_i) = m(w_j) = 2\} + \theta_{12} \sum_{w_i \sim w_j} \mathbf{1}\{m(w_i) + m(w_j) = 3\}. \end{aligned} \quad (3.15)$$

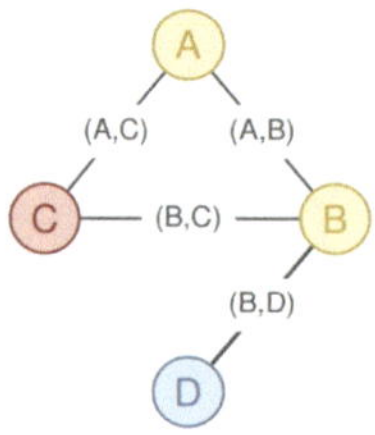

(a) Collaboration graph

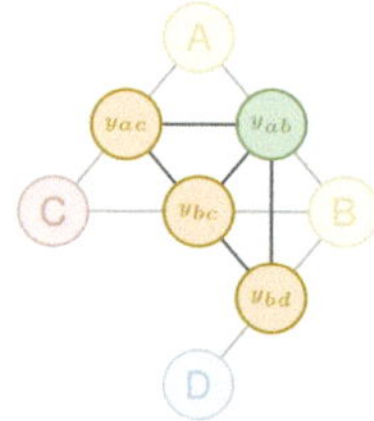

(b) The corresponding line graph

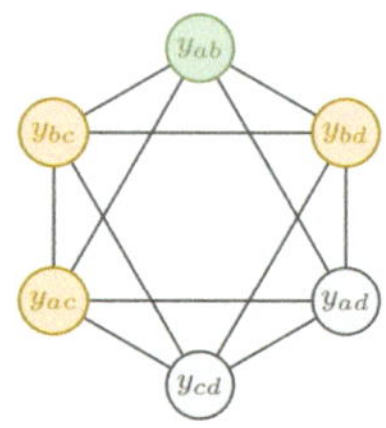

(c) Augmented line graph representation *taking nil edges into account*

FIGURE 3.21
An example of collaboration graph and its line graph representation.

where $\sim$ is the neighbourhood relation induced by the presence of an edge between two nodes.

The condition $\mathbf{1}\{m(w_i) = m(w_j) = 1\}$ is verified if a researcher works with two persons from the same team. Similarly, the condition $\mathbf{1}\{m(w_i) = m(w_j) = 2\}$ is valid if a researcher cooperates with two persons belonging two both to other research teams. Finally, the condition $\mathbf{1}\{m(w_i)+m(w_j) = 3\}$ is verified if a researcher works with a person from his own team and another from a different team.

The model acts similarly to the Potts model [121, 176, 318, 167]. Depending on the values of the parameters θ, if important patches of $(1,1)$ may appear this indicates a tendency towards inter-organisational cooperation. Similarly, the presence of patches of $(2,2)$ underlines the tendency towards intra-organisational cooperation, while the appearance of patches of $(1,2)$ shows a mixed cooperational behaviour.

3.5 Landscape modelling using T–tessellations

The development of genetics has encouraged studies of pollen propagation in different landscape structures. To this end, we are interested in modelling agricultural parcel landscapes. Here we propose a model based on T–tessellations. The different characteristics of a landscape are described by the geometric properties of the tessellations. An observed T–tesselation can be considered as a realisation of a random set of interacting segments and cells. This allows to describe the properties of such a tesselation in a probabilistic framework, similar to the patterns described by marked point processes. However, this also raises some very interesting mathematical problems, since T–tessellations are quite constrained geometric objects. The main ideas and results of our work are presented below [159, 1]. To make this section easier to read, the proofs of the results presented are omitted. These proofs, together with the necessary technical details, can be found in [159, 1].

3.5.1 Space of T–tessellations

Only tessellations defined on a compact domain $W \subset \mathbb{R}^2$ are considered here. This domain is assumed to be convex and polygonal. Let $a(W)$, $l(W)$, $n_e(W)$ and $n_v(W)$ be the area, perimeter, number of edges and vertices of W, respectively.

A *polygonal tessellation* of W is a finite partition of W into polygons. Such a polygon is called a *cell*. The vertices of the tessellation are the vertices of the cells. Its *edges* are pieces of cell boundaries bounded by adjacent vertices. The set of edges in a tessellation can be seen as a set of segments, some of which are unions of aligned and connected edges. A vertex of the tessellation is a *T–vertex* if it is the intersection of exactly three edges, two of which are aligned. The tessellation of W is called *T–tessellation* if all its vertices, with the exception of those of W, are T–vertices, and there is no pair of segments lying on the same line. Figure 3.22 shows an example of T–tesselation.

The space of T–tesselations on the domain W is denoted by $\mathcal{T}$. For any T–tesselation $T \in \mathcal{T}$, let $C(T)$ be the set of its cells, $V(T)$ the set of its vertices, $E(T)$ the set of its edges and $S(T)$ the set of its segments. The cardinals of these sets are denoted by $n_c(T)$, $n_v(T)$, $n_e(T)$ and $n_s(T)$ respectively. Vertices, edges and segments are said to be *internal* if they do not lie on the boundary of the domain W. Sets and measures corresponding to internal objects are denoted by the same symbol with a circle on top. For example, $\overset{\circ}{V}(T)$ and $\overset{\circ}{n}_v(T)$ denote the set and number of internal vertices.

A tessellation induces a configuration of lines $L(T)$, which is the set of lines supporting the segments. Conversely, a finite configuration of lines touching W gives rise to a set of tessellations $\mathcal{T}(L)$, such that $L(T) = L$ for all $T \in \mathcal{T}(L)$.

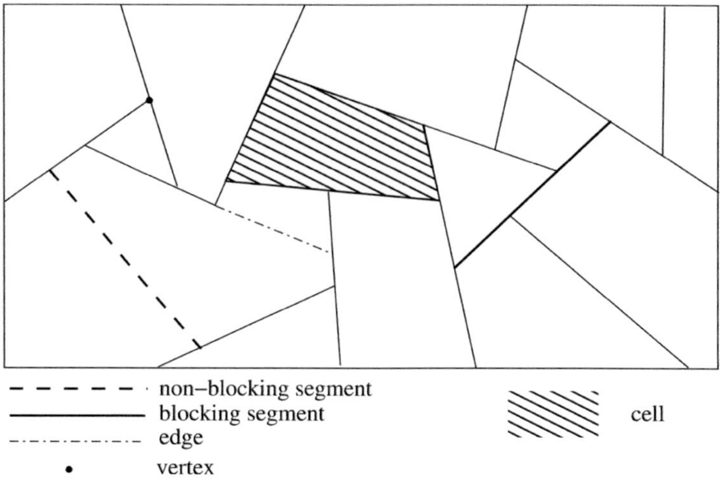

FIGURE 3.22
Example of a T–tesselation with its components.

The cardinality of this set is called $n_t(L)$. The following inequality holds:

$$n_t(L) \leq C^k \left(\frac{k}{(\log k)^{1-\epsilon}} \right)^{k-k/\log k}, \tag{3.16}$$

with $\epsilon > 0$, $k \geq 2$ and C depending on ϵ. This result was obtained by [147] and is used to prove the integrability of the proposed models for T–tesselations.

Definition 3.1. A function $\phi : \mathcal{T} \to \mathbb{R}^+$ is said to be *stable* if there exists a real positive constant K such that

$$\phi(T) \leq K^{\overset{\circ}{n}_s(T)},$$

for all $T \in \mathcal{T}$.

The previous definition is a tentative connection with the stability conditions for point processes [108, 229, 269]. However, the analogy with the local stability of point processes is not trivial. This is because the addition or deletion of a segment must preserve the structure of a T–tesselation.

In the following we show that it is possible to transform one T–tesselation into another T–tesselation, using three operations we call split, merge, and flip.

A *split* is the operation that cuts a cell in two (see Figures 3.23a and 3.23b). A split S of T is represented by $S = (c, l)$, where c is a cell and l is a line intersecting the interior of c. Dividing T by S adds the edge $c \cap l$ to T. The

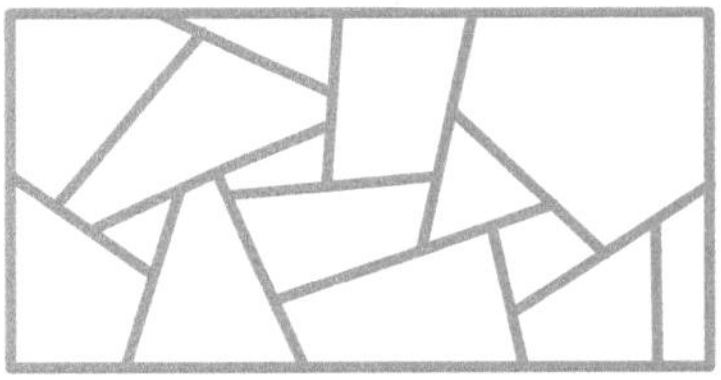

a) A T–tesselation

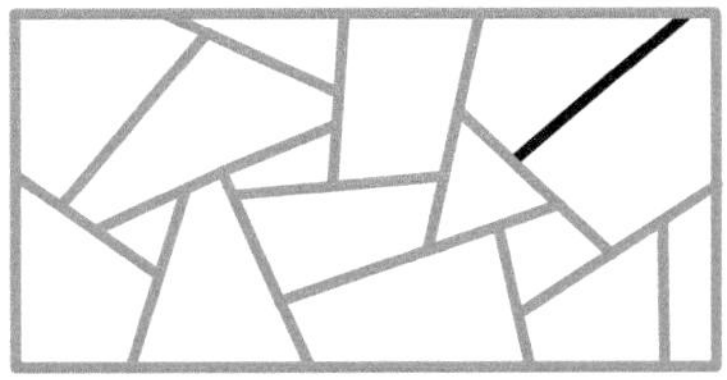

b) Split: the new edge (black) is blocked by two existing segments.

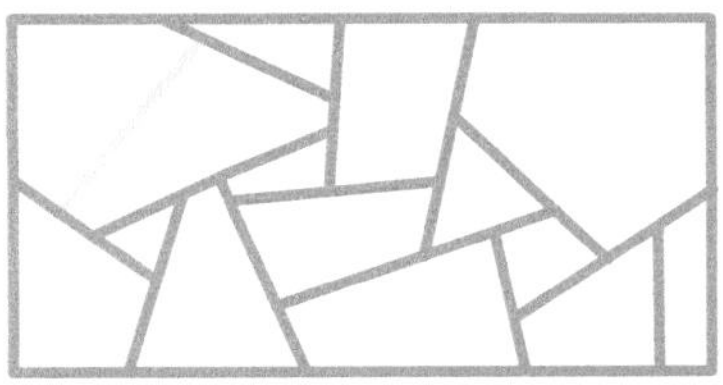

c) Merge: a non-blocking segment is removed (in grey)

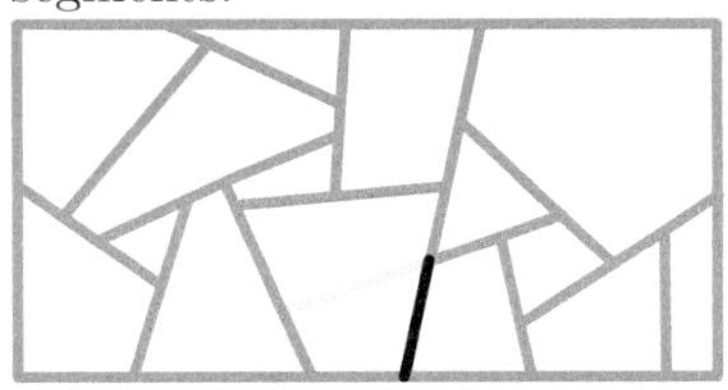

d) Flip: two segments are modified; one is shortened (suppressed edge in grey) the other is extended (added edge in black).

FIGURE 3.23
Local transformations a T–tesselation a): split b), merge c) and flip d).

new tessellation is $ST = T \cup (c \cap l)$. A *merge* is an operation that removes a segment containing a single edge (cf. Figures 3.23a and 3.23c). Segments that can be affected by a merge are called non-blocking, and the others are called blocking. A merge in a T–tesselation is identified by a non-blocking segment of T. For any T–tesselation, the number of possible merges is $\mathring{n}_{\mathrm{s,nb}}(T)$, the number of non-blocking segments in T. The tesselation after the merge is denoted by MT. A *flip* is an operation that removes a final edge from a blocking segment and then adds a new edge. This new edge extends the previously blocked segment (see Figure 3.23d). A flip is identified with the final edge of a blocking internal segment. For a tessellation there are $2\mathring{n}_{\mathrm{s,b}}(T)$ possible flips. The tessellation obtained after a flip is denoted by FT.

The effect of a split can be compensated by a merge, and vice versa. For any $S \in \mathcal{S}_T$, S^{-1} is the merge which compensates S. Any flip F can be compensated by another flip F^{-1}. The empty tesselation T_W is the T–tesselation which has the domain W as its only cell.

Proposition 3.1. *Let T be a non-empty T–tesselation. There is a finite sequence of splits, merges and flips that transforms T into an empty tessellation. Conversely, there is a finite sequence of splits and flips that transforms the empty tessellation T_W into T.*

Corollary 3.1. *Given any pair of distinct T–tesselations, there is a finite sequence of split, merge, and flip operations that transform one into the other.*

Let $\mathcal{S}_T$, $\mathcal{M}_T$ and $\mathcal{F}_T$ be the set of all splits, merges and flips that can be applied to T. We will construct probability measures on each of these spaces. The set $\mathcal{M}_T$ is a finite discrete set. Consequently, if this set is not empty, we can associate a uniform distribution with it. If it is not possible to perform a merge, the probability of a merge will be zero. For the same reasons, we proceed in a similar way for the set $\mathcal{F}_T$. For the probability measure on $\mathcal{S}_T$, we take two steps, since this set is not discrete. First, we select a cell c of the tessellation T with a weight proportional to its perimeter, then we generate a uniform line that intersects c. This construction allows us to choose a uniform split in the set $\mathcal{S}_T$. All the details of these constructions are given in [159].

3.5.2 Completely random T–tessellations

The space $\mathcal{T}$ is equipped with the $\sigma-$algebra $\sigma(\mathcal{T})$, which is generated by the events

$$\left\{ T \in \mathcal{T} : \left(\bigcup_{e \in E(T)} e \right) \cap K \neq \emptyset \right\}$$

where K is a compact subset of W [207]. A T–tesselation is seen as a closed random set.

Our construction of the completely random T–tesselation is based on the Poisson line process, denoted here as $\mathbf{L}$. This process is assumed to be of unit intensity. This means that the average length of the pattern of lines observed across a unit area is one. Here, we consider the T–tesselations of a bounded domain W. Thus, only the restriction of $\mathbf{L}$ to W is relevant to our construction. For simplicity, we denote this restriction $\mathbf{L}$. All the details are given in [159]. For the definition of the Poisson line process, we recommend [54, 300].

Let be the probability measure μ defined by

$$\mu(A) = Z^{-1} \mathbb{E} \sum_{T \in \mathcal{T}(\mathbf{L})} \mathbf{1}_A(T), \quad A \in \sigma(\mathcal{T}), \tag{3.17}$$

where $Z = \mathbb{E} \sum_{T \in \mathcal{T}(\mathbf{L})} \mathbf{1}_{\mathcal{T}}(\mathbf{L})(T)$ is the normalisation constant. For μ to be well-defined, we need to check that Z is finite. This is a direct consequence of the following result.

Theorem 3.2. *Let $\mathbf{L}$ be the restriction to the domain W of the Poisson line process of unit intensity and let $\phi : \mathcal{T} \to \mathbb{R}^+$ be a stable function. Then*

$$\mathbb{E} \sum_{T \in \mathcal{T}(\mathbf{L})} \phi(T)^\alpha$$

is finite for all $\alpha \in \mathbb{R}$.

By applying Theorem 3.2 with $\phi = 1$, we obtain that the measure μ defined by (3.17) is finite. The normalisation constant Z has no known analytic expression. Another consequence of this theorem is that for $\mathbf{T} \sim \mu$ and ϕ a stable function, all moments of $\phi(\mathbf{T})$ are finite.

To characterise the μ measure (3.17), we will use Campbell measures and Papangelou kernels. Used with point processes, these tools allow us to study the behaviour of a point with respect to the rest of the observed configuration. Extending these tools to T–tesselations is not trivial. This is because the components of T–tesselations – vertices, edges and segments – are much more constrained than a point in a configuration of points.

In a T–tesselation, the components that can be easily added or removed are the non-blocking segments. In the following, each split represented by the pair (c, l) formed by a cell c touched by a line l is identified with the segment $c \cap l$. Consequently, $\mathcal{S}_T$ is identified with the space of these segments and dS the uniform measure on $\mathcal{S}_T$ induces a measure on the space of non-blocking segments, denoted ds in the sequel. Similarly, $\mathcal{M}_T$ is identified with the number of segments that can be removed from the tessellation T.

Definition 3.2. Let $\mathcal{C}_\mathrm{s}$ be the space,

$$\mathcal{C}_\mathrm{s} = \{(s, T) : s \in \mathcal{S}_T, T \in \mathcal{T}\}.$$

The split (reduced) Campbell measure of a random T–tesselation is defined by

$$C_\mathrm{s}(\phi) = \mathbb{E} \sum_{m \in \mathcal{M}_{f\mathbf{T}}} \phi\left(m, \mathbf{T} \setminus \{m\}\right), \quad \phi : \mathcal{C}_\mathrm{s} \to \mathbb{R}. \tag{3.18}$$

The split Papangelou kernel is the P_s kernel, characterised by the identity

$$C_\mathrm{s}(\phi) = \mathbb{E} \int_{\mathcal{S}_\mathbf{T}} \phi(s, \mathbf{T})\, P_\mathrm{s}(\mathbf{T}, ds), \quad \phi : \mathcal{C}_\mathrm{s} \to \mathbb{R}. \tag{3.19}$$

It should be noted that for a point process, the reduced Campbell measure is defined by replacing the sum in (3.18) by a sum on all points of the process. In the case of a T–tessellation, the aforementioned sum is reduced to segments that are non-blocking.

Proposition 3.2. *Let* $\mathbf{T}$ *be a complete random T–tesselation with distribution* μ *given by* (3.17). *Then we have*

$$C_s(\phi) = \mathbb{E} \int_{\mathcal{S}_\mathbf{T}} \phi(s, \mathbf{T})\, ds, \quad \phi : \mathcal{C}_s \to \mathbb{R}. \tag{3.20}$$

As a direct consequence of this result, we obtain the following corollary.

Corollary 3.2. *The split Papangelou kernel of a complete random T–tesselation $\mathbf{T} \sim \mu$ is*

$$P_s(T, ds) = ds. \tag{3.21}$$

If we compare these results with similar results for homogeneous Poisson point processes, a possible intuitive interpretation would be that the cost of adding a non-blocking segment does not depend on the existing tessellation, suggesting the absence of interaction in a T–tessellation distributed according to μ.

In what follows, the reasoning applied to split and merge operations is extended to flips.

Definition 3.3. Let $\mathcal{C}_{\mathrm{f}}$ be the space

$$\mathcal{C}_{\mathrm{f}} = \{(F, T) : F \in \mathcal{F}_T, T \in \mathcal{T}\}.$$

The flip Campbell measure of a random T–tesselation is defined by

$$C_{\mathrm{f}}(\phi) = \mathbb{E} \sum_{F \in \mathcal{F}_{\mathbf{T}}} \phi\left(F^{-1}, F\mathbf{T}\right), \quad \phi : \mathcal{C}_{\mathrm{f}} \to \mathbb{R}. \tag{3.22}$$

The flip Papangelou kernel is the kernel P_{f} given by

$$C_{\mathrm{f}}(\phi) = \mathbb{E} \sum_{F \in \mathcal{F}_{\mathbf{T}}} \phi(F, \mathbf{T})\, P_{\mathrm{f}}(\mathbf{T}, F), \quad \phi : \mathcal{C}_{\mathrm{f}} \to \mathbb{R}. \tag{3.23}$$

Proposition 3.3. *Let $\mathbf{T}$ be a complete random T–tesselation with distribution μ given by* (3.17). *Then we have*

$$C_f(\phi) = \mathbb{E} \sum_{F \in \mathcal{F}_{\mathbf{T}}} \phi(F, \mathbf{T}), \quad \phi : \mathcal{C}_f \to \mathbb{R}. \tag{3.24}$$

Corollary 3.3. *The flip Papangelou kernel of a complete random T–tesselation $\mathbf{T} \sim \mu$ is*

$$P_f(T, F) \equiv 1. \tag{3.25}$$

Continuing the analogy with stationary Poisson point processes, the flip Papangelou kernel of a T–tesselation distributed according to μ shows that the cost of the associated transformation is constant and independent of the initial T–tesselation.

All these results lead us to consider the μ generated T–tesselations as completely random structures with minimal spatial dependence. For these reasons, this model is called the CRTT (Completely Random T Tesselation) model in [159].

3.5.3 Gibbs models for T–tessellations

In this section we present some models for T–tesselations. These models are described by probability densities constructed with respect to the reference measure μ given by (3.17). They allow the construction of T–tesselations with structured geometric properties, close to what might be encountered in practical applications.

Let $h : \mathcal{T} \to \mathbb{R}^+$ be a stable function. A random *Gibbs T–tesselation* is described by the unnormalised density h and has the distribution

$$\mathbb{P}(dT) \propto h(T)\, \mu(dT), \tag{3.26}$$

with $\int_{\mathcal{T}} \mathbb{P}(dT) = 1$. The integrability of the probability density h is ensured by the stability condition. In the terminology used for Gibbs processes,

$$U(T) = -\log h(T)$$

is called the energy function of the tesselation.

The results obtained for the Papangelou kernels of completely random T–tesselations can be extended to Gibbs T–tesselations, under an assumption similar to heredity for point processes.

Definition 3.4. A Gibbs T–tesselation with probability density h is *hereditary* if the following conditions are satisfied

- for all (T, S) such that $T \in \mathcal{T}$ and $S \in \mathcal{S}_T$, $h(ST) = 0 \Rightarrow h(T) = 0$.
- for all (T, F) such that $T \in \mathcal{T}$ and $F \in \mathcal{F}_T$, $h(FT) = 0 \Rightarrow h(T) = 0$.

Proposition 3.4. *Let* $\mathbf{T}$ *be a hereditary Gibbs T–tesselation. Then the split and flip Papangelou kernels are*

$$P_s(T, ds) = \frac{h\,(T \cup \{s)}{h(T)}\, ds, \quad P_f(T, F) = \frac{h\,(FT)}{h(T)}, \tag{3.27}$$

and the following Georgii-Nguyen-Zessin type formulae are verified

$$\begin{aligned} \mathbb{E} \sum_{m \in \mathcal{M}_{\mathbf{T}}} \phi\,(m, \mathbf{T} \setminus \{m\}) &= \mathbb{E} \int_{s \in \mathcal{S}_{\mathbf{T}}} \phi(s, \mathbf{T}) \frac{h\,(\mathbf{T} \cup \{s\})}{h(\mathbf{T})}\, ds, \\ \mathbb{E} \sum_{F \in \mathcal{F}_{\mathbf{T}}} \phi(F^{-1}, F\mathbf{T}) &= \mathbb{E} \sum_{F \in \mathcal{F}_{\mathbf{T}}} \phi(F, \mathbf{T}) \frac{h\,(F\mathbf{T})}{h(\mathbf{T})}, \end{aligned}$$

while considering $0/0 = 0$.

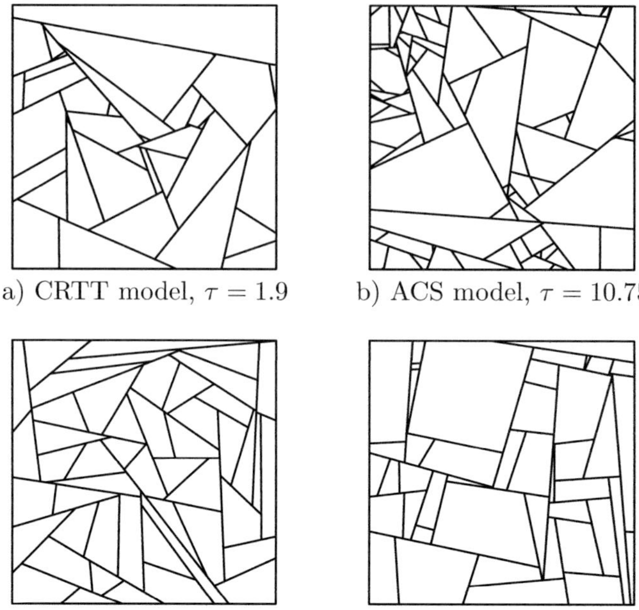

a) CRTT model, $\tau = 1.9$ b) ACS model, $\tau = 10.75$

c) Control of the cells area, $\tau = 0.043$, $\alpha = 10$ d) Control of the cells angles, $\tau = 12.1$, $\beta = 2.5$

FIGURE 3.24
Realisations of the four models that are presented as examples.

Example 3.4 (CRTT modified model). Let the energy function be

$$U(T) = -\mathring{n}_{\mathrm{s}}(T) \log \tau, \tag{3.28}$$

which associates the potential $\log \tau$ with each internal segment of the configuration. The parameter τ controls the density of internal segments or lines on which the T–tesselation is constructed. If $\tau = 1$, we have the CRTT model. Indirectly, the τ parameter can also be used to control the number of cells in a T–tesselation. The Papangelou kernels associated with the energy function (3.28) are

$$P_{\mathrm{s}}(T, ds) = \tau \, ds, \quad P_{\mathrm{f}}(T, F) \equiv 1.$$

A realisation of this model is shown in Figure 3.24a.

Example 3.5 (ACS model). The energy function

$$U(T) = \frac{\tau}{\pi} l(T) + \mathring{n}_{\mathrm{v}}(T) \log 2 - \mathring{n}_{\mathrm{s}}(T) \log \tau \tag{3.29}$$

is the T–tesselation model proposed in [2]. This model is called the *ACS model*, after the authors of the paper, Arak, Clifford and Surgailis. A realisation of the ACS model is shown in Figure 3.24b. The τ parameter has been adjusted so that the average number of cells is close to the one obtained by the CRTT model in Figure 3.24a.

The model introduced by [2] has remarkable properties. Depending on its parameters, it is able to generate different types of random sets, such as segments, broken lines, polygons or random T–tesselations. The normalisation constant of the probability density h has a precise analytical expression. The parameter h is equal to the linear intensity of this T–tesselation. The ACS model is Markovian in a sense similar to that expressed by Theorem 2.8. A simulation algorithm for this model is also given in [2]. The distributions of the lengths of the segments and edges, and the distributions of the areas and perimeters of the cells have been studied by [213].

The split and flip Papangelou kernels for the ACS model can be calculated from the relations (3.27) and (3.29). For the split with a segment s we get

$$\log \frac{P_s(T, ds)}{ds} = -\frac{\tau}{\pi} l(s) - 2\log 2 + \log \tau,$$

where $l(s)$ is the length of the segment. Compared to CRTT, the ACS model tends to divide cells using short segments, producing cells with a small surface area. This aspect has been numerically verified by simulation. The result of this comparison is shown in Figure 3.25.

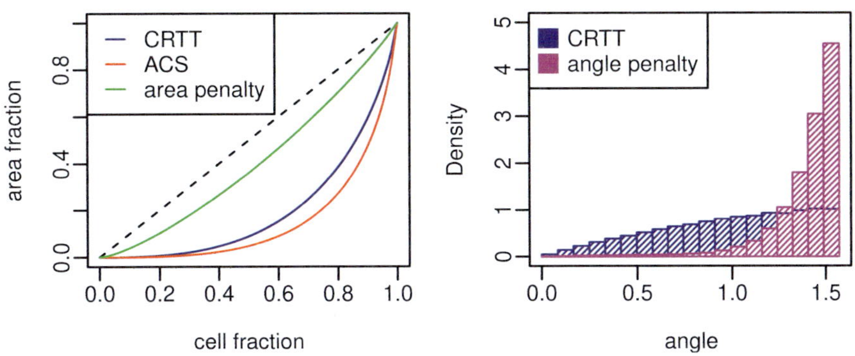

FIGURE 3.25
Comparisons of some statistics for the presented Gibbs models. Left: cell area distribution of the CRTT, ACS and cell area control models shown as Lorenz curves (x-axis: fraction of smallest cells, y-axis: covered area fraction, discontinuous: theoretical curve for cells with uniformly distributed areas). Right: distribution of acute angles between incident segments of the CRTT and angles control models.

Example 3.6 (Cell area control). To obtain T–tesselations with cells with less dispersed areas, we can consider the following energy function

$$U(T) = -\overset{\circ}{n}_{\mathrm{s}}(T)\log\tau + \alpha a^2(T), \tag{3.30}$$

where $a^2(T)$ is the sum of the squares of the areas of the cells in the tessellation. The $a^2(T)$ statistic is minimal for cells of the same size. Noticing that $a^2(T) \leq a(W)^2$, it easily follows that the energy function (3.30) is stable, so the model is well-defined. Figure 3.24c shows a realisation of the model. The model parameters have been set so that the mean number of cells is approximately the same as for the CRTT and ACS model realisations shown in Figures 3.24a and 3.24b respectively. Figure 3.25 shows the result of a simulation study which confirms that with these parameters this model produces a much smaller number of small cells than the CRTT and ACS models.

Example 3.7 (Angles control). To control the angle between two intersecting segments, the following energy function is considered

$$U(T) = -\overset{\circ}{n}_{\mathrm{s}}(T)\log\tau - \beta\sum_{v\in V(T)}\phi(v), \tag{3.31}$$

where $\phi(v)$ is the smallest angle between two segments touching at vertex v. With $\beta > 0$ we prefer configurations with angles $\phi(v)$ close to $\pi/2$. Since $\phi(v)$ is a finite quantity, we can easily prove that the energy function (3.31) is stable, so the model is well-defined. A realisation of this model is shown in Figure 3.24d. By simulation we have compared the angles of the T–tesselations obtained by this model with those obtained by the CRTT model. The histograms in Figure 3.25 show that the angles tend to be sharper in the CRTT model.

Part II

Build the pattern: Markov chains Monte Carlo simulation

4

Markov chains: notions, properties and simulation algorithms

Most marked point processes with interactions, as well as many other models for spatial structures, are described by probability distributions with normalisation constants that cannot be calculated analytically. To simulate such processes, it is often necessary to use Markov Chain Monte Carlo (MCMC) techniques. The key principle of these techniques is to simulate a Markov chain whose unique equilibrium distribution is the law of the process of interest.

The following two chapters are devoted to the development and analysis of MCMC techniques for the simulation of the models previously presented. Two families of MCMC methods are considered: the Metropolis-Hastings and the exact simulation algorithms.

The Metropolis-Hastings algorithms are MCMC techniques that can be considered relatively easy to implement. The theoretical construction of these algorithms guarantees the necessary convergence properties. Their convergence time is unknown. The samples generated are asymptotically distributed according to the distribution of interest. However, the Metropolis-Hastings algorithms allow the use of transition kernels adapted to the model in question, which can lead to improved convergence properties. The use of customised transition kernels is particularly recommended when sampling models with complex interactions.

The exact simulation algorithms are MCMC techniques, which themselves indicate when convergence is reached. However, their implementation is not straightforward. The samples generated will follow the distribution of interest, provided that the random number generators used by these algorithms are also exactly simulated. When these methods are used to sample models with strong interactions, the convergence time can be prohibitively long.

This chapter discusses Metropolis-Hastings and exact simulation algorithms from a theoretical perspective. First, Markov chains on general state spaces

DOI: 10.1201/9781003380382-4

and their properties are introduced. These concepts are then used to construct a Metropolis-Hastings algorithm for marked point processes, with a focus on its convergence properties. The chapter continues with the presentation of the principle of the Coupling From the Past perfect sampler and its implementation for the simulation of locally stable marked point processes. The numerical performance of all the algorithms presented is analysed and discussed in the next chapter.

4.1 "Classical" MCMC simulation

A Markov chain can be defined as a sequence of random variables indexed by a discrete time, such that the realisation of the variable at time t depends only on its realisation at time $t-1$. A very general algorithm for simulating a Markov chain is as follows:

Algorithm 4.1. General MCMC sampler. *Data: π probability distribution of interest, $\boldsymbol{x}_0$ the initial condition, T the discrete time horizon.*

1. *For $i = 1$ to T, do*

   {

   $\boldsymbol{x}_i = Update(\boldsymbol{x}_{i-1})$

   }

2. *Return $\boldsymbol{x}_T$.*

From a mathematical perspective, the *Update* procedure is described by the transition kernel of the chain that is being simulated. Its construction is directly related to the distribution of interest π. The algorithm is considered to work properly if the transition kernel has the following properties: irreducibility, recurrence and ergodicity. These properties guarantee that the infinite iteration of the transition kernel converges in total variation to π which is the unique invariant distribution of the chain. Moreover, the convergence occurs independently of the initial state of the chain.

Obviously, Algorithm 4.1 cannot be used for infinite time. However, for sufficiently large values of T, the samples obtained almost follow the invariant distribution of the chain, thus allowing the use of important results such as the law of large numbers and the central limit theorem. This justifies the success of this approach, provided that the updating mechanism is sufficiently simple with respect to the distribution one desires to sample.

The chapter continues with the presentation of the mathematical tools needed to construct MCMC algorithms for simulating point processes or other models of random geometrical patterns. For a deeper understanding of these notions, we warmly recommend to the reader some of the references we have followed for this presentation [5, 25, 101, 109, 108, 139, 152, 165, 181, 186, 212, 229, 238, 262, 267, 311, 312, 318].

Let $(\Omega, \mathcal{F}, \mu)$ be a probability space and $X_0, X_1, X_2, \ldots, X_n, \ldots$ a sequence of random variables defined on this space.

A *transition kernel* is a function $P : \Omega \times \mathcal{F} \to [0, 1]$ such that

i) for any $\boldsymbol{x} \in \Omega$, $P(\boldsymbol{x}, \cdot)$ is a probability measure,

ii) for all $A \in \mathcal{F}$, $P(\cdot, A)$ is measurable.

If the Ω space is discrete, the transition kernel is given by the transition matrix P

$$P_{\boldsymbol{x}\boldsymbol{y}} = \mathbb{P}(X_n = \boldsymbol{y} | X_{n-1} = \boldsymbol{x}).$$

If Ω is continuous, the transition kernel is given by the conditional density $P(\boldsymbol{x}, \boldsymbol{x}')$ of the transition kernel $P(\boldsymbol{x}, \cdot)$, that is

$$\mathbb{P}(X \in A | \boldsymbol{x}) = \int_A P(\boldsymbol{x}, d\boldsymbol{x}').$$

Let P be a transition kernel. The sequence of random variables $X_0, X_1, \ldots, X_n, \ldots$ is a *Markov chain* denoted (X_n), if for all t the conditional distribution of X_t knowing $\boldsymbol{x}_{t-1}, \boldsymbol{x}_{t-2}, \ldots \boldsymbol{x}_0$ is the same as that of X_t knowing $\boldsymbol{x}_{t-1}$:

$$\mathbb{P}(X_t \in A | \boldsymbol{x}_0, \boldsymbol{x}_1, \ldots, \boldsymbol{x}_{t-1}) = \mathbb{P}(X_t \in A | \boldsymbol{x}_{t-1}) = \int_A P(\boldsymbol{x}_{t-1}, d\boldsymbol{x}).$$

A Markov chain is said to be *homogeneous* if

$$\mathbb{P}(X_{t_1}, X_{t_2}, \ldots, X_{t_n} | \boldsymbol{x}_{t_0}) = \mathbb{P}(X_{t_1 - t_0}, X_{t_2 - t_0}, \ldots, X_{t_n - t_0} | \boldsymbol{x}_0)$$

for all $n \in \mathbb{N}^{\star}$.

As in the discrete case, the probability of going from $\boldsymbol{x}$ to A in n steps is defined iteratively:

$$P^n(\boldsymbol{x}, A) = \int_\Omega P(\boldsymbol{x}, d\boldsymbol{y}) P^{n-1}(y, A) \quad \boldsymbol{x} \in \Omega, A \in \mathcal{F}.$$

We set $P^0(\boldsymbol{x}, A) = \delta_{\boldsymbol{x}}(A)$ which is the Dirac measure in x:

$$\delta_{\boldsymbol{x}}(A) = \begin{cases} 1 & \boldsymbol{x} \in A \\ 0 & \boldsymbol{x} \notin A. \end{cases}$$

Theorem 4.1. The Chapman-Kolmogorov equations. *For any pair of integers m and n, we have*

$$P^{m+n}(\boldsymbol{x}, A) = \int_\Omega P^m(\boldsymbol{x}, d\boldsymbol{y}) P^n(\boldsymbol{y}, A), \quad \boldsymbol{x} \in \Omega, A \in \mathcal{F}.$$

Intuitively, this result indicates that to go from $\boldsymbol{x}$ to A in $m+n$ steps, the chain must transit in a certain $\boldsymbol{y}$ state at step m.

In the discrete case, the Chapman-Kolmogorov equations can be interpreted as a matrix product. In the general case, the kernel P^n is an operator which acts both on bounded measurable functions and on $\sigma-$finite measures on $\mathcal{F}$ via the formulae

$$P^n f(\boldsymbol{x}) = \int_\Omega P^n(\boldsymbol{x}, d\boldsymbol{y}) f(\boldsymbol{y}), \quad \mu P^n(A) = \int_\Omega \mu(d\boldsymbol{x}) P^n(\boldsymbol{x}, A).$$

So we can also write

$$P^n f(\boldsymbol{x}) = \int_\Omega P^n(\boldsymbol{x}, d\boldsymbol{y}) f(\boldsymbol{y}) = \delta_{\boldsymbol{x}} P^n f.$$

A $\sigma-$finite measure on $(\Omega, \mathcal{F})$ is *invariant* if

$$\pi(A) = \int_\Omega \pi(d\boldsymbol{x}) P(\boldsymbol{x}, A), \quad A \in \mathcal{F}.$$

The invariant measure guarantees the stationarity of the Markov chain as follows: if an initial state is chosen with an invariant distribution π, then the action of the transition kernel is to produce a new state following the same law π. This aspect is illustrated by the following calculation:

$$\begin{aligned}
\pi(A) &= \int_\Omega \pi(d\boldsymbol{x}) P(\boldsymbol{x}, A) \\
&= \int_\Omega \left[\int_\Omega \pi(d\boldsymbol{y}) P(\boldsymbol{y}, d\boldsymbol{x})\right] P(\boldsymbol{x}, A) \\
&= \int_\Omega \pi(d\boldsymbol{y}) \left[\int_\Omega P(\boldsymbol{y}, d\boldsymbol{x}) P(\boldsymbol{x}, A)\right] \\
&= \int_\Omega \pi(d\boldsymbol{y}) P^2(\boldsymbol{y}, A) \\
&\;\;\vdots \\
&= \int_\Omega \pi(d\boldsymbol{y}) P^n(\boldsymbol{y}, A),
\end{aligned}$$

for all $n \in \mathbb{N}^\star$ and $A \in \mathcal{F}$.

Let μ be any initial distribution. If there exists a measure $\gamma(\mu)$ that verifies

$$\gamma_\mu(A) = \lim_{n\to\infty} \int_\Omega \mu(d\boldsymbol{x}) P^{(n)}(\boldsymbol{x}, A)$$

for a certain topology defined on the space of probability measures and for all $A \in \mathcal{F}$, then

$$\begin{aligned}\gamma_\mu(A) &= \lim_{n\to\infty} \int_\Omega \mu(d\boldsymbol{x}) \int_\Omega P^{n-1}(\boldsymbol{x}, d\boldsymbol{y}) P(\boldsymbol{y}, A) \\ &= \int_\Omega \gamma_\mu(d\boldsymbol{y}) P(\boldsymbol{y}, A).\end{aligned}$$

It follows that γ_μ is an invariant probability measure.

Therefore, if the limit γ_μ exists, it is an invariant measure. Moreover, if the chain has a single invariant measure, in this case γ_μ does not depend on μ and it represents the *equilibrium distribution* of the chain.

A transition kernel $P(\boldsymbol{x}, A)$ is *reversible* with respect to the π measure, if the integral

$$\int_A \int_B \mathbf{1}_A(\boldsymbol{x}) \mathbf{1}_B(\boldsymbol{y}) \pi(d\boldsymbol{x}) P(\boldsymbol{x}, d\boldsymbol{y}) \tag{4.1}$$

is symmetrical by substituting A and B.
Applying the previous definition and the formula (4.1), we obtain

$$\int_A \pi(d\boldsymbol{x}) P(\boldsymbol{x}, B) = \int_B \pi(d\boldsymbol{x}) P(\boldsymbol{x}, A),$$

which means that the probabilities of transiting from A to B or from B to A are equal.

Let's take $B = \Omega$. Since $P(\boldsymbol{x}, \Omega) = 1$, it immediately follows that

$$\pi(A) = \int_\Omega \pi(d\boldsymbol{x}) P(\boldsymbol{x}, A),$$

It thus appears that a π-reversible kernel has π as its invariant distribution.

From a practical and computing point of view, the *Update* procedure represents and simulates a reversible transition kernel with respect to the distribution of interest. The *Update* mechanism, that has only this property, guarantees the convergence of the algorithm, only if the initial conditions are chosen according to the equilibrium distribution. But this is precisely the law one wishes to simulate. Therefore, it is necessary that the *Update* procedure "push" the simulated chain towards the equilibrium regime from any initial condition. This is what happens if the constructed chain is irreducible, aperiodic and recurrent. These concepts are described below. Details and proofs regarding the presented results can be found in [5, 108, 212, 229, 262, 312].

For any set $A \in \mathcal{F}$, the variable

$$\tau_A = \min\{n \geq 1 : X_n \in A\}$$

represents the first time the chain reaches the set A. It is also called *first return* time on A. If $X_n \notin A$ for all n, we pose by convention $\tau_A = +\infty$.

For Markov chains defined on a discrete state space, irreducibility means that all the states communicate, that is

$$\mathbb{P}_{\boldsymbol{x}}(\tau_{\boldsymbol{y}} < \infty) > 0 \quad \forall \boldsymbol{x}, \boldsymbol{y} \in \Omega.$$

But if the state space is continuous we always have $\mathbb{P}_{\boldsymbol{x}}(\tau_{\boldsymbol{y}}) = 0$. To properly define irreducibility for chains on a general state space, an auxiliary measure ϕ is needed.

Let ϕ be a measure on $\mathcal{F}$. The Markov chain (X_n) is said to be *$\phi-$irreducible* if $\phi(A) > 0$ implies $\mathbb{P}_{\boldsymbol{x}}(\tau_A < \infty) > 0$ for all $\boldsymbol{x} \in \Omega$.

An equivalent way of defining $\phi-$irreducibility is to say that there exists $n > 0$ such that $P^n(\boldsymbol{x}, A) > 0$. This condition guarantees that all the sets A that are "seen" by ϕ can be reached by the chain from any initial state $\boldsymbol{x} \in \Omega$. One may also notice, that the $\phi-$irreducibility for chains defined on a discrete space with ϕ the counting measure is just the usual irreducibility.

A set C is *small* if there exists $m \in \mathbb{N}^\star$ and a positive measure ν_m such that

$$P^m(\boldsymbol{x}, A) \geq \nu_m(A), \tag{4.2}$$

for all $\boldsymbol{x} \in C$ and $A \in \mathcal{F}$. The set C which verifies (4.2) is called ν_m-small.

The following result shows the link between $\phi-$irreducibility and small sets.

Theorem 4.2. *Let (X_n) be a $\phi-$irreducible chain. For any $A \in \mathcal{F}$ such that $\phi(A) > 0$, there exists $m \in \mathbb{N}^\star$ and a small set $C \subset A$ such that $\nu_m(C) > 0$. Moreover, Ω is a countable union of small sets.*

Thus, small sets can be seen as a tool for verifying non-irreducibility.

Irreducibility, or the accessibility of any state from any other state, is a desirable property when constructing Markov chains to simulate probability distributions. However, for the same purpose, cyclic or periodic behaviour of chains is an undesirable property. Such behaviour does not guarantee convergence to the equilibrium distribution.

Let us consider a $\phi-$irreducible chain (X_n) and C a ν_M-small set such that for $\boldsymbol{x} \in C$, $\nu_M(C) > 0$. Hence, for the chain started in C there is a positive probability that it will return to C at time M. The following set

$$T_C = \{m : \exists \epsilon_m > 0 \quad \text{such that} \quad P^m(\boldsymbol{x}, \cdot) \geq \nu_m(\cdot) = \epsilon_m \nu_M(\cdot)\}$$

contains all the time points m for which C is small with ν_m proportional to ν_M. It can be observed that for $A \subseteq C$, $n, m \in T_C$ one gets

$$\begin{aligned} P^{n+m}(\boldsymbol{x}, A) &\geq \int_C P^n(\boldsymbol{x}, d\boldsymbol{y}) P^m(\boldsymbol{y}, A) \\ &\geq [\epsilon_n \epsilon_m \nu_M(C)] \nu_M(A) \\ &\geq \epsilon_{n+m} \nu_M(A), \quad \boldsymbol{x} \in C. \end{aligned}$$

Hence, T_C is closed under addition. Thus, the "period" for the set C is naturally defined by the greatest common divisor of T_C. The chain is called *aperiodic* if

$$\gcd(T_C) = 1$$

Algorithm 4.1 simulates an aperiodic Markov chain if the *Update* procedure allows to remain in the same state with positive probability.

Theorem 4.3. *Let (X_n) be a Markov chain with transition kernel P. Assume the chain is $\pi-$irreducible, aperiodic and reversible with respect to π. Then π is the unique invariant distribution of the chain. Furthermore, there exists a set $\Omega' \subseteq \Omega$ such that $\pi(\Omega') = 1$ and*

$$\sup_{A \in \mathcal{F}} |P^n(\boldsymbol{x}, A) - \pi(A)| \to 0 \tag{4.3}$$

for all $\boldsymbol{x} \in \Omega'$.

The previous result allows the construction of a Markov chain which converges to its equilibrium distribution π from almost any initial state. Indeed, according to Theorem 4.3, the chain can be in a state $\boldsymbol{x} \in \Omega''$ with $\pi(\Omega'') = 0$, for which the limit (4.3) is non-zero.

Recurrence is the property that protects against the existence of such a set of zero measure Ω''. It thus guarantees the independence of the initial conditions for a MCMC-type algorithm. Recurrence also ensures us that any set $A \in \mathcal{F}$ is sufficiently visited from any initial state.

For any set $A \in \mathcal{F}$, the number of visits after time zero by the chain (X_n) to A is

$$\eta_A = \sum_{n=1}^{\infty} \mathbf{1}\{X_n \in A\}$$

and it is called the *occupancy time* of the set A.

A set A is Harris recurrent if $\mathbb{P}_{\boldsymbol{x}}(\eta_A = \infty) = 1$ for all $\boldsymbol{x} \in A$. The chain (X_n) is *Harris recurrent* if it is $\phi-$irreducible and if any set $A \in \mathcal{F}$ such that $\phi(A) > 0$ is Harris recurrent.

Proposition 4.1. *If for every set $A \in \mathcal{F}$, $\mathbb{P}_{\boldsymbol{x}}(\tau_A < \infty) = 1$ for all $\boldsymbol{x} \in A$, then $\mathbb{P}_{\boldsymbol{x}}(\eta_A = \infty) = \mathbb{P}_{\boldsymbol{x}}(\tau_A < \infty)$ for all $\boldsymbol{x} \in \Omega$ and (X_n) is Harris recurrent.*

This result shows that Harris recurrence is a stronger property than ϕ-irreducibility. Harris recurrence requires $\mathbb{P}_{\boldsymbol{x}}(\tau_A < \infty) = 1$, whereas $\phi-$irreducibility requires only that $\mathbb{P}_{\boldsymbol{x}}(\tau_A < \infty) > 0$. Recurrence for general state space Markov chains requires $\mathbb{E}_{\boldsymbol{x}}[\eta_A] = \infty$. The Harris recurrence $\mathbb{P}_{\boldsymbol{x}}(\eta_A = \infty) = 1$ is a stronger property [262, 212].

Often, the *Update* procedures representing transition kernels that are $\phi-$irreducibles are leading to Markov chains that are also Harris recurrent. The following result gives a tool for establishing Harris recurrence. This result uses the *drift condition.*

Theorem 4.4. *Let (X_n) be a $\phi-$irreducible Markov chain. Suppose there exists a small set $C \in \mathcal{F}$ and a function $V : \Omega \to (0, \infty)$ such that the level sets*

$$C_V(\alpha) = \{\boldsymbol{x} \in \Omega : V(\boldsymbol{x}) \leq \alpha\}$$

are small. The chain is Harris recurrent if the drift $\triangle V(\boldsymbol{x}) = PV(\boldsymbol{x}) - V(\boldsymbol{x})$ is negative for all $\boldsymbol{x} \notin C$, that is:

$$PV(\boldsymbol{x}) = \mathbb{E}\left[V(X_{t+1})|X_t = \boldsymbol{x}\right] = \int_\Omega P(\boldsymbol{x}, d\boldsymbol{y})V(\boldsymbol{y}) \leq V(\boldsymbol{x}), \quad \boldsymbol{x} \notin C. \tag{4.4}$$

An intuitive interpretation of these results might be as follows. From any initial configuration, the transition kernel moves the chain towards those configurations where the average behaviour of the function tends to decrease. The iteration of the transition kernel moves the chain towards an equilibrium region where $\Delta V = 0$.

The main interest of MCMC algorithms is that the samples obtained can be used within strong results such as the Law of Large Numbers or the Central Limit Theorem. An algorithm that simulates a recurrent Harris chain does not allow such an application directly, unless the equilibrium regime has been established. Except for the perfect simulation methods, this is impossible in practice. However, this problem is solved if the chain is also ergodic.

A Markov chain (X_n) is *ergodic* if it is both Harris recurrent and aperiodic. In this case, the limit (4.3) converges independently of the initial conditions. Moreover, we know the convergence speed of this limit, i.e. the speed with which the chain reaches its equilibrium regime. So even if we only have a limited number of samples, we have a legitimate hope of obtaining results that are all the more accurate the larger the number of samples.

The *total variation norm* between two probability measures, ν_1 and ν_2 is

$$\| \nu_1 - \nu_2 \| = \sup_{A \in \mathcal{F}} |\nu_1(A) - \nu_2(A)|.$$

Let (X_n) be an ergodic chain with invariant distribution π. The chain is said to be *geometrically ergodic* if there exists a function $M : \Omega \to \mathbb{R}^+$ with $\pi|M| < \infty$ and a constant $\rho \in (0, 1)$ such that

$$\| P^n(\boldsymbol{x}, \cdot) - \pi \| \leq M(\boldsymbol{x})\rho^n$$

for all $\boldsymbol{x} \in \Omega$. The chain is said to be *uniformly ergodic* if there exist constants $M > 0$ and $\rho in(0, 1)$ such that

$$\sup_{\boldsymbol{x} \in \Omega} \| P^n(\boldsymbol{x}, \cdot) - \pi \| \leq M\rho^n.$$

Geometric ergodicity means that repeated iteration of the transition kernel approaches the equilibrium distribution of the chain at a geometric speed weighted by a constant depending on the initial state of the chain. The following result allows to establish this property.

Theorem 4.5. *Let (X_n) be a $\phi-$irreducible and aperiodic chain. The chain is geometrically ergodic if there exists a function $V : \Omega \to [1, \infty)$, constants $a < 1$ and $b < \infty$, and a small set $C \in \mathcal{F}$ such that*

$$PV(\boldsymbol{x}) \leq aV(\boldsymbol{x}) + b\mathbf{1}_C(\boldsymbol{x}), \quad \forall \boldsymbol{x} \in \Omega. \tag{4.5}$$

It can be shown that if V satisfies (4.5), then for all $\alpha > 0$, the level sets $C_V = \{\boldsymbol{x} \in \Omega : V(\boldsymbol{x}) \leq \alpha\}$ are small . This implies that the drift condition for ergodicity (4.5) implies the drift condition for recurrence (4.4). It follows that if a chain is geometrically ergodic, then it is Harris recurrent ([108] pp. 106).

4.2 Metropolis Hastings algorithm for marked point processes

4.2.1 Algorithm presentation

Historically, the Metropolis-Hastings algorithm was introduced to simulate physical systems with a fixed number of particles [211]. The generalisation of this approach to Markov chains with discrete states was achieved by [129]. The adaptation of this framework for Markov chains with general states has been done by [116, 119]. The simulation of point processes using a Metropolis-Hastings algorithm has been proposed by [109]. The presentation in this section is based on the following references [108, 181, 186, 192, 195, 229, 283].

Let X be a hereditary marked point process given by the probability density p with respect to μ the reference measure (2.17). To simulate this process we construct an `Update` procedure which represents the kernel of a Markov chain with equilibrium distribution

$$\pi(A) = \int_A p(\boldsymbol{x})\mu(d\boldsymbol{x}), \quad A \in \mathcal{F}.$$

In the context of Metropolis-Hastings algorithms, the principle of such a procedure is as follows: starting from an initial configuration $\boldsymbol{x}$, another configuration $\boldsymbol{y} \in A$ is proposed. This other state is either accepted as the new initial state, or rejected. If the $\boldsymbol{y}$ state is rejected, the chain remains in its $\boldsymbol{x}$ initial state. This idea is described by the transition kernel

$$P(\boldsymbol{x}, A) = \int_A \alpha(\boldsymbol{x}, \boldsymbol{y})Q(\boldsymbol{x}, d\boldsymbol{y}), \quad \boldsymbol{x} \in \Omega, A \in \mathcal{F}, \boldsymbol{x} \notin A \tag{4.6}$$

and

$$P(\boldsymbol{x}, \{\boldsymbol{x}\}) = \int_{\boldsymbol{y} \neq \boldsymbol{x}} [1 - \alpha(\boldsymbol{x}, \boldsymbol{y})]Q(\boldsymbol{x}, d\boldsymbol{y}),$$

where Q is the proposal kernel and $\alpha(\cdot, \cdot)$ is the acceptance probability.

The transition kernel (4.6) is constructed using three possible transformations or moves applied to an initial object configuration $\boldsymbol{x}$. These transformations are: add or birth of a new object to $\boldsymbol{x}$, delete or death of an existing object from $\boldsymbol{x}$ and change the properties of a selected object in $\boldsymbol{x}$.

The moves add, delete and change of an object are proposed with probabilities p_b, p_d and p_c, respectively. A new object η to be added is produced using the probability density $b(\boldsymbol{x}, \eta)$ on $W \times M$. An existing object $\eta \in \boldsymbol{x}$ is selected to be deleted with probability $d(\boldsymbol{x}, \eta)$. And finally, an object $\eta \in \boldsymbol{x}$ is first selected with probability $q(\boldsymbol{x}, \eta)$ and then replaced by a new object ζ built with the probability density $c(\boldsymbol{x}, \eta, \zeta)$.

A transition kernel that implements these transformations is

$$\begin{aligned} P(\boldsymbol{x}, A) &= p_b \int_{W \times M} b(\boldsymbol{x}, \eta)\alpha(\boldsymbol{x}, \boldsymbol{y} := \boldsymbol{x} \cup \{\eta\})\mathbf{1}\{\boldsymbol{y} \in A\}d\sigma(\eta) \\ &+ \quad p_d \sum_{\eta \in \boldsymbol{x}} d(\boldsymbol{x}, \eta)\alpha(\boldsymbol{x}, \boldsymbol{y} := \boldsymbol{x} \setminus \{\eta\})\mathbf{1}\{\boldsymbol{y} \in A\} \\ &+ \quad p_c \sum_{\eta \in \boldsymbol{x}} q(\boldsymbol{x}, \eta) \int_{W \times M} c(\boldsymbol{x}, \eta, \zeta)\alpha(\boldsymbol{x}, \boldsymbol{y} := \boldsymbol{x} \setminus \{\eta\} \cup \{\zeta\})\mathbf{1}\{\boldsymbol{y} \in A\}d\sigma(\zeta) \\ &+ \quad \mathbf{1}\{\boldsymbol{x} \in A\} \Bigg[1 - p_b \int_{W \times M} b(\boldsymbol{x}, \eta)\alpha(\boldsymbol{x}, \boldsymbol{x} \cup \{\eta\})d\sigma(\eta) \end{aligned}$$

$$- \quad p_d \sum_{\eta \in \boldsymbol{x}} d(\boldsymbol{x}, \eta)\alpha(\boldsymbol{x}, \boldsymbol{x} \setminus \{\eta\})$$

$$- \quad p_c \sum_{\eta \in \boldsymbol{x}} q(\boldsymbol{x}, \eta) \int_{W \times M} c(\boldsymbol{x}, \eta, \zeta)\alpha(\boldsymbol{x}, \boldsymbol{x} \setminus \{\eta\} \cup \{\zeta\}) d\sigma(\zeta) \Bigg], \quad (4.7)$$

with $d\sigma(\eta) = d\sigma((w,m)) = d\nu(w) \times d\nu_M(m)$ and $0 < p_b + p_d + p_c \leq 1$.

If the initial state is the empty set, the transition kernel cannot propose to delete or to change an existing object. In this case, the kernel (4.7) writes as

$$\begin{aligned} P(\emptyset, A) &= p_b \int_K b(\emptyset, \eta)\alpha(\emptyset, \{\eta\}) \mathbf{1}\{\{\eta\} \in A\} d\sigma(\eta) \\ &+ \mathbf{1}\{\emptyset \in A\} \left[1 - p_b \int_K b(\emptyset, \eta)\alpha(\emptyset, \{\eta\}) d\sigma(\eta)\right]. \end{aligned} \quad (4.8)$$

It can be observed that the transition kernel (4.7) is a linear combination of two kernels. The first kernel adds or deletes an object from the initial configuration. This kernel concentrates on the subspaces $\Omega_{n-1} \cup \Omega_n \cup \Omega_{n+1}$ (or $\Omega_0 \cup \Omega_1$ if $n = 0$). The second kernel changes the characteristics of an object in a configuration. This kernel is limited to the subspace Ω_n, as this transformation does not change the number of objects in configuration.

The distribution given by $p(\boldsymbol{x})$ is the invariant law of the chain if the associated transition kernel is reversible. It is not difficult to check that if two kernels are reversible, then so is their linear combination [108].

Establishing the reversibility of a Markov chain directly may be difficult. As alternative we may consider a two-step strategy as explained in [229, 181, 186]. First, choose the proposal distribution and the acceptance probability of a kernel so that the detailed balance condition is respected. Next, within this context, prove the reversibility of the chain.

Suppose that the chain is in the state $\boldsymbol{x} \in A \subseteq \Omega_n$ and that an "add" is proposed so that the chain may go into the state $\boldsymbol{x} \cup \{\eta\} \in B \subseteq \Omega_{n+1}$. The only move that compensates for the add is "delete" of an object. In this case, the detailed balance condition is written as

$$p(\boldsymbol{x}) p_b b(\boldsymbol{x}, \eta)\alpha(\boldsymbol{x}, \boldsymbol{x} \cup \{\eta\}) = p(\boldsymbol{x} \cup \{\eta\}) p_d d(\boldsymbol{x} \cup \{\eta\}, \eta)\alpha(\boldsymbol{x} \cup \{\eta\}, \boldsymbol{x}). \quad (4.9)$$

In what follows, it will be shown how the condition (4.9) implies the reversibility of the transition kernel constructed from the "add" and "delete" moves.

By using (2.17), (4.7) and (4.1), the probability of being in A and then in B at the next iteration is equal to

$$\begin{aligned} &\tfrac{\exp[-\nu(W)]}{n!} \int_{(W \times M)^{(n+1)}} p_b p(\boldsymbol{x}) b(\boldsymbol{x}, \eta)\alpha(\boldsymbol{x}, \boldsymbol{x} \cup \{\eta\}) \times \\ &\times 1\{\boldsymbol{x} \in A, \boldsymbol{x} \cup \{\eta\} \in B\} d\sigma(x_1) \dots d\sigma(x_n) d\sigma(\eta), \end{aligned} \quad (4.10)$$

whereas the probability of being in B and then in A at the next iteration is

$$\frac{\exp[-\nu(W)]}{(n+1)!} \int_{(W\times M)^{(n+1)}} \sum_{y_i \in \boldsymbol{y}} p_d p(\boldsymbol{y}) d(\boldsymbol{y}, y_i) \alpha(\boldsymbol{y}, \boldsymbol{y} \setminus \{y_i\}) \times \\ \mathbf{1}\{\boldsymbol{y} \in B, \boldsymbol{y} \setminus \{y_i\} \in A\} d\sigma(y_1) \dots d\sigma(y_n) d\sigma(y_{n+1}). \tag{4.11}$$

The swap of the sum and integral signs in the last equation gives $(n+1)$ equal integrals, since the order of the objects in a configuration does not matter. This allows to write (4.11) as follows:

$$\frac{\exp[-\nu(W)]}{n!} \int_{(W\times M)^{(n+1)}} p_d p(\boldsymbol{y}) d(\boldsymbol{y}, y_1) \alpha(\boldsymbol{y}, \boldsymbol{y} \setminus \{y_1\}) \times \\ \mathbf{1}\{\boldsymbol{y} \in B, \boldsymbol{y} \setminus \{y_1\} \in A\} d\sigma(y_1) d\sigma(y_2) \dots d\sigma(y_{n+1}).$$

Let $\boldsymbol{x} = \boldsymbol{y} \setminus y_1\}$ and $\eta = y_1$ in (4.10). Then, the condition (4.9) provides reversibility, i.e. the equality of the probabilities (4.10) and (4.11).

The acceptance probability can be obtained directly from the detailed balance conditions (4.9). Several solutions exist [23, 211, 181]. Here the classical choice presented by [109, 108, 181, 186, 229] was adopted.

Let $r(\boldsymbol{x}, \eta)$ denote the acceptance ratio

$$r(\boldsymbol{x}, \eta) = \frac{p_d d(\boldsymbol{x} \cup \{\eta\}, \eta)}{p_b b(\boldsymbol{x}, \eta)} \frac{p(\boldsymbol{x} \cup \{\eta\})}{p(\boldsymbol{x})} \tag{4.12}$$

and let's assume that it is always well-defined, that is the quantities $d(\cdot, \cdot)$, $b(\cdot, \cdot)$, $p(\cdot)$ are always strictly positive.

The acceptance probability of an "add" move writes as

$$\alpha(\boldsymbol{x}, \boldsymbol{x} \cup \{\eta\}) = \min\{1, r(\boldsymbol{x}, \eta)\}, \tag{4.13}$$

whereas the acceptance probability of a "delete" is obtained as the inverse of the acceptance ratio (4.12), which gives

$$\alpha(\boldsymbol{x}, \boldsymbol{x} \setminus \{\eta\}) = \min\{1, 1/r(\boldsymbol{x} \setminus \{\eta\}, \eta)\}. \tag{4.14}$$

To avoid any pathological situation, we assume that $\alpha(\boldsymbol{x}, \boldsymbol{x} \cup \{\eta\}) = 0$ when $p(\boldsymbol{x} \cup \{\eta\}) = 0$.

The acceptance probability of a "change" move is deduced in a similar way. The detailed balance condition for this type of transition is written as

$$p(\boldsymbol{x}) q(\boldsymbol{x}, \eta) c(\boldsymbol{x}, \eta, \zeta) \alpha(\boldsymbol{x}, \boldsymbol{x} \setminus \{\eta\} \cup \{\zeta\}) = \\ p(\boldsymbol{x} \setminus \{\eta\} \cup \{\zeta\}) q(\boldsymbol{x} \setminus \{\eta\} \cup \{\zeta\}, \zeta) c(\boldsymbol{x} \setminus \{\eta\} \cup \{\zeta\}, \zeta, \eta) \alpha(\boldsymbol{x} \setminus \{\eta\} \cup \{\zeta\}, \boldsymbol{x}).$$

Under these assumptions, the reversibility of the associated transition kernel is obtained by introducing (4.7) and (2.17) in (4.1).

As previously, pathological cases are eliminated using very weak assumptions. The result is that the acceptance probability for a "change" is

$$\alpha(\boldsymbol{x}, \boldsymbol{x} \setminus \{\eta\} \cup \{\zeta\}) = \tag{4.15}$$
$$= \min\left\{1, \frac{q(\boldsymbol{x} \setminus \{\eta\} \cup \{\zeta\}, \zeta)c(\boldsymbol{x} \setminus \{\eta\} \cup \{\zeta\}, \zeta, \eta)}{q(\boldsymbol{x}, \eta)c(\boldsymbol{x}, \eta, \zeta)} \frac{p(\boldsymbol{x} \setminus \{\eta\} \cup \{\zeta\})}{p(\boldsymbol{x})}\right\}.$$

The following algorithm is an example of an `Update` procedure that can be implemented to simulate marked point processes.

Algorithm 4.2. Metropolis-Hastings update procedure for simulating marked point processes. *Data: initial state $\boldsymbol{x}$, probability density of the process p, the moves selection probabilities p_b,p_d,p_c and the associated moves distributions $b(\cdot,\cdot)$, $d(\cdot,\cdot)$, $q(\cdot,\cdot)$ and $c(\cdot,\cdot,\cdot)$.*

$\boldsymbol{y}$ = *Update*($\boldsymbol{x}$)

1. *Set $\boldsymbol{y} = \boldsymbol{x}$.*
2. *Choose a move based on the probabilities p_b,p_d and p_c.*
3. *If "add" is chosen, generate a new object η via the probability density $b(\boldsymbol{x}, \eta)$. Accept the new configuration $\boldsymbol{y} = \boldsymbol{x} \cup \{\eta\}$ with probability $\alpha(\boldsymbol{x}, \boldsymbol{y})$ given by (4.26).*
4. *If "delete" is chosen, select an object to be deleted $\eta \in \boldsymbol{x}$ with probability $d(\boldsymbol{x}, \eta)$. The new configuration $\boldsymbol{y} = \boldsymbol{x} \setminus \{\eta\}$ is accepted with probability $\alpha(\boldsymbol{x}, \boldsymbol{y})$ given by (4.14).*
5. *If "change" is chosen, with probability $q(\boldsymbol{x}, \eta)$ select an object $\eta \in \boldsymbol{x}$ and transform it into ζ using the probability density $c(\boldsymbol{x}, \eta, \zeta)$. Accept the new configuration $\boldsymbol{y} = \boldsymbol{x} \setminus \{\eta\} \cup \{\zeta\}$ with probability $\alpha(\boldsymbol{x}, \boldsymbol{y})$ given by (4.15).*
6. *Return $\boldsymbol{y}$.*

The type of move is chosen using the following mechanism. First, u a random uniform variable u is generated on $[0, 1]$. Next, the "add", "delete" or "modify" moves are performed depending on whether $u \leq p_b$, $p_b < u \leq p_b + p_d$ or $p_b + p_d < u \leq p_b + p_d + p_c$. Similarly, the acceptance of a new configuration $\boldsymbol{y}$ from $\boldsymbol{x}$ is done if $w \leq \alpha(\boldsymbol{x}, \boldsymbol{y})$ where w is a random uniform variable over $[0, 1]$. Otherwise, the chain remains in the initial $\boldsymbol{x}$ configuration. Note that u and w are two different independent random variables.

4.2.2 Convergence properties

The previous algorithm allows a wide choice for the proposal laws. For instance, we can choose any positive value for p_b, p_d, p_c provided that $0 < p_b + p_d + p_c \leq 1$. Here, these probabilities are constant. It is possible to choose them according to the current configuration of the chain, while preserving its convergence properties [109, 116, 108]. In order to avoid pathological case, the densities $b(\boldsymbol{x}, \eta)$, $d(\boldsymbol{x}, \eta)$, $q(\boldsymbol{x}, \eta)$ and $c(\boldsymbol{x}, \eta, \zeta)$ must be chosen strictly positive [192, 290, 292].

To add a new object to the current configuration, the simplest choice is the uniform density

$$b(\boldsymbol{x}, \eta) = \frac{\mathbf{1}\{\eta \in W\}}{\nu(W)} \tag{4.16}$$

with respect to $d\sigma(\eta)$. By this choice, the proposed object has a uniformly distributed position in W while its mark is chosen independently according to ν_M.

Similarly, to delete an object, the simplest choice is

$$d(\boldsymbol{x}, \eta) = \frac{\mathbf{1}\{\eta \in \boldsymbol{x}\}}{n(\boldsymbol{x})} \tag{4.17}$$

with $\boldsymbol{x} \neq \emptyset$. By this uniform choice, all objects have the same chance of being proposed for elimination. If $\boldsymbol{x} = \emptyset$ no object can be proposed.

The choice offered by (4.16) and (4.17) is not the only one. For the time being, we wish to show that this simple choice leads to convergent algorithms for a large family of models. However, let us say that intuitively, the dynamics based on such uniform proposals are better suited to models with weak interactions.

There are several ways of modifying the characteristics of an object in a configuration [210, 263, 264]. Here, we once again apply a simple strategy which consists of uniformly choosing the object to be modified

$$q(\boldsymbol{x}, \eta) = \frac{1}{n(\boldsymbol{x})}, \quad \boldsymbol{x} \neq \emptyset.$$

Once the object has been selected, its properties are modified using the density $c(\boldsymbol{x}, \eta, (w, m))$. Let $C(\eta) = C_W(w_\eta) \times C_M(m_\eta)$ be a neighbourhood of the marked point $\eta = (w_\eta, m_\eta)$ such that $\nu(C_W(w_\eta))$ and $\nu_M(C_M(m_\eta))$ are strictly positive. Then the density $c(\boldsymbol{x}, \eta, (w, m))$ is

$$c(\boldsymbol{x}, \eta, (w, m)) = \frac{\mathbf{1}\{w \in C_W(w_\eta)\}}{\nu(C_W(w_\eta))} \times \frac{\mathbf{1}\{m \in C_M(m_\eta)\}}{\nu_M(C_M(m_\eta))}.$$

To guarantee reversibility, we must have $\zeta \in C(\eta)$ if and only if $\eta \in C(\zeta)$. Typically, $C(\eta)$ is a relatively small neighbourhood which gives a "local" character to the modification of an object. Such "local" modifications have an acceptance ratio (4.15) close to 1, which means that the proposed transition is often accepted. This is a very positive point. However, this property introduces a significant correlation between samples. From a practical point of view, there is a trade-off between the desired result, the result obtained and the computational cost.

The presented choices, together with the detailed balance conditions, ensure us that the simulated Markov chain is reversible. The following result shows that the simulated chain is $\phi-$irreducible, Harris recurrent and geometrically ergodic.

Theorem 4.6. *Let b, d, q and c be the functions described earlier. Suppose that $b(\boldsymbol{x}, \eta)$ and $d(\boldsymbol{x}, \eta)$ are strictly positive on their respective domains of definition and that*

$$\lim_{n\to\infty} u_n = \lim_{n\to\infty}\left[\sup_{\eta\in W\times M, \boldsymbol{x}\in\Xi_n} \frac{d(\boldsymbol{x}\cup\{\eta\},\eta)}{b(\boldsymbol{x},\eta)}\right] \to 0.$$

Let us also assume that $p_b, p_d, p_c \in (0,1)$ with $p_b + p_d + p_c = 1$ and that $p(\boldsymbol{x})$ is the probability density of a marked point process on $W \times M$. The point process is locally stable and $p(\boldsymbol{x})$ is constructed with respect to the reference measure μ, represented by the standard marked Poisson process. Then, the Metropolis-Hastings sampler for the density p defined by Algorithm 4.2 simulates a Markov chain that is $\phi-$irreducible, Harris recurrent and geometrically ergodic.

Proofs of this result may be found in [109, 108, 229, 181, 192, 186]. For pedagogical reasons we present the one given by [192].

Proof. Let $\Lambda > 0$ be the upper bound of the likelihood ratio $p(\boldsymbol{x}\cup\{\eta\})/p(\boldsymbol{x})$ and $V(\boldsymbol{x}) = A^{n(\boldsymbol{x})}$ with $A > 1$.

A new object $\eta \notin \boldsymbol{x}$ is added to the configuration $\boldsymbol{x}$ with the probability (4.26)

$$\min\left\{1, \frac{p_d d(\boldsymbol{x}\cup\{\eta\},\eta)p(\boldsymbol{x}\cup\{\eta\})}{p_b b(\boldsymbol{x},\eta)p(\boldsymbol{x})}\right\} \leq \frac{p_d\Lambda}{p_b} \times u_{n(\boldsymbol{x})} \leq \epsilon.$$

Since $u_{n(\boldsymbol{x})} \to 0$, when $n(\boldsymbol{x})$ is large enough the sequence cannot exceed a certain pre-fixed quantity.

Similarly, a selected object $\eta \notin \boldsymbol{x}$ is deleted from the configuration $\boldsymbol{x} \cup \{\eta\}$ with the probability

$$\begin{aligned} & \min\left\{1, \frac{p_b b(\boldsymbol{x},\eta)p(\boldsymbol{x})}{p_d d(\boldsymbol{x}\cup\{\eta\},\eta)p(\boldsymbol{x}\cup\{\eta\})}\right\} \\ \geq & \min\left\{1, \inf\left[\frac{b(\boldsymbol{x},\eta)}{d(\boldsymbol{x}\cup\{\eta\},\eta)}\right] \times \frac{p_b}{p_d \Lambda}\right\} \geq 1, \end{aligned}$$

because according to the hypothesis we have

$$\inf_{\eta \in W\times M, \boldsymbol{x}\in \Xi_n} \frac{b(\boldsymbol{x},\eta)}{d(\boldsymbol{x}\cup\{\eta\},\eta)} \to \infty$$

when $n(\boldsymbol{x}) \to \infty$.

It follows that for configurations $\boldsymbol{x}$ with a large number of objects $n(\boldsymbol{x}) > N_\epsilon$, we have $\alpha(\boldsymbol{x}, \boldsymbol{x}\cup\{\eta\}) \leq \epsilon$ and $\alpha(\boldsymbol{x}, \boldsymbol{x}\setminus\{\eta\}) = 1$. As a result, by applying the transition kernel (4.7) to the function $V(\boldsymbol{x})$, we obtain

$$\begin{aligned} PV(\boldsymbol{x}) &= \int_\Omega P(\boldsymbol{x}, d\boldsymbol{y})V(\boldsymbol{y}) \\ &= p_b A^{n(\boldsymbol{x})} \int_{W\times M} b(\boldsymbol{x},\eta)(A-1)\alpha(\boldsymbol{x},\boldsymbol{x}\cup\{\eta\})d\sigma(\eta) \\ &+ p_d A^{n(\boldsymbol{x})} \sum_{\eta\in\boldsymbol{x}} d(\boldsymbol{x},\eta)(A^{-1}-1)\alpha(\boldsymbol{x},\boldsymbol{x}\setminus\{\eta\}) + A^{n(\boldsymbol{x})} \\ &\leq \left[p_b(A-1)\epsilon + p_d(A^{-1}-1)+1\right] V(\boldsymbol{x}). \end{aligned} \tag{4.18}$$

Taking the limit in (4.18), it follows that

$$p_d(A^{-1}-1)+1 = p_b + p_c + \frac{p_d}{A} < 1,$$

whenever ϵ goes to zero. It is therefore possible to choose ϵ such that

$$PV(\boldsymbol{x}) = \int_\Omega P(\boldsymbol{x}, d\boldsymbol{y})V(\boldsymbol{y}) \leq aV(\boldsymbol{x}) \tag{4.19}$$

for a value $a < 1$.

Let C be the set of configurations containing less than N_ϵ objects, i.e.

$$C = \{\boldsymbol{x} \in \Omega : n(\boldsymbol{x}) \leq N_\epsilon\}.$$

This is a small set. To verify this statement, we first note that the probability of acceptance for removing an object is always greater than

$$\triangle := \min\left\{\frac{p_b}{u_n p_d \Lambda} : n(\boldsymbol{x}) \leq N_\epsilon\right\}.$$

Obviously, we have $\triangle < 1$. Then using the equations Chapman-Kolmogorov equations and also the fact that $P(\emptyset, \Xi_0) \geq p_d$, it follows that

$$P^{N_\epsilon}(\boldsymbol{x}, \Xi_0) \geq P^{n(\boldsymbol{x})}(\boldsymbol{x}, \Xi_0) P^{N_\epsilon - n(\boldsymbol{x})}(\emptyset, \Xi_0) \geq (p_d \triangle)^{N_\epsilon}, \tag{4.20}$$

for any configuration $\boldsymbol{x}$ containing at most N_ϵ objects. The set C is therefore small with respect to the non-zero measure $(p_d \triangle)^{N_\epsilon} \delta_{\boldsymbol{x}}(\emptyset)$.

Having a positive probability of remaining in the same state, the chain is aperiodic. It is also ϕ−irreducible as a consequence of (4.20). The drift condition (4.19) then implies that it is also Harris recurrent. Defining the function $V(\boldsymbol{x})$ and the set C as before, and posing $b = A^{N_\epsilon + 1}$, we see that the drift condition for geometric ergodicity (4.5) is valid for all $\boldsymbol{x} \in \Omega$. □

The authors in [229] prove uniform ergodicity, which is stronger property than geometric ergodicity, for Metropolis-Hastings kernels simulating hereditary point processes with a bounded number of points. In a certain sense a bounded number of points is a modelling limitation. Nevertheless, for several practical applications such bounds can be found quite easily, using heuristic approaches. Two examples of such heuristics can be given. The first would be to bound the size of the configurations by the highest unsigned integer allowed by the computer on which the algorithm is implemented. The second is to assume that the objects forming the pattern of interest exhibit a pairwise hard-core interaction with ϵ radius. For finite point processes, this implies a bound of the total number of objects in a configuration which is easy to calculate.

4.3 "Perfect" MCMC simulation

The primary criticism of classical MCMC simulation algorithms is that it is not possible to ascertain when their convergence regime has been reached. The solution to this problem is provided by *perfect or exact simulation methods*. These algorithms are MCMC samplers that indicate by themselves when convergence is reached. The Coupling From The Past (CFTP) algorithm presented in the groundbreaking paper of [252] allows to obtain perfect samples for a large class of models. Among these, it is worth mentioning the well-known processes with interactions such as the Ising and the random cluster models. The generality and the cleverness of the presented approach provoked an extraordinary enthusiasm. Over the following two decades after the publication of the paper, a very important part of the statistics community worked to extend the exact simulation framework to more general interacting spatial processes. This section is structured in two parts. The first part introduces the general principles of exact simulation and of the CFTP algorithm. The second part presents the CFTP algorithm for sampling marked point processes. For further reading we recommend the dedicated chapters from the monographs [101, 139, 165, 181, 229, 262, 318]. The excellent book of M. L. Huber [139] is at our best knowledge, the first book entirely dedicated to perfect simulation. For writing this part we have followed closely [181, 195].

Let us suppose that we wish to sample from a distribution π defined on a finite state space $\Omega = \{\omega_1, \omega_2, \ldots, \omega_m\}$, using a Markovian approach. The Markov chain (X_n) is represented by its transition function $\phi(\cdot, \cdot)$

$$X_{n+1} = \phi(X_n, V_n), \tag{4.21}$$

where V_n are i.i.d. random variables.

Let us consider m chains $(X_n(\omega_i))$, all initialised with a different state in Ω, and evolving from time $-T < 0$ to time 0. These chains are coupled, in the sense that they use the same variables V_n. If at a certain moment $n \in \{-T, \ldots, 0\}$ all the chains are in the same state that is $X_n \equiv \boldsymbol{x}$, this phenomenon is called *coalescence*. After this moment, the chains will all continue to evolve in the same state until time 0. The influence of the initial conditions has been eliminated. We can start the simulation well before time $-T$, at infinity, the chains will all be in the same state at the same time. The result is that X_0 is a "perfect" sample of the distribution of interest π.

Even if finite, the cardinal of the state space Ω may have an extremely high value m. So, it is not always practical to run so many chains in parallel. However, if the state space is ordered, with a maximum $\omega_{\max}$ and a mininum $\omega_{\min}$, and if in addition the transition function respects this order relation, in the sense that $\omega \leq \omega'$ implies $\phi(\omega) \leq \phi(\omega')$, then it is not necessary to consider all the possible initial states. Only the states $\omega_{\min}$ and $\omega_{\max}$ are sufficient, as the behaviour of all the other chains will be sandwiched by those of the extremal chains $X_n(\omega_{\min})$ and $X_n(\omega_{\max})$.

This idea is referred to as *Coupling From The Past (CFTP)*. The following algorithm illustrates this principle. The procedure `Updates` describes the transition function (4.21). It depends on the state of the chain and a random variable, which serves to establish the coupling.

Algorithm 4.3. General CFTP sampler. *Data: the probability distribution π of a model defined on the partially ordered finite state space $\Omega = \{\omega_{\min}, \omega_{\max}\}$, the transition function ϕ that preserves the order relation.*

1. *Set $T = 1$.*
2. *Set the random variables $V_n, n = -1, -2, \ldots$ i.i.d. as in (4.21)*

3. Do

$\{$

$U_{-T}(-T) = \omega_{\max}, L_{-T}(-T) = \omega_{\min}$

For $n = -T$ *to* -1 *do*

$\{$

$$U_{-T}(n+1) = \mathit{Update}(U_{-T}(n), V_n)$$
$$L_{-T}(n+1) = \mathit{Update}(L_{-T}(n), V_n)$$

$\}$

If $(U_{-T}(0) = L_{-T}(0))$ *Then* $\boldsymbol{x} = U_{-T}(0) = L_{-T}(0)$

Else $T = 2T$

$\}$

While $(U_{-T}(0) \neq L_{-T}(0))$

4. Return $\boldsymbol{x}$ *and* T.

If more perfect samples are needed, it is possible to restart Algorithm 4.3 or just to use a classical MCMC sampler from the distribution of interest with initial condition the perfect sample, previously obtained. The application of this new transition kernel keeps the samples in the equilibrium regime, since the distribution of interest is invariant by construction.

It is a valid question to ask why we do not couple into the future instead of coupling from the past. That is to say, we should start Algorithm 4.3 at time 0, run it towards the future, and stop at time T when coalescence occurs. The rationale behind this approach is that it may introduce a sampling bias [150, 181].

The following result sets out the conditions under which the output of Algorithm 4.3 will provide an unbiased sample of the equilibrium distribution [181].

Theorem 4.7. *Let* π *be a strictly positive probability on the finite space provided with an order relation* $(\Omega, \leq)$ *with* $\omega_{\min}$ *and* $\omega_{\max}$ *the minimum and the maximum states, respectively. Let* ϕ *be the transition function of an irreducible, aperiodic Markov chain with* π *as its invariant distribution. The function* ϕ *respects the order relation on* Ω. *If Algorithm 4.3 finishes almost surely, then it returns an unbiased sample of* π.

4.3.1 Perfect simulation for marked point processes

The key idea for developing perfect sampling algorithms based on the CFTP principle is to create transition functions ϕ that is `Update` mechanisms that mimic classical MCMC dynamics and ensure coalescence. For instance, the

CFTP algorithm that exactly simulates the Ising model uses a transition function that reproduces the Gibbs sampler [252]. Nevertheless, it should be noted that not all transition functions that reproduce MCMC dynamics guarantee coalescence [181].

The CFTP algorithm for the simulation of marked point processes uses a transition function that replicates the simulation of spatial birth-and-death processes. In the following, spatial birth-and-death processes are first introduced. Then the CFTP algorithm for obtaining perfect samples from locally stable hereditary marked point processes is given together with a convergence proof [152, 181, 195, 229].

4.3.2 Spatial birth-and-death processes

The use of spatial birth-and-death processes is perhaps the oldest technique for simulating point processes [149, 181, 229, 251, 258]. A *spatial birth-and-death process* is a continuous-time stochastic process $(X_t)_{t\geq 0}$ such that $X_t = \boldsymbol{x}$ is the configuration of a point process on ω. This process evolves by jumps. A jump consists of adding or deleting a point. The number of jumps in a bounded time interval is finite.

The law of X_t is characterised by the functions $b(\boldsymbol{x}, \eta)$ and $d(\boldsymbol{x}, \eta)$, representing the birth and the death rates, respectively. The total birth rate is

$$B(\boldsymbol{x}) = \int_{W\times M} b(\boldsymbol{x}, \eta) d\sigma(\eta), \tag{4.22}$$

and the total death rate is

$$D(\boldsymbol{x}) = \sum_{\eta\in\boldsymbol{x}} d(\boldsymbol{x}, \eta). \tag{4.23}$$

The following algorithm simulates a spatial birth-death process.

Algorithm 4.4. Simulating a spatial birth-and-death process. *Data: b birth rate, d death rate and T the continuous time horizon.*

1. *Set* $t = 0$, $X^{(0)} = \boldsymbol{x}_0$ *with* $\boldsymbol{x}_0 \in \Omega$ *and* $n = 0$
2. *While* $(t < T)$

   {

   · $\boldsymbol{x} = X^{(n)}$

   · *Compute* $D(\boldsymbol{x})$ *et* $B(\boldsymbol{x})$

· *The sojourn time* $t^{(n)}$ *is chosen according to the exponential distribution with mean* $1/(D(\boldsymbol{x}) + B(\boldsymbol{x}))$ *independently of all the other sojourn times and the other previous states*

· *The next jump is birth with probability* $\frac{B(\boldsymbol{x})}{B(\boldsymbol{x})+D(\boldsymbol{x})}$*: a new object* η *is generated by means of the probability density*

$$\frac{b(\boldsymbol{x}, \eta)}{B(\boldsymbol{x})}$$

with respect to the reference measure σ *and then added to the initial configuration* $\boldsymbol{x} := \boldsymbol{x} \cup \{\eta\}$

· *The next jump is death with probability* $\frac{D(\boldsymbol{x})}{B(\boldsymbol{x})+D(\boldsymbol{x})}$*: the object* η *is deleted from the configuration with probability*

$$\frac{d(\boldsymbol{x}, \eta)}{D(\boldsymbol{x})},$$

and the new configuration becomes $\boldsymbol{x} := \boldsymbol{x} \setminus \{\eta\}$

· $t = t + t^{(n)}$, $n = n + 1$, $X^{(n)} = \boldsymbol{x}$

}

3. Return $\boldsymbol{x}, t$ *and* n.

In relation to this algorithm, the Metropolis-Hastings dynamics already presented can be seen as a "discretisation". This is because we are more interested in what happens at the moments when a jump occurs. In both algorithms, the transitions are births and deaths, i.e. the addition or deletion of points to the current configuration. However, in the case of birth and death processes, transitions are always accepted.

As with the Metropolis-Hastings algorithm, there are several possible choices for the birth and death rates. These choices must guarantee that the algorithm converges and does not explode. This means that the birth and death process has an invariant measure and that the number of jumps remains finite in a bounded time interval. The distribution of the birth and death process induces a distribution on the point processes observed at the time of the jumps. We want to control the distribution of this point process.

Let $p(\boldsymbol{x})$ be the probability density with respect to the standard Poisson process of a point process that we wish to simulate. Let's assume here that the detailed balance condition

$$b(\boldsymbol{x}, \eta)p(\boldsymbol{x}) = p(\boldsymbol{x} \cup \{\eta\})d(\boldsymbol{x} \cap \{\eta\}, \eta) \tag{4.24}$$

is respected.

Proposition 4.2. *If (4.24) is verified and if* $\mathbb{E}B(X) < \infty$ *with* $X \sim \pi = \int p(\boldsymbol{x})d\mu(\boldsymbol{x})$, *then* $X_t, t \geq 0$ *is reversible with respect to* π.

This means that if $X_0 \sim \pi$, then for all $t > 0$, the process $X_s, 0 \leq s \leq t$ and $X_{t-s}, 0 \leq s \leq t$ are identically distributed.

Theorem 4.8. *Let* $n \in \mathbb{N}$, $\beta_n = \sup_{n(\boldsymbol{x})=n} B(\boldsymbol{x})$ *and* $\delta_n = \inf_{n(\boldsymbol{x})=n} D(\boldsymbol{x})$. *Suppose the detailed balance condition (4.24) satisfied and* $\delta_n > 0$ *for all* $n > 1$. *It is also assumed that one of the following conditions is satisfied*
(C1) $\beta_n = 0$ *for sufficiently large* $n > 0$,
(C2) $\beta_n > 0$ *for* $n > 1$ *and*

$$\sum_{n=2}^{\infty} \frac{\beta_1 \cdots \beta_{n-1}}{\delta_1 \cdots \delta_n} < \infty, \qquad \sum_{n=1}^{\infty} \frac{\delta_1 \cdots \delta_{n-1}}{\beta_1 \cdots \beta_n} = \infty.$$

Then there exists a unique spatial birth and death process defined by the transition rates $b(\cdot,\cdot)$ *and* $d(\cdot,\cdot)$. *This process has a unique equilibrium distribution to which it converges independently of the initial state. If the condition (4.24) follows from the distribution of a point process, i.e.* $\pi = \int p(\boldsymbol{x})d\mu(\boldsymbol{x})$, *then*

$$\pi(F) = \lim_{t \to \infty} \mathbb{P}(X_t \in F | X_0 = \boldsymbol{x}).$$

Let $p(\cdot)$ be the probability density of a locally stable point process, i.e. $\lambda(\eta; \boldsymbol{x}) < \Lambda$ with $\Lambda > 0$. The spatial birth and death process constructed with $b(\boldsymbol{x}, \eta) = \lambda(\eta; \boldsymbol{x})$ and $d(\boldsymbol{x}, \eta) = 1$ satisfies the detailed balance condition (4.24). The total birth rate is

$$B(\boldsymbol{x}) = \int_{W \times M} b(\boldsymbol{x}, \eta) d\sigma(\eta) \leq \Lambda \nu(W)$$

with $\beta_n \leq \Lambda \nu(W)$, and the total death rate is $D(\boldsymbol{x}) = n$ with $\delta_n = n$.

If $\beta_n > 0$ for all $n > 1$, then the condition $C2$ from the previous result is verified since

$$\sum_{n=2}^{\infty} \frac{\beta_1 \cdots \beta_{n-1}}{\delta_1 \cdots \delta_n} \leq \sum_{n=2}^{\infty} \frac{[\Lambda \nu(W)]^{n-1}}{n!} < \infty,$$

and

$$\sum_{n=1}^{\infty} \frac{\delta_1 \cdots \delta_{n-1}}{\beta_1 \cdots \beta_n} \geq \sum_{n=1}^{\infty} \frac{n!}{[\Lambda \nu(W)]^{n-1}} = \infty.$$

Under these conditions, Theorem 4.8 ensures that there is a unique spatial birth and death process defined by the transition rates $b(\boldsymbol{x}, \eta) = \lambda(\eta; \boldsymbol{x})$ and $d(\boldsymbol{x}, \eta) = 1$, and the law of this process converges to $\pi = \int p d\mu$ from any initial configuration.

The Markov chain simulated by Algorithm 4.4 is uniformly ergodic [229]. This has been an additional argument used by the community to simulate point processes from birth and death processes. However, the use of this algorithm is limited by the need to be able to compute the total birth rate $B(\boldsymbol{x})$. A priori, this problem can be circumvented by a *thinning procedure.* Its principle is as follows: simulate a birth and death process with a total birth rate $B(\boldsymbol{x}) = \Lambda\nu(W)$, and accept the transition $\boldsymbol{x} \to \boldsymbol{x} \cup \{\eta\}$ with probability $\lambda(\boldsymbol{x};\eta)/\Lambda$. This technique is numerically efficient when the interactions between points are weak. In the opposite case, Λ can be very large, which leads to a very low acceptance rate of births and limits the practical use of the algorithm [181].

4.3.3 CFTP algorithms for marked point processes

As announced earlier, this section shows how the spatial birth and death processes are embedded in the CFTP framework to obtain perfect samples of the probability densities of marked point processes. Let $p(\cdot)$ be the probability density of a marked point process X on $W \times M$. The space of configurations Ω is ordered by the inclusion relation. The empty set is the minimum element, but there is no maximum element. It is replaced by a stochastic maximum.

The birth-and-death process is a continuous time process, whereas the CFTP algorithm is a discrete time algorithm. The role of the transition function $\phi(\cdot,\cdot)$ in (4.21) is then played by the transition rates $B(\boldsymbol{x})$ and $D(\boldsymbol{x})$ given by (4.22) and (4.23) respectively.

Let us assume that the point process X is locally stable, i.e. that there exists $\Lambda > 0$ such that

$$0 \leq \lambda(\eta;\boldsymbol{x}) = \frac{p(\boldsymbol{x} \cup \{\eta\})}{p(\boldsymbol{x})} \leq \Lambda,$$

for any $\eta \in W$ and any $\boldsymbol{x} \in \Omega$ such that $p(\boldsymbol{x}) > 0$. Let us also suppose that the conditional density is increasing with respect to the inclusion, i.e.

$$\lambda(\eta;\boldsymbol{x}) \leq \lambda(\eta;\boldsymbol{x}') \quad \text{if} \quad \boldsymbol{x} \subseteq \boldsymbol{x}'. \tag{4.25}$$

The point processes that verify (4.25) are called *monotonic* or *attractive.*

Under these assumptions, the application of the CFTP principles for getting perfect samples of marked point processes probability densities is described by the following algorithm [152, 181, 229, 195]:

Algorithm 4.5. CFTP for marked point processes. *Data: the probability density* p *of the locally stable marked point process of interest. Output:* $\boldsymbol{x}$ *a perfect sample,* T *a realisation of the coalescence time.*

1. *Let* $V_t, t \leq 0$ *be a family of random variables identically, independently uniformly distributed over* $[0,1]$.

2. *Let $D(0)$ be a realisation of the Poisson process with intensity Λ on $W \times M$. The points are distributed uniformly in W, while the marks are generated independently following ν_M.*

3. *Set $T = 1$.*

4. *The process $D(\cdot)$ is simulated from 0 to $-T$ by means of a spatial birth-and-death process with birth rate $b(\boldsymbol{x}, \eta) = \Lambda$ and death rate $d(\boldsymbol{x}, \eta) = 1$.*

5. *Generate $L^{-T}(\cdot)$ (lower process) and $U^{-T}(\cdot)$ (upper process) from $-T$ to 0 as follows:*

   - *$L^{-T}(-T) = \emptyset$ and $U^{-T}(-T) = D(-T)$,*
   - *if $D(\cdot)$ experienced a birth while going into the past, that is $D(t-) = D(t) \cup \{(w, m)\}$ where $D(t-)$ is the state just prior to the time t, then the point (w, m) is deleted from both $L^{-T}(t-)$ and $U^{-T}(t-)$,*
   - *if $D(\cdot)$ experienced a death while going into the past, that is $D(t-) = D(t) \setminus \{(w, m)\}$, then the point (w, m) is added to $L^{-T}(t-)$ if*

     $$V_t \leq \alpha_{\min}(U^{-T}(t-), L^{-T}(t-), (w, m)),$$

     and to $U^{-T}(t-)$ if

     $$V_t \leq \alpha_{\max}(U^{-T}(t-), L^{-T}(t-), (w, m)).$$

6. *If $U^{-T}(0) = L^{-T}(0)$ stop. Else, put $T = 2T$ and restart.*

7. *Return $\boldsymbol{x} = U^{-T}(0)$ and T.*

The acceptance probabilities for the births are given by

$$\begin{aligned} &\alpha_{\min}(U, L, (w, m)) := \min\{\lambda((w, m); \boldsymbol{x})/\Lambda : L \subseteq \boldsymbol{x} \subseteq U\} \\ &\text{and} \\ &\alpha_{\max}(U, L, (w, m)) := \max\{\lambda((w, m); \boldsymbol{x})/\Lambda : L \subseteq \boldsymbol{x} \subseteq U\}. \end{aligned} \tag{4.26}$$

It should be noted that the values of the generated random processes are reused when the time is doubled.

When the process to be simulated is attractive, then the probabilities (4.26) become

$$\begin{aligned} &\alpha_{\min}(U, L, (w, m)) := \lambda((w, m); L)/\Lambda \\ &\text{et} \\ &\alpha_{\max}(U, L, (w, m)) := \lambda((w, m); U)/\Lambda. \end{aligned}$$

The point processes that satisfy the following condition are said to be *anti-monotonic* or *repulsive*,

$$\lambda(\eta; \boldsymbol{x}) \leq \lambda(\eta; \boldsymbol{x}') \quad \text{si} \quad \boldsymbol{x}' \subseteq \boldsymbol{x}.$$

The previous algorithm can be used for these processes too, by posing

$$\alpha_{\min}(U, L, (w, m)) := \lambda((w, m); U)/\Lambda$$
and
$$\alpha_{\max}(U, L, (w, m)) := \lambda((w, m); L)/\Lambda.$$

The following result proves the convergence of Algorithm 4.5.

Theorem 4.9. *Let X be a locally stable marked point process on $W \times M$ defined by the probability $\pi = \int p d\mu$, where μ is the standard Poisson reference measure with i.i.d. marks distributed according to ν_M. Algorithm 4.5 finishes almost certainly and returns an unbiased sample from the probability density $p(\cdot)$.*

Proof. In the following we present the proof given by [181]. The reader may also consult the related results and proofs that can be found in [152, 229].

The dominant process $D(\cdot)$ is a stationary Poisson process with intensity Λ. For all $T > 0$, we have

$$\emptyset = L_{-T}(-T) \subseteq U_{-T}(-T) = D(-T).$$

By construction, the functions α_{min} and α_{max} respect the order relation induced by inclusion, i.e.

$$L_{-T}(t) \subseteq U_{-T}(t), \tag{4.27}$$

for all $t \in [-T, 0]$. Furthermore, the processes L and U exhibit the *funneling property*

$$L_{-T}(t) \subseteq L_{-S}(t) \subseteq U_{-S}(t) \subseteq U_{-T}(t), \tag{4.28}$$

whenever $-S \leq -T \leq t \leq 0$. The first inclusion can be verified by noting that $L_{-T}(t) = \emptyset \subseteq L_{-S}(t)$ and that the definition of the transitions respects the inclusion. For the last inclusion, we have $U_{-T}(t) = D(-T) \supseteq U_{-S}(-T)$. If $L_{-T}(t_0) = U_{-T}(t_0)$, then $L_{-T}(t) = U_{-T}(t)$ for all $t \in [t_0, 0]$ because the processes are coupled.

Let $X_{-T} = \emptyset$ and define the process $X_{-T}(\cdot)$ on $[-T, 0]$ in the same way as the processes L and U, with the difference that a birth is accepted if $V_t \leq \lambda(\eta; X(t-))/\Lambda$. The process $X_{-T}(\cdot)$ is a spatial birth-and-death process with a birth rate $\lambda(\cdot, \cdot)$ and unity death rate. By hypothesis $\lambda(\eta; \boldsymbol{x}) \leq \Lambda$, so the total birth rate is bounded by $\Lambda\nu(W)$. By virtue of Theorem 4.8, the process $X_{-T}(\cdot)$ has π as its equilibrium distribution.

By the monotonicity property of the construction, we have $L_{-T}(0) \subseteq X_{-T}(0) \subseteq U_{-T}(0)$. Assuming for the moment that the algorithm finishes almost surely, then the limit $lim_{T \to \infty} X_{-T}(0)$ is well-defined with probability 1. Since D is in equilibrium and reversible, $X_{-T}(0)$ has the same distribution as its extension into the future, always coupled with the process D, during a period of time of length T. It follows that $L_{-T}(0)$ is an unbiased sample of $p(\cdot)$.

It remains to prove that coalescence occurs almost surely. We know that $\pi_\Lambda(\emptyset) = \exp[-\Lambda\nu(W)] > 0$ where π_Λ is the distribution of a Poisson process with intensity Λ. For $n \in \mathbb{N}^\star$, let us define $E_n = \mathbf{1}\{D(-n) \neq \emptyset\}$. It turns out that (E_n) is an irreducible and aperiodic Markov chain on $\{0, 1\}$, and therefore recurrent since it is discrete [96]. The probability of the 0 state is $\pi_\Lambda(\emptyset)$ which is strictly positive. Consequently, the 0 state is reached with probability 1. This means that the process $D(t)_{t\leq 0}$ will almost surely be "empty", in which case the coupling implies that the algorithm almost certainly ends and the proof is complete. □

5

Applications

This chapter presents first the construction and a simulation study of a Metropolis-Hastings dynamics, designed specifically to simulate the Candy and the Bisous models [192, 292]. The chapter continues with the presentation of several exact simulation algorithms together with their convergence proofs. Their numerical performances for different marked point processes with interactions are also analysed [195].

5.1 Building tailored to the model simulation algorithms

The convergence of Metropolis-Hastings algorithms is affected by the choice of the proposal distributions. For the simulation of marked point processes, we have shown that the choice of uniform proposals $b(\boldsymbol{x}, \eta) = \frac{1}{\nu(W)}$ and $d(\boldsymbol{x}, \eta) = \frac{1}{n(\boldsymbol{x})}$ guarantees remarkable convergence properties of the dynamics thus constructed. However, such a choice seems less natural if the process to be simulated exhibits more structured patterns, with strong interactions between objects. In the case of processes such as Candy or Bisous, it is legitimate to think of proposals that "help" the model, for example, by proposing objects connected to the existing pattern.

Following this idea, a first attempt to build a tailored to the model simulation algorithm for the Candy model was done in [282, 290]. The authors construct an adapted reversible jumps Monte Carlo dynamics and numerical results show the performance of the proposed algorithm, without however giving proof of its theoretical convergence. This problem was subsequently treated and solved in the context of Metropolis-Hastings dynamics by [192]. Since then, the proposed solution has been generalised to the simulation of the Bisous model [292].

The adopted principle for implementing this solution is to construct a non-uniform probability density for the proposal to add an object to the

DOI: 10.1201/9781003380382-5

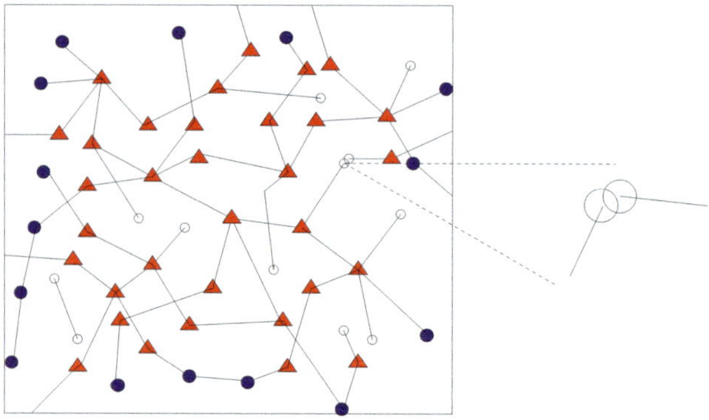

FIGURE 5.1
A random configuration of connected segments with attraction regions around their extremities.

configuration. Its expression is given by

$$b(\boldsymbol{x},\eta) = \frac{p_1 \mathbf{1}\{\eta \in W\}}{\nu(W)} + p_2 b_a(\boldsymbol{x},\eta), \tag{5.1}$$

where $p_1 + p_2 = 1$ and $b_a(\boldsymbol{x},\eta)$ is a probability density proposing an object in attraction and alignment with another object in the configuration. The role of the second component is to propose objects that are not penalised by the model and have more chances of being accepted by the simulation dynamics.

This strategy is illustrated in Figure 5.1. The white circles represent the extremities around which we want to connect other segments to extend the existing network. The red triangles represent the extremities that are already connected. Due to the definition of the model, a proposal to extend the network in these regions is likely to be rejected. The blue circles represent the free extremities that are too close to the edge of the W domain. Segments connected in these regions are unlikely to be part of the observed domain. Therefore, we propose to extend the network by regions where connectivity is allowed. Such a strategy must always take into account possible edge effects [282, 192, 292]. Sampling techniques are proposed in [13] to avoid these pitfalls.

The expression for the probability density $b_a(\boldsymbol{x},\eta)$ is given by

$$b_a(\boldsymbol{x},\eta) = \frac{1}{n(A(\boldsymbol{x}))} \sum_{x \in A(\boldsymbol{x})} \tilde{b}(x,\eta)$$

where $A(\boldsymbol{x})$ is the set of objects in the configuration $\boldsymbol{x}$ that do not have all their q extremities connected. First, an object x is uniformly selected from the set $A(\boldsymbol{x})$. Then a new object $\eta = (w_\eta, m_\eta)$ with parameters (w, m_1, m_2) is proposed according to the probability density function

$$\tilde{b}(w, m_1, m_2) = f(w|x)g(m_1)h(m_2). \tag{5.2}$$

The terms chosen to construct the density (5.2) are as follows:

$$\begin{aligned} f(w|x) &= \frac{\mathbf{1}\{w \in \tilde{a}(x)\}}{\nu(\tilde{a}(x) \cap W)} \\ g(m_1) &= 1 \\ h(m_2) &= \frac{2}{\tau}\mathbf{1}\{m_2 \in (1-\tau, 1)\}, \end{aligned}$$

where $\tilde{a}(x) = \bigcup_{u=1}^{\tilde{q}} b(e_u(x), r_a)$ is the union of all attracting spheres of x that do not contain an extremity of any other object in $\boldsymbol{x}$. Using m_1 and m_2, the vector m_η is obtained by one translation and two rotations to satisfy the alignment conditions with respect to m_x. By performing a translation according to the attraction rule $\mathcal{C}$, the point w becomes the centre of the new object w_η.

Using the proposal (5.1) to add objects and the uniform distribution to remove objects (4.17) leads to the construction of a convergent Metropolis-Hastings dynamics. The results presented earlier show that the Markov chain simulated by this algorithm is $\phi-$irreducible, Harris recurrent and geometrically ergodic, provided that

$$\sup_{\eta \in \Omega, n(\boldsymbol{x})=n} \frac{d(\boldsymbol{x} \cup \{\eta\}, \eta)}{b(\boldsymbol{x}, \eta)} \to 0$$

when $n \to \infty$. For more details on how to build a tailored to the model proposal we recommend [192, 292].

5.1.1 Simulation study of the tailored dynamics built for sampling the Candy model

This section presents a simulation study of a Metropolis-Hastings dynamics specifically tailored to simulate the Candy model [192]. This model generates random configurations of segments that connect and align to form a network. It can also be seen as a two-dimensional application of the Bisous model to build linear patterns [292]. The process takes place in $W = [0, 256] \times [0, 256]$. The segments are of uniform length between $[30, 40]$ and their orientation is uniform in $[0, \pi)$. The radius of attraction is $r_a = 1/\sqrt{\pi}$. The thresholds defining the repulsion and alignment interactions are $\delta = 0.05\pi$ and $\tau = 0.2\pi$ respectively. A realisation of the model with the interaction parameters and the observed sufficient statistics is shown in Figure 5.2. The parameters $\log \gamma_s$ control the number of segments n_s with s extremities connected to the rest of the network. Since a segment has two ends, $s = \{0, 1, 2\}$. The parameters

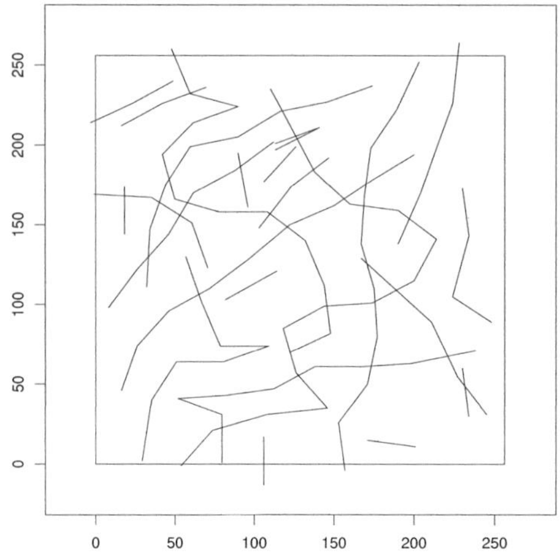

Interaction parameters	Sufficient statistics
$\log \gamma_2 = 2.5$	$n_2 = 59$
$\log \gamma_1 = -3.0$	$n_1 = 32$
$\log \gamma_0 = -8.5$	$n_0 = 7$
$\log \gamma_r = -2.5$	$n_r = 11$
$\log \gamma_o = -2.5$	$n_o = 12$

FIGURE 5.2
Realisation of the Candy model together with the interaction parameters and the observed sufficient statistics.

$\log \gamma_r$ and $\log \gamma_o$ control n_r and n_o, the number of pairs of segments that have repulsion and misalignment interactions respectively.

In a first experiment, the Metropolis-Hastings algorithm defined by (4.7) and the probabilities $p_b = 0.6$, $p_d = 0.2$ and $p_c = 0.2$ was run from an empty configuration for 2×10^7 iterations. Sufficient statistics were collected every 10^3 iterations. The probability density of adding a new object to the configuration $b(\boldsymbol{x}, \eta)$ is of the form (5.1), where $p_1 = 0.2$ and $p_2 = 0.8$. For a configuration of segments $\boldsymbol{x}$, the set $A(\boldsymbol{x})$ is the union of all attracting spheres with centres at a distance greater than $\frac{1}{2}l_{\max} + r_a$ from the edge of W. The probability of deleting an object is given by (4.17). An object is modified in two ways. First, with probability $p_{c1} = 0.1$ an object is uniformly selected in the configuration and new values are proposed for its parameters according to the law $c(\boldsymbol{x}, x_i, \eta) = b(\boldsymbol{x} \setminus \{x_i\}, \eta)$. Second, with probability $p_{c2} = 0.1$ an object is selected according to the same scheme, then a uniform modification of its orientation over $[0, \pi)$ is proposed.

Figure 5.3 illustrates the dependence of the morphology of networks simulated with the Candy model on the model parameters. The connectivity of the network is influenced by the parameters γ_0, γ_1 and γ_2, while its curvature is controlled by the parameters γ_o and γ_r.

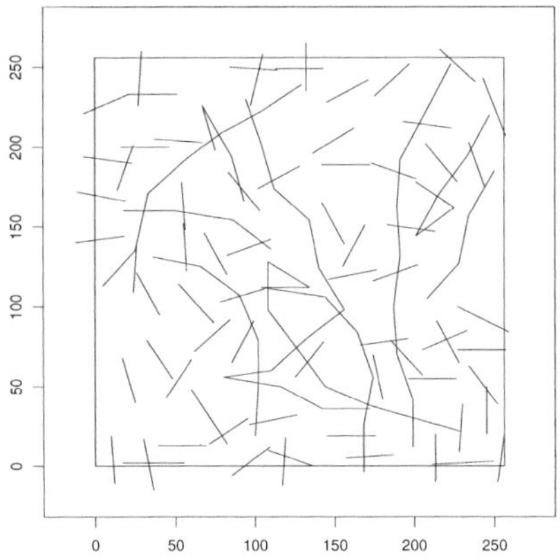

Interaction parameters	Sufficient statistics
$\log \gamma_2 = 2.5$	$n_2 = 39$
$\log \gamma_1 = -3.0$	$n_1 = 20$
$\log \gamma_0 = -5.0$	$n_0 = 68$
$\log \gamma_r = -2.5$	$n_r = 15$
$\log \gamma_o = -2.5$	$n_o = 4$

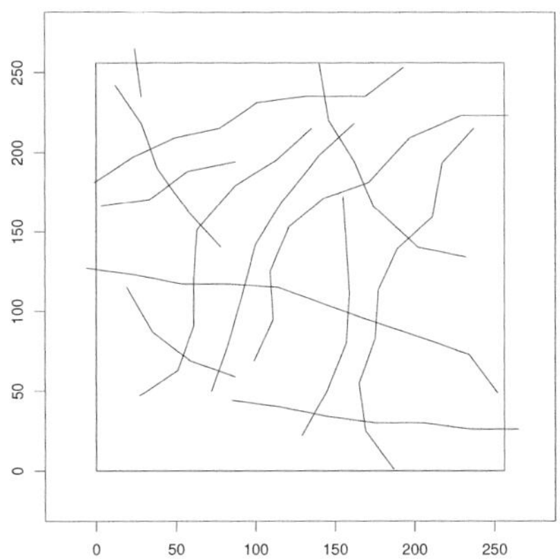

Interaction parameters	Sufficient statistics
$\log \gamma_2 = 2.5$	$n_2 = 47$
$\log \gamma_1 = -3.0$	$n_1 = 24$
$\log \gamma_0 = -8.5$	$n_0 = 1$
$\log \gamma_r = -6.5$	$n_r = 0$
$\log \gamma_o = -6.5$	$n_o = 0$

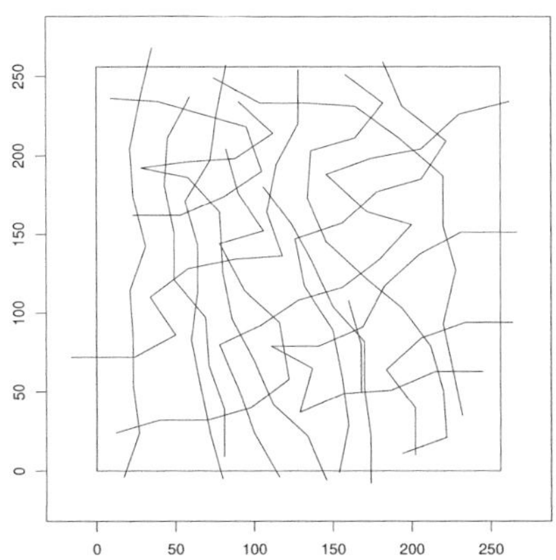

Interaction parameters	Sufficient statistics
$\log \gamma_2 = 4.0$	$n_2 = 107$
$\log \gamma_1 = -3.0$	$n_1 = 30$
$\log \gamma_0 = -8.5$	$n_0 = 0$
$\log \gamma_r = -2.5$	$n_r = 14$
$\log \gamma_o = -2.5$	$n_o = 23$

FIGURE 5.3
Control of the morphology of the network simulated with the Candy model via the interaction parameters. From up to bottom: fragmented network, well aligned network, dense network allowing repulsion and mis-alignment.

The second experiment examines the performance of the Metropolis-Hastings algorithm depending on the initial configuration and the influence of the different moves that make up the transition kernel on the speed of convergence.

Figure 5.4 shows two realisations of the reference Candy model (its parameters are given in Figure 5.2) obtained by the previous algorithm, where the initial configurations are the realisation of a binomial process and a random network, respectively. The binomial process is implemented for 200 segments. The random network is obtained by a Metropolis-Hastings algorithm using only "change" moves. The algorithm is defined by a transition kernel (4.7) with $p_b = p_d = 0$, $p_{c1} = p_{c2} = 0.5$ and proposal laws $c(\boldsymbol{x}, x_i, \eta)$ and $c_\theta(\boldsymbol{x}, x_i, \eta)$ defined as above. The initial state of this algorithm is a realisation of a binomial process with 200 segments. As in the first experiment, the algorithms were run for 2×10^7 iterations and sufficient statistics were taken every 10^3 iterations. It can be observed that the mean values of the observed statistics are close and their evolution during the simulation does not give any indication against convergence.

In the following the simulation conditions of the reference model in Figure 5.2 were kept exactly the same as before, while the probabilities of proposing different moves were varied. Each row in Figure 5.5 shows results obtained with a different set of values for the Metropolis-Hastings algorithm's proposal probabilities. The results shown for each simulation are the final configuration, the evolution of the cumulative mean of the total number of segments, and the mean of the other sufficient statistics. For the first two rows (top and middle), each algorithm considered uses a tailor-made proposal (5.1) while all moves are proposed with a different probability. The results obtained in these two situations are very similar and comparable to those of the first experiment. For the third row (bottom), the simulation was performed using only uniform births and deaths. It can be seen that the evolution of the cumulated mean does not indicate convergence and that the values of all the sufficient statistics are really different from those obtained in the previous simulations.

All these results are consistent with Theorem 4.6. Clearly, the choice of proposal probabilities affects the speed of convergence. Using a transition kernel tailored to the model improves the general performance of the algorithm. The price for this is computational time, since the evaluation of (5.1) requires more mathematical operations than the simple uniform proposal.

5.1.2 Monte Carlo dynamics for sampling T–tessellations

For the construction of a T–tesselation, we implemented the ideas used for the adapted dynamics simulating point processes with interaction. This time we had to take into account that the transition from one configuration to

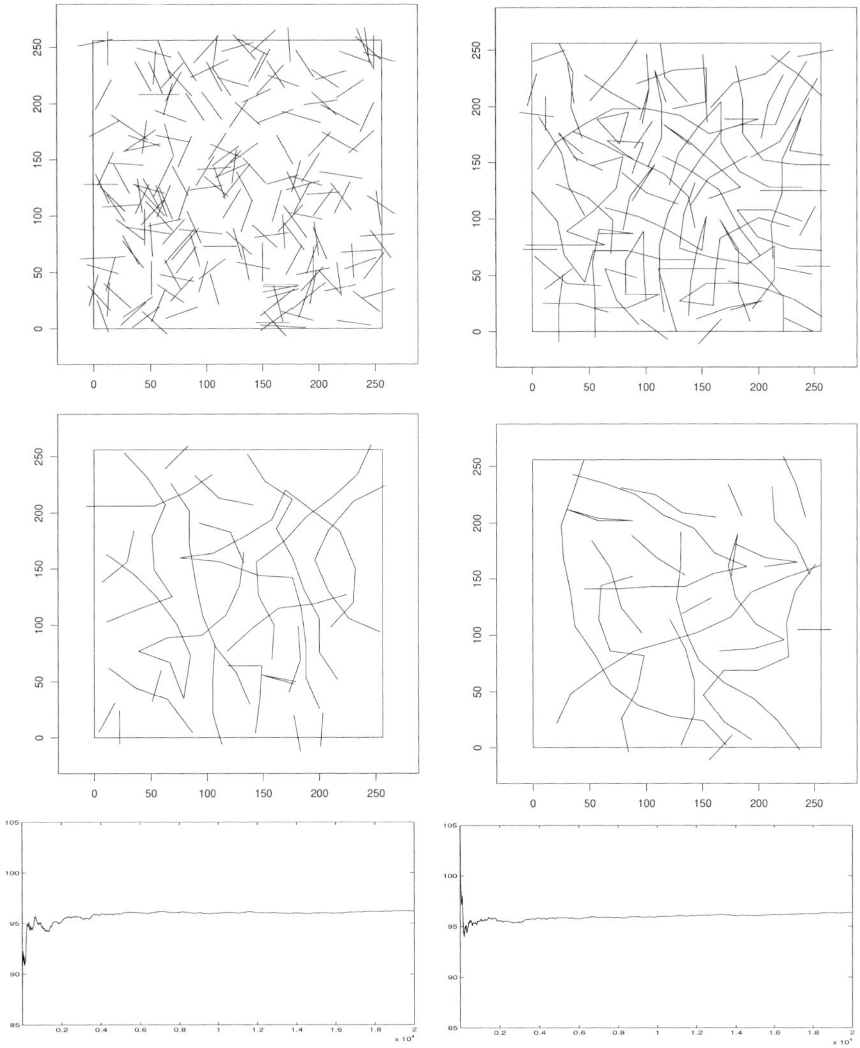

$\bar{n}_t = \bar{n}_2 + \bar{n}_1 + \bar{n}_0 = 96.23$
$\bar{n}_o = 8.94,\ \bar{n}_r = 7.84$

$\bar{n}_t = \bar{n}_2 + \bar{n}_1 + \bar{n}_0 = 96.39$
$\bar{n}_o = 8.93,\ \bar{n}_r = 8.19$

FIGURE 5.4
Initial conditions dependence of the Metropolis-Hastings dynamics for the simulation of the Candy model. Results of two simulations are presented in the left and right column, respectively. For each simulation we have: initial condition (up), final realisation (middle), evolution of the cumulated mean of the total number of segments during the simulation and the mean value of the other sufficient statistics (bottom).

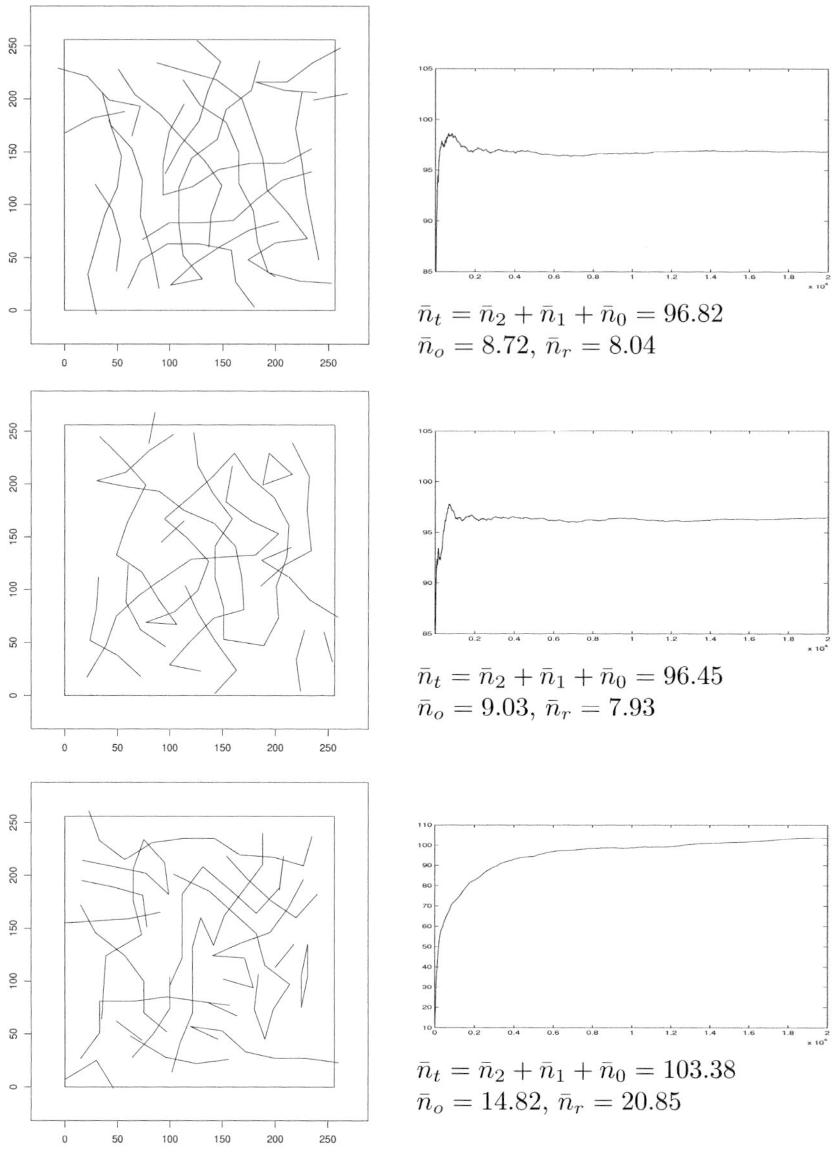

$\bar{n}_t = \bar{n}_2 + \bar{n}_1 + \bar{n}_0 = 96.82$
$\bar{n}_o = 8.72,\ \bar{n}_r = 8.04$

$\bar{n}_t = \bar{n}_2 + \bar{n}_1 + \bar{n}_0 = 96.45$
$\bar{n}_o = 9.03,\ \bar{n}_r = 7.93$

$\bar{n}_t = \bar{n}_2 + \bar{n}_1 + \bar{n}_0 = 103.38$
$\bar{n}_o = 14.82,\ \bar{n}_r = 20.85$

FIGURE 5.5
Three simulations results (final configuration and behaviour of the sufficient statistics) for three different sets of parameters for the simulation dynamics. Up: $p_b = 0.45$, $p_d = 0.15$, $p_{c1} = 0.3$, $p_{c2} = 0.1$. Middle: $p_b = 0.7$, $p_d = 0.1$, $p_{c1} = 0.1$ and $p_{c2} = 0.1$. For both these cases we have: $p_1 = 0.2, p_2 = 0.8$. as well as $p_M = p_{tildeM} = 0.5$. Bottom: $p_b = 0.75$,$p_d = 0.25$, $p_1 = 1.0$ and $p_2 = p_c = 0$.

another is done by three transformations that preserve the structure of a T–tesselation. The three transformations are split, merge and flip. The proposed algorithm is a tailored to the model Metropolis-Hastings-Green (MHG) dynamics [108, 109, 116]. Here we present only the key points in the construction of this simulation dynamics. All the details can be can be found in the paper [159, 1].

Algorithm 5.1. MHG update for T–tesselations. *Data: the probability density $p \propto h$ of the T–tesselation of interest, the initial configuration $\mathbf{T}_n$, the probabilities p_s, p_m, p_f, the proposal kernels $q_s(T,\cdot)$, $q_m(T,\cdot)$ and $q_f(T,\cdot)$. Output: the new T–tesselation $\mathbf{T}_{n+1}$.*

1. *Choose a split, merge or flip transformation with the corresponding probabilities p_s, p_m, p_f.*
2. *If the chosen transformation is not applicable, return $\mathbf{T}_n$.*
3. *Depending on the chosen transformation, propose a new tessellation $\mathbf{T}_{n+1}$, using the appropriate kernel ($q_s(T,\cdot)$, $q_m(T,\cdot)$ or $q_f(T,\cdot)$).*
4. *Calculate the corresponding acceptance ratio $r_t(\mathbf{T}_n, \mathbf{T}_{n+1})$, where $t = \{s, m, f\}$ is the chosen transformation.*
5. *Accept the new configuration $\mathbf{T}_{n+1}$ with the probability*
$$\alpha = \min\{1, r_t(\mathbf{T}_n, \mathbf{T}_{n+1})\}.$$
6. *If the new configuration $\mathbf{T}_{n+1}$ is accepted, then return $\mathbf{T}_{n+1}$, else return $\mathbf{T}_n$.*

Several realisations from the model of interest are obtained by simply iterating the previous algorithm.

A simple version of the algorithm is obtained by choosing the probabilities p_s, p_m and p_t independently of the current tessellation $\mathbf{T}_n$ and considering the following uniform proposals:

$$\begin{aligned} q_s(T,S) &= \frac{\pi}{2l(T) - l(D)}, \\ q_m(T,M) &= \frac{1}{\mathring{n}_{\mathrm{s,nb}}(T)}, \\ q_f(T,F) &= \frac{1}{2\mathring{n}_{\mathrm{s,b}}(T)}. \end{aligned}$$

The acceptance ratio of a proposed split S is

$$r_s(T, ST) = \frac{p(ST)}{p(T)} \frac{p_m}{p_s} \frac{2l(T) - l(W)}{\pi(\mathring{n}_{\mathrm{s,nb}}(T) + 1 - \xi)},$$

where $\xi \in \{0, 1, 2\}$ is the number of internal non-blocking segments of T that are formed by the new segment that creates the split of a cell.

Similarly, for a proposed merge M, we obtain

$$r_m(T, M) = \frac{p(MT)}{p(T)} \frac{p_s}{p_m} \frac{\pi \overset{\circ}{n}_{\mathrm{s,nb}}(T)}{2(l(T) - l(M)) - l(W)},$$

where $l(M)$ is the length of the non-blocking segment removed by the merge M.

Finally, for a proposed flip F, we have

$$r_f(T, F) = \frac{p(FT)}{p(T)} \frac{\overset{\circ}{n}_{\mathrm{s,b}}(T)}{\overset{\circ}{n}_{\mathrm{s,b}}(FT)},$$

$\overset{\circ}{n}_{\mathrm{s,b}}(FT) - \overset{\circ}{n}_{\mathrm{s,b}}(T)$ taking values between -2 and 2.

In [159] the authors prove that the chain constructed in this way is aperiodic, $\phi-$irreducible and converges to the equilibrium distribution $\mathbb{P}(dT) = p(T)\mu(dT)$. These results were obtained within the general state Markov chain framework given by [109, 108]. The authors in [159] also present a simulation study to analyse the speed of convergence and the mixing properties of the proposed algorithm.

5.2 Analysis of some exact simulation algorithms for marked point processes

CFTP is not the only algorithm for obtaining "perfect" samples of a marked point process. In [195], the authors carried out a theoretical and numerical study to compare the performances of different perfect simulation methods for marked point processes.

5.2.1 Clan of ancestors

If the law of the process to be simulated has a partial order structure, the CFTP algorithm is particularly efficient. However, many objects are generated unnecessarily and the coalescence time is determined iteratively. These observations motivated [94] to propose an algorithm that avoids the creation of objects that do not affect the final perfect realisation and that determines its convergence time in a single sweep. The application of this algorithm to the simulation of point processes is given in in [30, 94, 229]. Its adaptation to marked point processes is presented in [195].

Let r be the range of interaction of the probability density p of a marked point process, in the sense that

$$\lambda((w,m);\boldsymbol{x}) = \lambda((w,m);\boldsymbol{x} \cap B(w,r)),$$

and $B(w,r) = \{(u,i) \in W \times M, \| u - w \| \leq r\}$. The previous equation is of course satisfied for r equal to the diameter of W. However, the algorithm which is about to be described is efficient for values of r which are quite small compared with Λ [195].

Algorithm 5.2. Clan of ancestors for marked point processes. *Data: the probability density p of the locally stable marked point process of interest. Output: $\boldsymbol{x}$ a perfect sample, T a realisation of the convergence time.*

Let $D(0)$ be the realisation of a Poisson marked point process on $W \times M$ with intensity Λ and marks independently distributed according to ν_M. The clan of ancestors is initialised as $A = D(0)$.

1. *The process $D(\cdot)$ is simulated into the past using a spatial birth-and-death process with a birth rate $b(\boldsymbol{x},\eta) = \Lambda$ and a death rate $d(\boldsymbol{x},\eta) = 1$. The simulation takes place only if $A_t = A \cap D(t) \neq \emptyset$, otherwise it stops. For each death into the past, i.e. $D(t-) = D(t) \setminus \{(w,m)\}$ for $t < 0$, if $(w,m) \in A \cap D(t)$ then add the points in $B(w,r) \cap D(t)$ to the clan of ancestors A. The resulting stopping time is denoted $-T$.*
2. *Generate $X(\cdot)$ into the future as follows:*
   - *$X(-T) = \emptyset$;*
   - *if a birth occurs into the past, i.e. $D(t-) = D(t) \cup \{(w,m)\}$, then delete (w,m) from $X(t-)$*
   - *if a death occurs into the past i.e. $D(t-) = D(t) \setminus \{(w,m)\}$, and if $(w,m) \in A_t$, then add (w,m) to $X(t-)$ with probability $\lambda((w,m);X(t-))/\Lambda$*
3. *Return $\boldsymbol{x} = X(0)$ and T.*

Algorithm 5.2 is also based on the simulation of spatial birth-and-death processes, just like Algorithm 4.5. However, Algorithm 5.2 does not require monotonicity of the point process, and obtaining coalescence is not necessary. A simulation into the past followed by a return into the future is enough to build the clan of ancestors and to thin out the dominant process. The structure of the interaction is crucial. Here a fixed range interaction has been used. It can be replaced by the *incompatibility index*:

$$I((w,m),(u,n)) := \mathbf{1}\{\sup_{\boldsymbol{x}} |\lambda((w,m);\boldsymbol{x}) - \lambda((w,m);\boldsymbol{x} \cup \{(u,n)\})| > 0\}. \tag{5.3}$$

The convergence proof of Algorithm 5.2 is similar to that of unmarked point processes [94, 30, 229]. Here the proof presented in [193] is given.

To check that Algorithm 5.2 ends, note that the process D is reversible and in equilibrium. The hypotheses imply that by going backwards, the process D will almost certainly empty itself after a finite time T, at which point $A_T = \emptyset$ as well. When simulating into the future, the births depend on the conditional intensity. This guarantees that the resulting process respects the detailed balance condition and admits $p(\cdot)$ as its equilibrium distribution.

Note that for the births of the simulation into the future, a point added to $D(t)$ is also added to $Y(t)$ according to the conditional intensity, which depends only on its neighbours at a distance of at most r. Thus, points that are not in A_t can be ignored.

Furthermore, this algorithm can be used to simulate within a bounded domain, point processes defined on infinite spaces [94].

5.2.2 Exact Metropolis-Hastings

An exact Metropolis-Hastings algorithm for the simulation of locally stable point processes is constructed by [152]. This algorithm uses a discretisation of W space and proposals that depend on the current configuration. A simpler algorithm using fixed proposal distributions and adapted to marked point processes is proposed by [195]. It is the development of this algorithm that is presented below.

The Metropolis-Hastings algorithm and its properties were presented in a previous section (see Algorithm 4.2). We recall here that this method is a technique that makes proposals for updates that may or may not be accepted [109, 108, 181, 229]. If the current state $\boldsymbol{x}$ is a configuration of objects having a positive probability density $p(\boldsymbol{x})$, then a standard update is to choose a birth (add an object) with probability $p_b \in (0,1)$, or a death (delete an object) with probability $p_d = 1 - p_b$. For a birth, a new object η is proposed according to the density $\nu \times \nu_M/\nu(W)$ and accepted with probability

$$\min\left\{1, \frac{p_d \nu(W)\lambda(\eta;\boldsymbol{x})}{p_b(1+n(\boldsymbol{x}))}\right\}. \tag{5.4}$$

For a death, if the $\boldsymbol{x}$ configuration is empty, then the $\boldsymbol{x}$ state remains unchanged. Otherwise, an object η is selected uniformly in $\boldsymbol{x}$ and deleted with probability

$$\min\left\{1, \frac{p_b n(\boldsymbol{x})}{p_d \nu(W)\lambda(\eta; \boldsymbol{x} \setminus \{\eta\})}\right\}. \tag{5.5}$$

This dynamics converges towards an unbiased sample of p [109, 108, 181, 192].

Adapting these updates to an exact simulation algorithm requires a great deal of attention, especially for death transitions. The idea behind CFTP methods for locally stable point processes is to simulate a pair of processes that can coalesce. These processes are coupled to a dominant process. The dominant process is supposed to respect the order relation based on inclusion. For this, it would be appropriate for the probability of acceptance of a death to always be 1, regardless of the initial configuration. The acceptance rate (5.5) is clearly larger than $p_b/(p_d\nu(W)\Lambda)$, but it can be smaller than 1. To overcome this difficulty, we discretise W in the manner of [152] and construct updates by strips.

Let $\cup_{i=1}^{n_W} W_i$ be a finite partition of W such that $\nu(W_i) > 0$ for all cells W_i, $i = 1, \ldots, n_W$. If the current configuration is $\boldsymbol{x}$ and the visited strip is $W_i \times M$, then we choose to add a new object with probability p_b^i, or with probability $p_d^i = 1 - p_b^i$ to remove one. Let's assume that the last equality holds for all $i = 1, \ldots, n_W$. In the case of a birth, a new object is created. The position of the object is uniform on W_i and its label is distributed according to ν_M. In case of a death, an object is chosen uniformly among those of W_i, if possible. Metropolis Hastings rates similar to (5.4) and (5.5) are then computed to accept or reject the proposed transitions. This procedure is well-defined on the set of positive probability configurations and converges to p.

To describe the coupling based on this dynamics, we begin by defining the dominant process D. This process accepts all the proposed transitions. The D chain visits the strips with equal probability. If the strip $W_i \times M$ is visited, then a new object is generated with probability p_b^i. The position of the object is uniformly distributed in W_i and its mark is chosen according to ν_M. With probability $p_d^i = 1 - p_b^i$ an object is chosen uniformly among those in W_i and deleted from the current configuration. This operation is executed if the strip W_i is not empty. Otherwise, the D process remains unchanged.

Let us assume that $0 < p_b^i/p_d^i < 1$, i.e. $0 < p_b^i < 1/2$, and let $\pi_i(n) = \mathbb{P}(n \ \text{ points in } \ W_i \times M)$. The detailed balance conditions for the number of objects in each strip i are

$$\pi_i(n)\, p_b^i = \pi_i(n+1)\, p_d^i, \quad n = 0, 1, 2, ;$$

and they give the unique solution

$$\pi_i(n) = \left(\frac{p_b^i}{p_d^i}\right)^n \left(1 - \frac{p_b^i}{p_d^i}\right). \tag{5.6}$$

Consequently, the number of objects in $W_i \times M$ follows a skewed geometric distribution with parameter $1 - p_b^i/p_d^i$. Knowing that we have n objects in a strip $W_i \times M$, at equilibrium they are i.i.d. with the positions distributed according to $\nu(\cdot)/\nu(W_i)$ and the marks chosen according to ν_M. The chain is reversible and the configurations in the different strips $W_i \times M$ are independent.

Let us now consider the target dynamics. The Metropolis-Hastings rate for a transition from the state $\boldsymbol{x} \neq \emptyset$ to the state $\boldsymbol{x} \setminus \{x_j\}$ with $x_j \in (W_j \times M) \cap \boldsymbol{x}$ is given by

$$\frac{p_b^i n(\boldsymbol{x} \cap (W_i \times M))}{p_d^i \nu(K_i) \lambda(x_j; \boldsymbol{x} \setminus \{x_j\})}.$$

By considering the restriction of the chain on the set $\{\boldsymbol{x} : p(\boldsymbol{x}) > 0\}$ and using the fact that p is hereditary, the Metropolis-Hastings rate is well-defined. A sufficient condition for it to be greater than 1 is

$$\frac{p_b^i}{p_d^i} \geq \Lambda \nu(W_i)$$

where Λ is the upper bound of the conditional intensity. Under this condition, these transitions are always accepted.

Let $L \subseteq U$ be two configurations with a finite number of objects, and let

$$\begin{aligned} &\alpha_{\min}(U, L, (w,m), i) := \min\left\{ \tfrac{p_d^i \nu(W_i)\, \lambda((w,m);\boldsymbol{x})}{p_b^i (1+n(\boldsymbol{x} \cap (W_i \times M)))} : L \subseteq \boldsymbol{x} \subseteq U \right\} \\ &\text{and} \\ &\alpha_{\max}(U, L, (w,m), i) := \max\left\{ \tfrac{p_d^i \nu(W_i)\, \lambda((w,m);\boldsymbol{x})}{p_b^i (1+n(\boldsymbol{x} \cap (W_i \times M)))} : L \subseteq \boldsymbol{x} \subseteq U \right\} \end{aligned} \tag{5.7}$$

be the bounds of the Metropolis-Hastings ratios for the birth of an object $(w, m) \notin U$ in the strip i, for the patterns $\boldsymbol{x}$ sandwiched in between L and U by the inclusion.

We now have all the ingredients to propose the following algorithm. As with Algorithm 4.5, the random variables already constructed are reused when the time is doubled.

Algorithm 5.3. Exact Metropolis-Hastings for marked point processes. *Data: the probability density p of the locally stable marked point process of interest. Output: $\boldsymbol{x}$ a perfect sample, T a realisation of the convergence time.*

Let W be partitioned such that $0 < \Lambda\nu(W_i) < 1$ and choose probabilities p_b^i and p_d^i such that $p_b^i/p_d^i \in [\Lambda\nu(W_i), 1)$ for all $i = 1, \ldots, n_W$. Let V_t, $t = -1, -2, \ldots$ be a family of independent uniform random variables on $\{1, \ldots, n_W\}$ and let

U_t be a family of independent uniform random variables on $(0,1)$. Initialise $T = 1$ and let $D(0)$ be a realisation of the marked point process with the points independently distributed by strips as follows: for each strip a number n of objects is chosen with respect to (5.6), then the positions of the objects are independently chosen in W_i according to the $\nu-$uniform distribution and finally an independent mark is associated to each point position according to ν_M.

1. *Simulate $D(\cdot)$ into the past till time $-T$: with probability $p_d^{V_t}$ remove a randomly chosen object from W_{V_t}; otherwise add an object ξ_t with a $\nu-$uniform position in W_{V_t} and a mark given by ν_M, independently of the other variables.*
2. *Generate $L^{-T}(\cdot)$ and $U^{-T}(\cdot)$ into the future as follows:*
   - *$L^{-T}(-T) = \emptyset$ and $U^{-T}(-T) = D(-T)$*
   - *If $D(\cdot)$ experienced a birth in the past, i.e. $D(t) = D(t+1) \cup \{(w,m)\}$, then an object is deleted from $L^{-T}(t)$ and $U^{-T}(t)$ by using the permutation mechanism [151, 152, 195, 284]*
   - *If $D(\cdot)$ experience a death in the past, i.e. $D(t) = D(t+1) \setminus \{(w,m)\}$, the object (w,m) is added to $L^{-T}(t)$ if*

     $$U_t \leq \alpha_{\min}(U^{-T}(t), L^{-T}(t), (w,m), V_t)$$

     and to $U^{-T}(t)$ if

     $$U_t \leq \alpha_{\max}(U^{-T}(t), L^{-T}(t), (w,m), V_t)$$
3. *If $U^{-T}(0) = L^{-T}(0)$ stop. Else $T = 2T$ and re-start.*
4. *Return $\mathbf{x} = U^{-T}(0)$ and T.*

If p is repulsive, the equations (5.7) become

$$\alpha_{\min}(U, L, (w,m), i) = (p_d^i \nu(W_i)\, \lambda((w,m); U))/p_b^i(1 + n(U \cap (W_i \times M)))$$

and respectively

$$\alpha_{\max}(U, L, (w,m), i) = (p_d^i \nu(W_i)\, \lambda((w,m); L))/p_b^i(1 + n(L \cap (W_i \times M))).$$

For the attractive case, the following bounds may be used

$$\alpha_{\min}(U, L, (w,m), i) = (p_d^i \nu(W_i)\lambda((w,m); L))/p_b^i(1 + n(U \cap (W_i \times M)))$$

and

$$\alpha_{\max}(U, L, (w,m), i) = (p_d^i \nu(W_i)\lambda((w,m); U))/p_b^i(1 + n(L \cap (W_i \times M))).$$

The convergence of the algorithm given by [195] is based on the verification of the conditions of Theorem 4.9.

First, the dynamics of the dominant process and the associated random variables must be specified. Apart from V_t and U_t, each $D(t)$ is associated with a permutation $\Sigma(t)$ of the marked points lying within the strip V_t. This permutation is used to determine the order in which the points can die in the dominant process and in the corresponding sub-processes in the simulation towards the future. For example, if a death is proposed at time t, then $D(t) \cap (W_{V_t} \times M) = \boldsymbol{x} = \{x_1, \ldots, x_n\}$ and $\Sigma(t) = (x_{j_1}, \ldots, x_{j_n})$ for a permutation $(j_1, \ldots, j_n)$ of the set $\{1, \ldots, n\}$, the removed object of the dominant process is x_{j_1}. The law of the Σ process is defined conditionally on D and V as follows. If $D(t+1)$ is obtained from $D(t)$ by a death in the strip indexed by V_t, the rank 1 is assigned to the object to be removed, while the ranks of the objects in the strip V_t of the $D(t+1)$ process are given by a uniform permutation independent of the other variables. When $D(t+1)$ is obtained from $D(t)$ by a birth in the indexed by V_t, then $\Sigma(t)$ is a uniform permutation of the points marked in $D(t). \cap (W_{V_t} \times M)$. Consequently, $\Sigma(t)$ depends on V_t, $D(t)$ and $D(t+1)$. The reversibility of the process $D(\cdot)$, the independence of U_t and V_t and the construction of Σ imply the joint stationarity of the process.

$$\{(D(t), V_t, \Sigma(t), U_t) : -\infty < t < \infty\}.$$

Note that for $D(t) = \boldsymbol{x}$ and $V_t = i$ the permutation Σ of $\{1, \ldots, n(\boldsymbol{x} \cap (W_i \times M))\}$ has marginal conditional probability

$$\begin{aligned}
\mathbb{P}(\sigma|\boldsymbol{x}, i) &= \\
&= p_b^i \frac{1}{n(\boldsymbol{x} \cap (W_i \times M))!} + p_d^i \frac{1}{n(\boldsymbol{x} \cap (W_i \times M))} \frac{1}{(n(\boldsymbol{x} \cap (W_i \times M)) - 1)!} \\
&= \frac{1}{n(\boldsymbol{x} \cap (W_i \times M))!},
\end{aligned}$$

which is a uniform distribution whenever $n(\boldsymbol{x} \cap (W_i \times M)) > 0$ or a degenerate distribution otherwise.

The dominant process can be simulated as follows. In $D(0)$, the strip i contains a number of objects of shifted geometric distribution with parameter $1 - p_b^i/p_d^i$. The objects are independent. The points are uniformly distributed in W_i and the marks are chosen according to the ν_M distribution. The random number U_0 is uniform on $(0, 1)$, while V_0 is uniform on $\{1, \ldots, n_W\}$, independent of any other source of randomness. Thanks to reversibility, D can be simulated from 0 to $-T$, and from 0 to T. The same applies to the independent quantities U_t and V_t. Conditional on the trajectories of D and V, the realisations of $\Sigma(-T), \ldots \Sigma(-1), \Sigma(0), \ldots, \Sigma(T-1)$ are generated as explained above.

When a death occurs, the object with the lowest rank assigned by σ is eliminated from the U and L processes. We need to check that this choice is equivalent to the uniform removal of a point in the strip selected for updating. To do this, let Σ be a uniform permutation of $A = \{x_1, \dots, x_n\}$ and let its restriction to the subset $B = \{x_{i_1}, \dots, x_{i_m}\}$ of A. We have $n!$ possible permutations for the set A. For the restriction to the subset B, the number of possibilities is calculated as follows. First we choose m indices, which gives $\binom{n}{m}$ possibilities. Then the $n-m$ free places give $n-m$ possible permutations. The probability of each permutation is the ratio between the number of favourable permutations and the number of possible permutations and we get

$$\mathbb{P}(\sigma(B)) = \frac{\binom{n}{m} \times (n-m)!}{n!} = \frac{1}{m!}$$

as it should be. It is easy to check that selecting the lowest rank point in a configuration after a uniform permutation is equivalent to directly selecting that point uniformly in the configuration. Furthermore, the death transition implemented in this way respects the order by inclusion: by removing the point with the lowest σ-rank, the modified B set is a subset of the modified A set. In [151, 152] the authors propose a procedure called *MHDeath* to compute only the part of the permutations necessary for the simulation (see Appendix A).

In conclusion, the sub-processes $U^{-T}(\cdot)$ and $L^{-T}(\cdot)$ are adapted functionals of the dominant marked process $\{(D(t), V_t, \Sigma(t), U_t) : -\infty < t < \infty\}$. By construction, the inclusion and funneling properties [(2.4)–(2.6) in [152]] are respected. These properties are quite similar to those required by the CFTP simulation via spatial birth-and-death processes described by (4.27) and (4.28).

Second, since the shifted geometric distribution assigns a positive probability to 0, and since the dominant process is in equilibrium, the extension of the process into the past almost certainly reaches the state $\emptyset$, hence Algorithm 5.3 almost certainly stops.

It remains to ensure the convergence of the target process $X^{-T}(\cdot)$, defined as follows. Let $X^{-T}(-T) = \emptyset$ and we apply to it the same dynamics as to $U^{-T}(\cdot)$ and $L^{-T}(\cdot)$, with the difference that if $X^{-T}(t) = \boldsymbol{x}$ and if the birth of an object (w, m) is proposed in the strip indexed by V_t, then it is accepted if

$$U_t \leq \frac{p_d^{V_t} \nu(W_{V_t}) \lambda((w,m); \boldsymbol{x})}{p_b^{V_t}(1 + n(\boldsymbol{x} \cap (W_{V_t} \times M))}.$$

Of course, $L^{-T}(t) \subseteq X^{-T}(t) \subseteq U^{-T}(t)$ for any integer $0 \geq t \geq -T$.

The process $X(t)$ with $t = 0, 1, 2, \dots$ and $X(0) = \emptyset$ is defined in an analogous way, from a simulation into the future of the dominant process and its associated random variables. The process $X(t)$ evolves according to a Metropolis Hastings dynamics. If the current state is $\boldsymbol{x}$, a strip $W_i \times M$ is uniformly chosen. With probability p_b^i a new marked point η is generated in $W_i \times M$ according to the law $\nu \times \nu_M / \nu(W_i)$ and accepted with probability $\left(p_d^i \nu(W_i)\, \lambda(\xi; \boldsymbol{x})\right) / \left(p_b^i n(\boldsymbol{y} \cap K_i \times M)\right) \leq 1$. With the complementary probability $p_d^i = 1 - p_b^i$ a uniform point of $W_i \times M$ is removed, unless the strip i is empty, in which case the configuration remains unchanged. The procedure is well-defined on the set of configurations with positive probability density.

By construction, p is an invariant probability density of the transition kernel. Since deaths are always accepted and since $p_d^i > 1/2$, the probability of reaching the empty configuration from any configuration $\boldsymbol{x}$ is at least

$$(2n_W)^{-n(\boldsymbol{x})} > 0.$$

It follows that the Metropolis-Hastings dynamics is δ_0-irreducible, where δ_0 is the Dirac measure of $\emptyset$. If the current state is empty, the probability for it to remain empty is at least

$$\sum_i p_d^i / n_W > 1/2 \geq (2n_W)^{-1}.$$

Therefore, sets of the form $\{\boldsymbol{x} : n(\boldsymbol{x}) \leq n\}$, as well as their measurable subsets, are small with respect to $(2n_W)^{-n}\delta_0$. Since the probability of remaining in the same state is positive, the simulated chain is aperiodic. The result is that $X(t)$ converges weakly in total variation to the law p.

Convergence is guaranteed for all p-positive states, but $X(0) = \emptyset$ is an atom of the limit distribution. To eliminate the influence of null measure sets, the drift condition can be used.

Let V be the function defined by

$$V(\boldsymbol{x}) = c(\boldsymbol{x}) A^{m(\boldsymbol{x})}$$

where $m(\boldsymbol{x}) = \max_i n(\boldsymbol{x} \cap (W_i \times M))$ and $c(\boldsymbol{x})$ the multiplicity of the maximum and $A > n_W$. We can see that V is strictly positive and that its level sets $\{\boldsymbol{x} : V(\boldsymbol{x}) \leq c\}$ for $c > 0$ are small sets. This is because $V(\boldsymbol{x}) \leq c$ implies $n(\boldsymbol{x}) \leq n_W \log(c)/\log(A)$ or 0, depending on whether $c > A$ or not. Now, for any $\epsilon > 0$, we can choose an integer $N_\epsilon > 0$ such that the probability of accepting a birth in a strip i that satisfies $n(\boldsymbol{x} \cap (W_i \times M)) \geq m(\boldsymbol{x}) - 1$ does not exceed ϵ for any $\boldsymbol{x}$ such that $n(\boldsymbol{x}) \geq N_\epsilon$. Changes in strips with less than $m(\boldsymbol{x}) - 1$ objects do not affect $V(\cdot)$. For the configurations $\boldsymbol{x}$ with $n(\boldsymbol{x}) \geq N_\epsilon$ we have

$$\mathbb{E}\left[V(X(t+1)) | X(t) = \boldsymbol{x}\right] - V(\boldsymbol{x}) \leq$$

$$\frac{\max_i p_b^i}{n_W} (A + n_W - 2)\, \epsilon\, V(\boldsymbol{x}) - \frac{\min_i p_d^i}{n_W} (1 - \frac{n_W}{A})\, V(\boldsymbol{x}).$$

It is possible to choose ϵ such that $0 < \max_i p_b^i(A + n_W - 2)\epsilon < \min_i p_d^i(1 - n_W/A)$, so that the expected increase in V is negative. The chain induced by X is Harris recurrent, which eliminates the influence of null measure sets.

5.2.3 Exact Gibbs sampler

The Gibbs sampler is a technique for simulating the probability distributions of several variables. The principle of the technique is to simulate the law of one variable conditionally on the others. The densities of marked point processes do not necessarily fit the framework of multi-variable distributions. For example, the study of multi-type marked point processes shows that simulating the marginal distribution of a single component is just as difficult as simulating the joint law of all components [181][1]. However, the approximation of a point process by a process on a grid is quite common, particularly in the context of parameter estimation by pseudo-likelihood [35, 114]. In this context, the Gibbs sampler naturally applies. The approach used by [195] is developed below.

Discretisation of marked point process

As in Section 5.2.2, let $W = \bigcup_{i=1}^{n_W} W_i$ be a finite partition of the position space such that $0 < \nu(W_i) < \infty$ for all $1 \leq i \leq n_W$. Let us also assume that the space of marks is partitioned, i.e. $M = \bigcup_{j=1}^{n_M} M_j$ with $0 < \nu_M(M_j) \leq 1$ for all $1 \leq j \leq n_M$. Then the induced partition on the product space is $W \times M = \bigcup_{i=1}^{n_W} \bigcup_{j=1}^{n_M} (W_i \times M_j)$. Each cell $W_i \times M_j$ has an associated representative point ξ_{ij}, for example its centre. The set of all representative points is called Ξ.

According to [35, 114], the probability density p of X is approximated by an autologistic distribution. More precisely, let N be a random field on the lattice induced by Xi taking binary values as follows. The state space is finite and consists of the vectors (n_{ij}), $i = 1, \ldots, n_W$, $j = 1 \ldots n_M$ such that $n_{ij} \in \{0, 1\}$ and $\sum_{j=1}^{n_M} n_{ij} \leq 1$ for all i. If n_{ij} is interpreted as an indicator of the presence/absence of a marked point in $W_i \times M_j$, the condition on $\sum_j n_{ij}$ ensures that each $W_i \times M$ band contains at most a single marked point. Thus the random vector $N_i = (N_{i1}, \ldots, N_{in_M})$ almost certainly takes its values in $S := \{\mathbf{0}, \boldsymbol{e}_1, \cdots, \boldsymbol{e}_{n_M}\}$, the set containing the null vector and the unit vectors.

In order to specify the probability function $\mathbb{P}$ of the random field N, we recall that p is hereditary by hypothesis and that by the Hammersley-Clifford

[1]There is, however, an exception with the Widom-Rowlinson model [127]

Theorem 2.9, the density can be factorised in the form

$$p(\boldsymbol{x}) = \prod_{\boldsymbol{z} \subseteq \boldsymbol{x}} \phi(\boldsymbol{z}) \tag{5.8}$$

with ϕ the interaction functions.

The probability distribution of N is obtained from (5.8) by restricting the product to subsets of Ξ and correcting for the mass of a cell [114]. The result is

$$\mathbb{P}(\boldsymbol{n}) \propto \prod_{\{i,j\}} [\nu(W_i)\nu_M(M_j)]^{n_{ij}} \prod_{\emptyset \neq \boldsymbol{z} \subseteq \Xi} \varphi(\boldsymbol{z})^{\eta(\boldsymbol{n};\boldsymbol{z})}$$

where $\boldsymbol{n}$ is a configuration in the state space and $\eta(\boldsymbol{n};\boldsymbol{z}) = \prod_{\xi_{ij} \in \boldsymbol{z}} n_{ij}$ with values 0 or 1. By convention, $0^0 = 1$.

Local characteristics

The *local characteristics* of a random field are given by the laws of a point of the field conditionally on the rest [318]. They describe the local behaviour of the field. Let $\boldsymbol{n}_{-i}$ be the vector formed by $n_{i'j}$ with $i' \neq i$. The local characteristics of the strip i are given by

$$\mathbb{P}(N_i = \boldsymbol{e}_j | \boldsymbol{n}_{-i}) = \frac{\nu(K_i)\nu_M(M_j) \prod_{\boldsymbol{z}} \varphi(\boldsymbol{z} \cup \{\xi_{ij}\})^{\eta(\boldsymbol{n}_{-i};\boldsymbol{z})}}{1 + \nu(W_i) \sum_{j'=1}^{n_M} \nu_M(M_{j'}) \prod_{\boldsymbol{z}} \varphi(\boldsymbol{z} \cup \{\xi_{ij'}\})^{\eta(\boldsymbol{n}_{-i};\boldsymbol{z})}} \tag{5.9}$$

whenever $j \neq 0$, provided that the event on which we condition has a strictly positive mass under $\mathbb{P}$. The product is performed on the sets $\boldsymbol{z} \subseteq \Xi \setminus \{\xi_{ij'} : j' = 1, \ldots, n_M\}$, and $\eta(\boldsymbol{n}_{-i};\boldsymbol{z})$ is defined as before. If the point processes we are discretising are Markovian, then the equation (5.9) uses only a reduced subset of points around ξ_{ij}. In this case the used set of points $\xi_{i'j}$ is called the neighbourhood of ξ_{ij} and the characteristics given by (5.9) are truly local.

The random field N is said to be *monotone* with respect to a given order relation $\leq$ on S, if for all $s \in S$,

$$\mathbb{P}(N_i \leq s | \boldsymbol{n}_{-i}) \geq \mathbb{P}(N_i \leq s | \tilde{\boldsymbol{n}}_{-i})$$

if $\boldsymbol{n}_{-i} \leq \tilde{\boldsymbol{n}}_{-i}$ occurs strip by strip, $\mathbb{P}(\boldsymbol{n}_{-i}) > 0$ and $\mathbb{P}(\tilde{\boldsymbol{n}}_{-i}) > 0$.

Example 5.1 (Monotonicity of the discrete Widom-Rowlinson and bivariate point processes). The discrete counterpart of the Widom-Rowlinson model (2.30) is monotone with respect to the order relation on S given by $\boldsymbol{e}_1 < \boldsymbol{0} < \boldsymbol{e}_2$ [126]. Similarly for the bivariate pairwise interaction point process (2.31), for which $gamma_{ii} \equiv 1$, $\gamma_{ij}(t) > 0$ for all $t > 0$ and $i, j \in \{1, 2\}$.

Let us verify this last assertion. After the discretisation of the process, the monotonicity with respect to the order relation $\boldsymbol{e}_1 < \mathbf{0} < \boldsymbol{e}_2$ on S, requires

$$\mathbb{P}(N_i = \boldsymbol{e}_1 | \boldsymbol{n}_{-i}) \geq \mathbb{P}(N_i = \boldsymbol{e}_1 | \tilde{\boldsymbol{n}}_{-i})$$

and

$$\mathbb{P}(N_i = \boldsymbol{e}_2 | \boldsymbol{n}_{-i}) \leq \mathbb{P}(N_i = \boldsymbol{e}_2 | \tilde{\boldsymbol{n}}_{-i})$$

for every i and all the pairs of configurations such that $\boldsymbol{n}_{-i} \leq \tilde{\boldsymbol{n}}_{-i}$. By posing $\xi_{i1} = (w_i, 1)$, $\xi_{i2} = (w_i, 2)$ for the representative points of the cell W_i and $\alpha_i = \nu(W_i)/2$, then the probability $\mathbb{P}(N_i = \boldsymbol{e}_1 | \boldsymbol{n}_{-i})$ is

$$\frac{\alpha_i \beta_1 \prod_{w \in n_{-i}(2)} \gamma_{12}(||w_i - w||)}{1 + \alpha_i \beta_1 \prod_{w \in n_{-i}(2)} \gamma_{12}(||w_i - w||) + \alpha_i \beta_2 \prod_{w \in n_{-i}(1)} \gamma_{12}(||w_i - w||)} =$$

$$\frac{1}{\left[\frac{1}{\alpha_i \beta_1} + \frac{\beta_2}{\beta_1} \prod_{w \in n_{-i}(1)} \gamma_{12}(||w_i - w||)\right] \prod_{w \in n_{-i}(2)} \gamma_{12}(||w_i - w||)^{-1} + 1}.$$

Here we have denoted by $n_{-i}(m)$ the positions of the representative points $\xi_{i'j}$, $i' \neq i$, which have the mark $m \in \{1, 2\}$ in $\boldsymbol{n}$. Since by hypothesis $\gamma_{12}(t)$ takes values in $(0, 1]$ and since $\tilde{n}_{-i}(1) \subset n_{-i}(1)$ while $\tilde{n}_{-i}(2) \supset n_{-i}(2)$, the local features satisfy the required order relation. The same reasoning applies to the conditional probabilities of label 2.

The discretisation presented above applies to the Candy and Bisous models. However, we have been not able to order the local characteristics [195].

The Gibbs sampler for discretised marked point processes

The exact Gibbs sampler for random fields on finite grids introduced by [126] is based on the CFTP principles [252]. To avoid the simultaneous coupling of too many chains, the authors propose to keep track of all possible values in each grid cell at each moment. The state space of the resulting Markov chain is built from subsets of the set of values of the random field distribution to be simulated. In our case, the grid is represented by Ξ and each n_{ij} takes values in $\{0, 1\}$.

A straightforward implementation of this method is likely to be inefficient if it does not exploit the constraint $\sum_{j=1}^{n_M} n_{ij} \leq 1$ for all i. For this reason, we propose to implement strip updates. More precisely, suppose the current state is $\boldsymbol{n} = (\boldsymbol{n}_1, \ldots, \boldsymbol{n}_W)$, where $\boldsymbol{n}_i \in S$ is the strip vector i of the components n_{ij}, $j = 1, \ldots, n_M$. Of course, each $\boldsymbol{n}$ belongs to the product space $S^{\otimes n_W}$. At each moment t, a strip $W_i \times M$ is randomly selected and its current value $\boldsymbol{n}_i$ is replaced by a sample of local characteristics (5.9), leaving the values of the other strips unchanged. Since S is finite, a single pseudo-random number u is sufficient to perform the update once the strip has been selected. Noting ψ the transition kernel, the new state is $\psi(\boldsymbol{n}, u, i)$.

If CFTP is applied to all chains initialised by all possible states with positive probability, and if we track the union of possible states at each grid point over time, rather than the evolution of the individual trajectories of each chain, the following algorithm is obtained. As before, the random variables generated are reused when the time is doubled.

Algorithm 5.4. Exact Gibbs sampler for discretised marked point processes. *Data: the probability density p of the locally stable marked point process of interest and its discrete counterpart as presented in Section 5.2.3. Output: $\boldsymbol{x}$ a perfect sample, T a realisation of the convergence time.*

Let V_t, $t = -1, -2, \ldots$ be a family of uniform independent random variables on $\{1, \ldots, n_W\}$ and let U_t be a family of uniform independent random variables on $(0,1)$. Suppose $\psi : S^{\otimes n_W} \times (0,1) \times \{1, \ldots, n_W\}$ is a transition kernel for the Gibbs sampler. This means that $\psi(\boldsymbol{n}, U, v)$ modifies the state $\boldsymbol{n}$ in the strip v in such a way that the new state follows the $\mathbb{P}-$conditional law knowing $\boldsymbol{n}_{-v}$. Initialise $T = 1$.

1. *Let $\mathcal{X}_i^{-T}(-T) = S$ for all $i = 1, \ldots, n_W$.*
2. *Simulate into the future starting from $-T$ by setting*
$$\mathcal{X}_{V_t}^{-T}(t+1) = \{s \in S : \exists \chi \in \mathcal{X}^{-T}(t) : \psi(\chi, U_t, V_t) = s\}$$
and leaving unchanged the sets of the other strips.
3. *If for all the strips $W_i \times M$, cardinal of $\mathcal{X}_i^{-T}(0)$ equals 1, then stop. Otherwise, set $T = 2T$ and restart.*
4. *Return $\boldsymbol{x} = \mathcal{X}^{-T}(0)$ and T.*

The result ([126] Thm. 1 pp. 400) implies that if the probability of $N_i = \mathbf{0}$ knowing the values of the other strips is uniformly minorised by a value of $\delta > 0$, then the algorithm will almost certainly terminate and return an unbiased sample of $\mathbb{P}$. Under the assumption that (5.8) is locally stable, for all $\boldsymbol{n}_-i$ such that $\mathbb{P}(\boldsymbol{n}_{-i}) > 0$ we have

$$\begin{aligned}
\mathbb{P}(N_i = \mathbf{0}|\boldsymbol{n}_{-i}) &= \frac{1}{1 + \nu(W_i)\sum_{j=1}^{n_M} \nu_M(M_j) \prod_{\boldsymbol{z}} \varphi(\boldsymbol{z} \cup \{\xi_{ij}\})^{\eta(\boldsymbol{n}_{-i};\boldsymbol{z})}} \\
&\geq \frac{1}{1 + \nu(W_i)\sum_{j=1}^{n_M} \nu_M(M_j)\Lambda} \geq \frac{1}{1 + W^*\Lambda} > 0
\end{aligned}$$

where $W^* = \max\{\nu(W_i) : i = 1, \ldots, n_W\}$, so the lower bound condition is satisfied.

The following are some points related to the implementation. Algorithm 5.4 is particularly interesting when monotonicity can be established with respect to a suitably chosen order relation [252, 125]. In this case, if the transition kernel respects this order, for example

$$\psi(\boldsymbol{n}, u, i) = \max\{j = 0, \ldots, n_M : \mathbb{P}(N_i \geq \boldsymbol{e}_j | \boldsymbol{n}_{-i}) \geq u\},$$

Algorithm 5.4 is the original algorithm of Propp and Wilson in [252], or again, according to our presentation, Algorithm 4.3 with function *Update* given by ψ.

It is often possible to save time by explicitly computing a new state $\mathcal{X}_{V_t}^{-T}(t+1)$ only if it is known in advance that it will be a singleton, or by setting $\mathcal{X}_{V_t}^{-T}(t+1) = S$ otherwise [126, 153, 232]. In fact, since

$$\mathbb{P}(N_i = \boldsymbol{0} | \boldsymbol{n}_{-i}) \geq \frac{1}{1 + \Lambda \max_i \nu(W_i)} > 0$$

is true for all i and all $\boldsymbol{n}_{-i}$, we can choose the transition kernel such that $\psi(\chi, u, v) = \boldsymbol{0}$ whenever $u \leq 1/(1 + \Lambda \max_i \nu(W_i))$. Thus, for small values of U_t, the value $\boldsymbol{0}$ is assigned to the strip V_t, regardless of $\mathcal{X}^{-T}(t)$. If $U_t > 1/(1 + \Lambda \max_i \nu(W_i))$, then if a set $\mathcal{X}_v^{-T}(t)$ of a neighbour v of V_t is not a singleton, then no update is performed, i.e. $\mathcal{X}_{V_t}^{-T}(t+1) = S$. In the following, this approach will be referred to as the "multigamma" coupler.

5.2.4 Simulation study

This section presents a simulation study to demonstrate the applicability and effectiveness of the exact simulation algorithms studied. The models considered here are Widom-Rowlinson (2.30) and Strauss bivariate (2.31). The full study, including the Cauchy and Candy models, can be found in [195].

For each of the four algorithms considered, we generate 25 independent samples and plot the evolution of the convergence time, or more precisely the number of "jumps", as a function of the model parameters.

For the Widom-Rowlinson model, the intensity parameters verify $\beta_1 = \beta_2$. The same is true for the bivariate Strauss model. Its interaction functions are $\gamma_{11}(t) = \gamma_{22}(t) = 1$ and $\gamma_{12}(t) = 0.5\mathbf{1}\{t \leq r\}$. For these two models the intensities vary from 1 in steps of 1 up to a maximum, depending on the algorithm. The range parameter r evolves from 0.005 in steps of 0.005 for both models, except for the Gibbs sampler where the discretisation is slightly different.

The algorithms are written in C++ and are available on simple request [281].

CFTP

For Algorithm 4.5, a jump is defined as a birth or death in the dominant chain. Note that the number of operations is actually greater, due to the doubling of time T. If the coalescence time is $T = -4$, the jumps between $T = -4$ and $T = -2$ are thinned once, those between $T = -2$ and $T = -1$ twice, and those after $T = -1$ three times.

Figure 5.6 shows the evolution of the average number of jumps as a function of the intensity and range parameters. The upper bounds are 150 and 0.15. We can see that the stronger the interaction, the longer it takes for the algorithm to converge. Indeed, for the Widom-Rowlinson model, the interaction is all-or-nothing and is said to be hard-core. For the Strauss model, the interaction is specified by a positive constant and is called soft-core.

Clan of ancestors

For Algorithm 5.2, a jump is defined as a birth or death in the simulation towards the past. The total number of jumps represents the computational load of this algorithm because there is no doubling of time.

The average number of jumps is shown in Figure 5.7(a) as a function of the model parameters. Here the maximum intensity is 100 and the maximum range is 0.08. The evolution towards the past is only determined by Λ and r. Consequently, the structure of the interaction function is irrelevant and all models obtained by thinning the same dominant process require the same simulation time. With respect to the parameters considered, the clan of ancestors is faster than coupling from the past. However, the coupling from the past is effective for a larger family of parameters.

We then used the incompatibility index (5.3). This index describes the neighbourhood structure better than a sphere of fixed radius. In fact, for both the Widom-Rowlinson model and the bivariate Strauss model, the conditional intensity of a point (w, m) depends only on points of type of m that are less than r away from w. Again, 25 independent samples were taken to approximate the average number of jumps for each pair of parameters. The results are shown in Figure 5.7(b). The obvious gain in efficiency is due to the fact that the number of ancestors is less than the number of neighbours for a fixed size neighbourhood system.

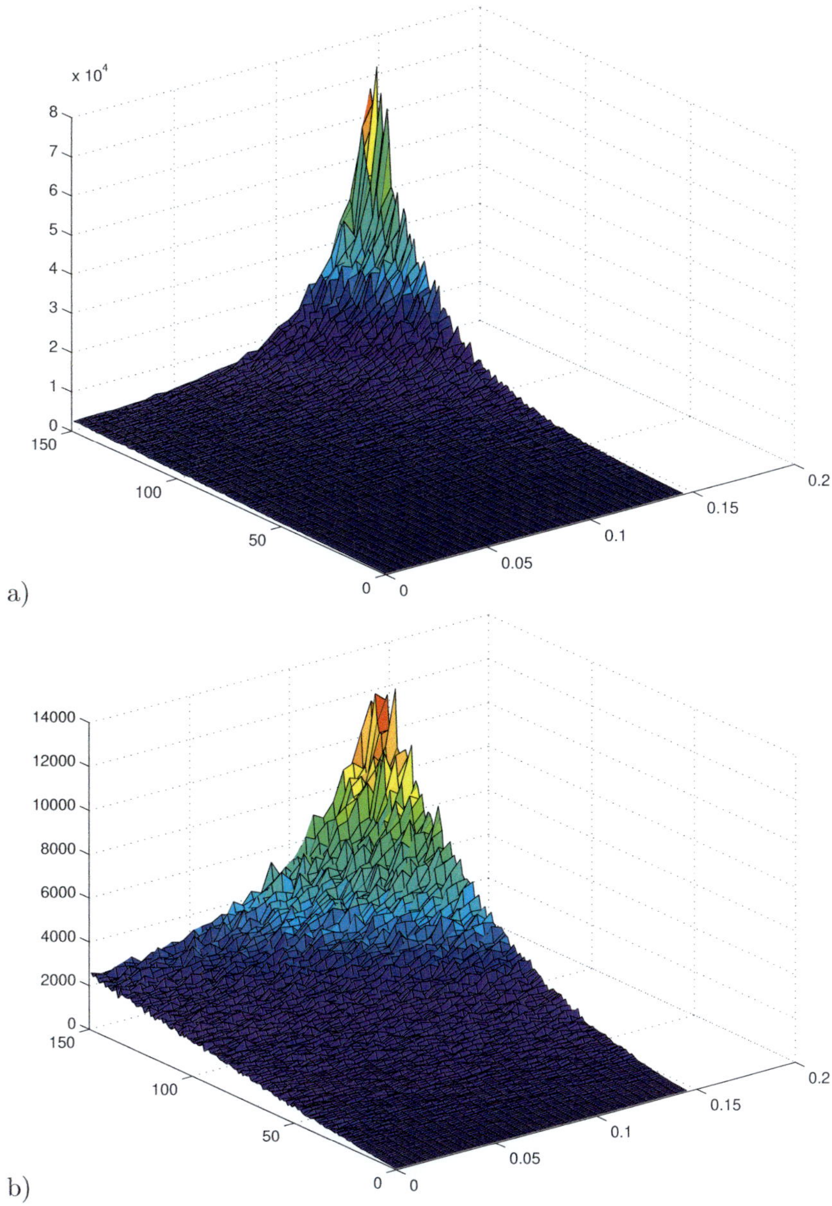

FIGURE 5.6
Average number of jumps as a function of intensity and range parameters, for 25 independent samples obtained with the CFTP algorithm. The models are: a) Widom-Rowlinson, b) Strauss bi-variate.

TABLE 5.1
Average coalescence time and mean number of jumps as a function of p_b, obtained from 500 simulations of a Poisson process with intensity $\beta = 50$.

p_b	Coalescence time	Number of jumps
0.01	34471.94	690.80
0.10	4816.90	965.64
0.25	4048.90	2023.12
0.45	24707.07	22233.11

TABLE 5.2
Average coalescence time and mean number of jumps as function of the grid size while $p_b = 0.1$. The computations were performed using 500 samples of a Poisson process with $\beta = 50$.

Cell size	Coalescence time	Number of jumps
25	6520.83	1304.47
30	9437.18	1890.16
35	13828.10	2768.14
40	17580.03	3518.75

Exact Metropolis–Hastings

Algorithm 5.3 allows some freedom in the choice of the domain partition and the birth probabilities. The effect of these choices will be analysed in the following experiments. We assume that all birth probabilities are equal $p_b^i \equiv p_b$ and that the partitioning of the domain $W = [0, 1] \times [0, 1]$ is done in $m \times m$ cells of equal volume $1/m^2$.

In the first experiment, the parameter m controlling the size of a cell is adjusted to p_b by taking $\lceil \sqrt{(\Lambda/p)} \rceil$ where $p = p_b/(1 - p_b)$. Using Algorithm 5.3, we simulate a Poisson point process with intensity $\Lambda = 50$ for different values of p_b. It can be observed that if the points are independent, the choice of their distribution does not affect the convergence time of the algorithm. For each value of p_b we simulated 500 independent realisations and calculated the average convergence time. The average number of jumps was also calculated by discarding deletions in empty strips. The results are shown in Table 5.1: small and large values of p_b increase the coalescence time, while intermediate values ($p_b = 0.10$ or 0.25) give rather good results.

The influence of the cell size m is studied with $p_b = 0.1$. The results are given in Table 5.2. As before, the mean coalescence time and the mean number of jumps were approximated from 500 samples of a Poisson process with intensity

50. For $p_b = 0.10$ we get $\lceil \sqrt{(\Lambda/p)} \rceil = 22$. Table 5.2 shows that grid refinement increases the coalescence time.

Figure 5.8 shows the average coalescence time as a function of the intensity and range parameters. It has been calculated from 25 simulations using Algorithm 5.3. The computational cost is increased by doubling the time. The upper bounds of the density and the range are 150 and 0.15 respectively. The probability of birth is $p_b = 0.1$. The domain W is divided into m^2 square cells of dimension $m = \lceil \sqrt{(\Lambda/p)} \rceil$. The results obtained are comparable to those of Algorithm 4.5: a strong interaction requires longer coalescence times.

Exact Gibbs Sampler

The average coalescence time of the Exact Gibbs Sampler Algorithm 5.4 as a function of the intensity and range parameters of the simulated models is shown in Figure 5.9. As for the coupling from the past, the computation time is larger due to the doubling of the time. For the bivariate Widom-Rowlinson and Strauss models, monotonicity relations are used for updating. All models were simulated on a W grid of size 20×20. Due to the discretisation, only a few range values can be considered. Here we used $r = 0.005$, 0.055, 0.075, 0.105, 0.115 and 0.155. The upper limit for the intensity is 100. For the Widom-Rowlinson and bivariate Strauss models, the mark space is discrete.

For the models studied, Figure 5.9 shows results comparable to those obtained by coupling from the past: the stronger the interaction, the longer the coalescence time. The authors in [195] have also studied the performance of this sampler on a version of the Candy model using the multigamma coupler. It appears that Gibbs algorithms, which can exploit the properties of monotonicity, are much faster.

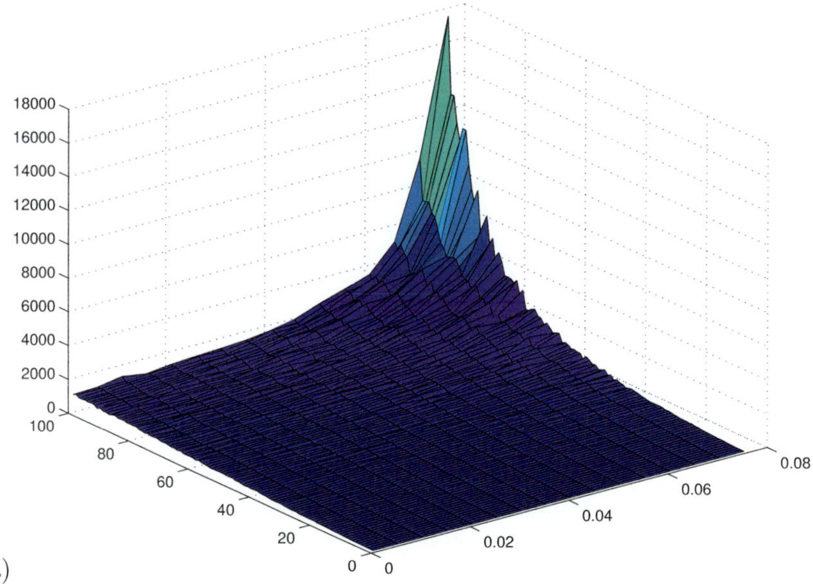

a)

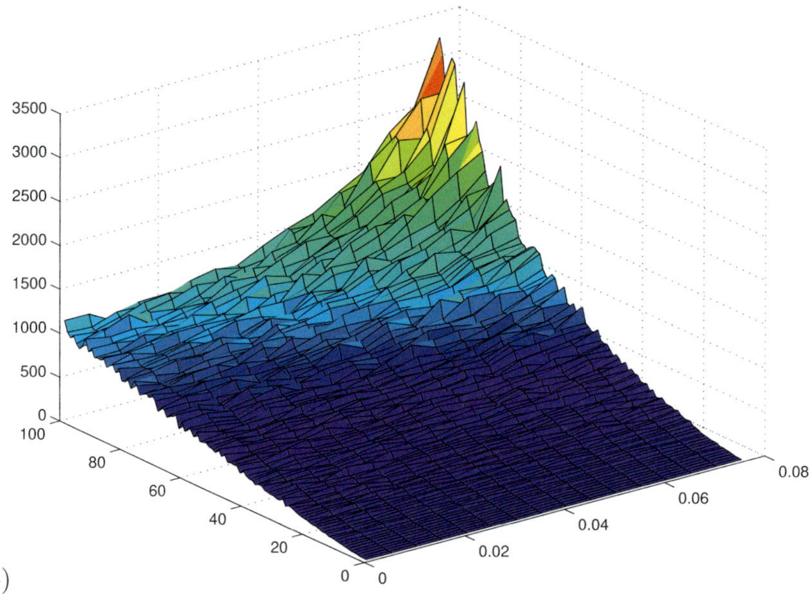

b)

FIGURE 5.7
Average number of jumps as a function of intensity and range, parameters, for 25 independent samples obtained with the clan of ancestors algorithm. The models are the Widom-Rowlinson and the bi-variate Strauss. a) The algorithm uses the fixed size neighbourhood, b) The algorithm uses the incompatibility index.

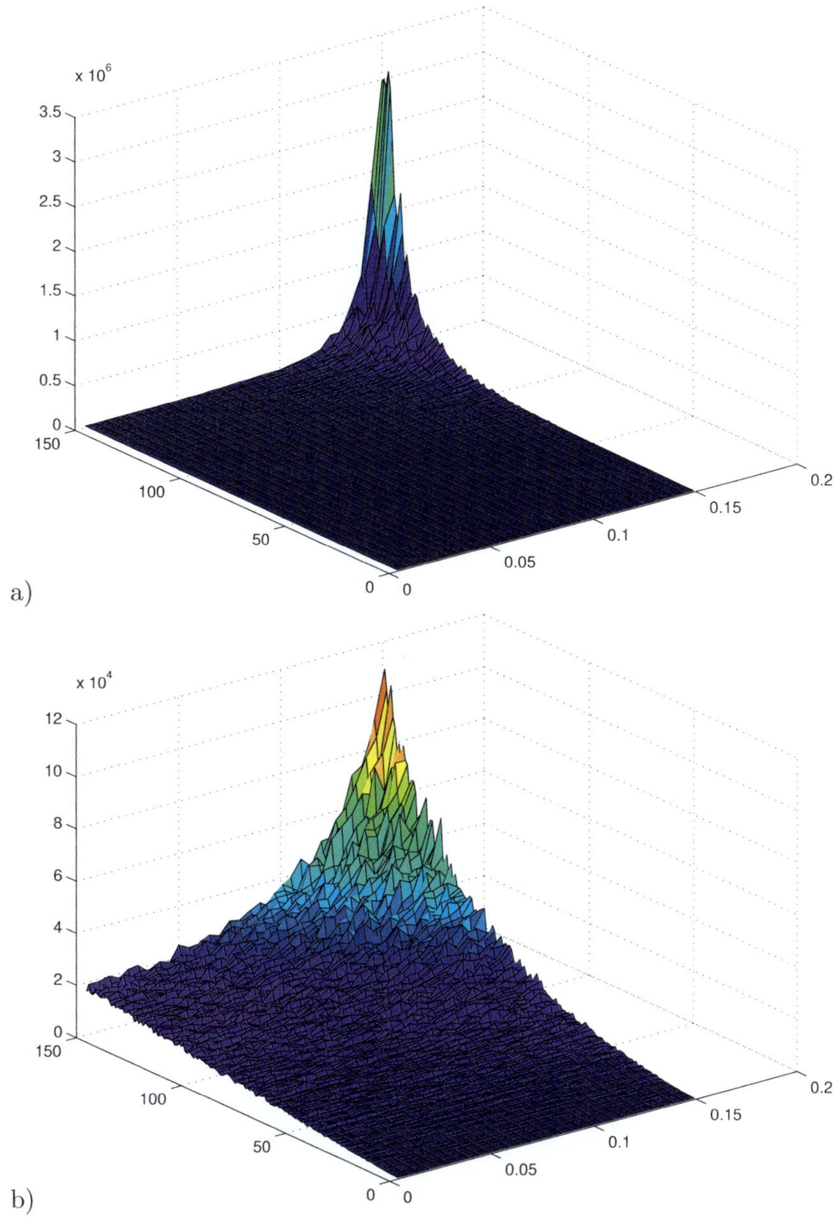

FIGURE 5.8
Average coalescence time as a function of intensity and range parameters estimated from 25 independent samples simulated by the Exact Metropolis-Hastings algorithm. The models are: a) Widom-Rowlinson, b) Strauss bivariate. In both cases, the grid was adapted to the birth probability $p_b = 0.1$.

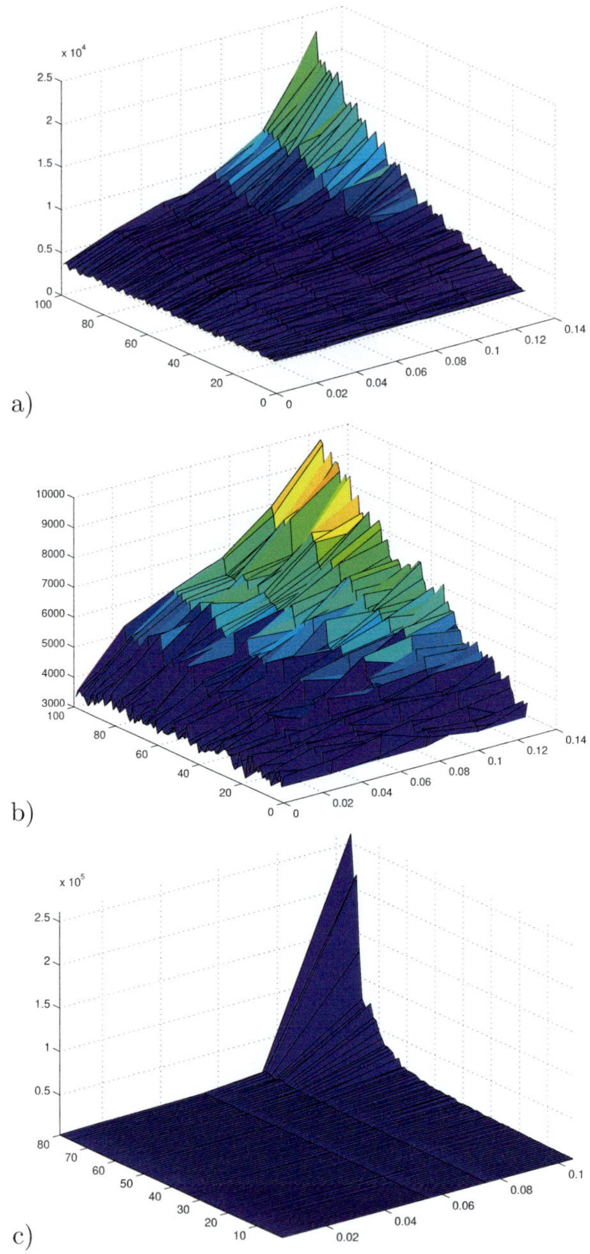

FIGURE 5.9
Average coalescence time as a function of the intensity and the range parameters estimated from 25 independent samples obtained with the Exact Gibbs Sampler. The models are: a) Widom-Rowlinson, b) bivariate Strauss, c) Candy.

Part III

Describe the pattern: statistical inference

6

Mathematical tools for statistical pattern detection and characterisation

6.1 Questions related to the pattern

By hypothesis, the pattern of interest is a set of interacting objects. Let us now assume that we have a well-defined model of the pattern, together with a suitable algorithm for simulating it. It is now time to address the issues raised in the Introduction by answering the following questions:

Q1: **pattern detection** – *given a spatial data field* $\mathbf{d}$ *and knowledge of the model parameters* θ*, what is the pattern* $\boldsymbol{x}$ *hidden in the data?*

Q2: **pattern characterisation** – *given an observed pattern* $\boldsymbol{x}$*, some covariates* $\mathbf{d}$*, and assuming that the pattern is the realisation of a specific statistical parametric model, what are the model parameters that best reproduce the observed pattern?*

Q3: **results verification** – *does the detected pattern really exist or is it just an artefact due to the general setup of the methodology used? how close are the simulations obtained by a fitted model to the observed pattern?*

The first question treats the data as an external field. To model the pattern within it, two components must be considered. The first, the data term, locates the objects that form the pattern in the data field. The second controls the interaction between the objects. The degree of knowledge of the model parameters determines how this question is answered. The parameters can be known in advance, or they can be modelled by a priori knowledge of their behaviour. Once this choice has been made, a probability distribution can be constructed. This distribution controls the pattern, taking into account the data field and the model parameters. The answer to the formulated question

DOI: 10.1201/9781003380382-6

is provided by the object configuration that maximises this distribution. This maximisation can be done using a simulated annealing algorithm. This algorithm is based on MCMC dynamics specifically designed to simulate the proposed model.

The second question is formulated after observing the pattern and assuming that it is the realisation of a stochastic parametric model. The statistical descriptors of the observed pattern are sufficient statistics of the model under consideration. These statistics may be fully or partially observed. In the first situation, i.e. the case of complete data, the model parameters can be estimated using MCMC algorithms that approximate maximum likelihood (MCMCML). Sampling strategies for the posterior distribution are also available. These include: auxiliary variable methods, exchange algorithms, Approximate Bayesian Computation strategies. When the statistics are only partially observed, i.e. in the case of incomplete data, the model parameters can be estimated again using MCMC algorithms. In this context, these algorithms are based on the Expectation Maximisation (EM) principle to approximate a likelihood function based on incomplete observations.

The last question aims at developing tools to verify the results obtained. Once a pattern has been detected in a data field, it is necessary to check whether the result obtained is due to the presence of the pattern in the data or to the general parametric setup of the method used. Strategies such as circular permutations or calculations of the mean pattern characteristics can be implemented. The first allows us to calculate a $p-$ value of the result obtained, i.e. to validate the existence of a structure in the data. The second allows us to infer pattern properties from average quantities rather than from a single realisation.

After fitting a parametric model to some observed data, it is necessary to check the quality of the parameter estimation and the adequacy of the model. Under certain assumptions, it is possible to calculate estimation errors in the parameter estimation of a model. First, it is possible to approximate the standard error, which is the difference between the maximum likelihood estimator and the true value of the parameters. We can then calculate the Monte Carlo standard error, which approximates the difference between the maximum likelihood estimator and its Monte Carlo counterpart. The calculation of these quantities is used to assess the accuracy of the estimation [110, 108, 181, 186, 229]. Validation or selection of fitted models can be done using Monte Carlo tests based on the calculation of selected residuals or summary statistics [19, 12, 142, 234, 235].

The chapter continues with the presentation of tools that are able to formulate answers to the previously formulated questions. First, tools for pattern detection are presented. These tools are the simulated annealing algorithm and the level set estimators. Then, pattern characterisation is addressed via two major families of methods: pointwise parameter estimation and posterior sampling. Answers to the third question are more application oriented and are therefore presented in the Chapter 7.

6.2 Some tools for pattern detection

Let us recall the pattern detection problem presented in the Introduction while adding few more details. The spatial data set $\boldsymbol{d}$ is observed within a compact window $W \subseteq \mathbb{R}^d$, we are interested in detecting the pattern exhibited by the data. In the following, the pattern of interest is assumed to be the realisation of a marked point process described by the probability density function

$$p(\boldsymbol{x}|\theta) = \frac{\exp[-(U_{\boldsymbol{d}}(\boldsymbol{x}|\theta) + U_i(\boldsymbol{x}|\theta))]}{Z(\theta)}. \tag{6.1}$$

where $U_{\boldsymbol{d}}(\boldsymbol{x}|\theta)$ and $U_i(\boldsymbol{x}|\theta)$ are the data and the interaction energies respectively. The probability density (6.1) is specified with respect to μ, the unit intensity Poisson process (2.17). Examples of such models were given in the Chapter 3. They represented hidden filaments in cosmological catalogues, clusters in epidemiological data, or faults in seismic images.

Within these considerations, assuming available knowledge about the parameters θ, the unknown pattern is estimated by

$$\hat{\boldsymbol{x}} = \arg\max_{\boldsymbol{x}\in\Omega} p(\boldsymbol{x}|\theta) = \arg\min_{\boldsymbol{x}\in\Omega}[U_{\boldsymbol{d}}(\boldsymbol{x}|\theta) + U_i(\boldsymbol{x}|\theta)]. \tag{6.2}$$

The convexity of this optimisation problem is not always guaranteed. In this context, global optimisation techniques such as the *simulated annealing* algorithms can be considered [105, 180, 318, 292, 176].

6.2.1 Simulated annealing: global maximisation of point process probability densities

The principle of simulated annealing is to sample the distributions

$$p_n(\boldsymbol{x}|\theta) = [p(\boldsymbol{x}|\theta)]^{\frac{1}{T_n}}, \tag{6.3}$$

while the temperature T_n slowly decreases towards 0. At the beginning of the algorithm, the initial temperature is very high and the probability density (6.3) is close to the uniform distribution on Ω. At lower temperatures, the law $p_n(\boldsymbol{x}|\theta)$ becomes more and more concentrated on configurations that maximise the law (6.1). Finally, at convergence, only the global optima are selected by the algorithm.

Simulated annealing requires two ingredients: a method for sampling $p(\boldsymbol{x}|\theta)$ and a cooling schedule for the temperature [105, 122, 123, 128, 157, 180, 292]. A first convergence result has been obtained in the case of models defined on discrete state spaces with a finite number of states [105]. In this case, simulated annealing is based on a Gibbs sampler and requires a logarithmic cooling scheme. Similar results are obtained by [128], in a slightly more general context, but always for models defined on a finite state space.

The case of marked point processes is more difficult because the configuration space is infinite. The convergence of simulated annealing on general configuration spaces has been studied by [122, 123]. The case of point processes simulated with spatial birth-and-death processes is treated by [180]. Here we present the construction and convergence proof of simulated annealing for point processes simulated with adapted Metropolis-Hastings dynamics [292, 283].

Lemma 6.1. *Let $\Omega_\star$ be the set of all configurations maximising $p(\boldsymbol{x}|\theta)$ given by* (6.1) *and suppose $\mu(\Omega_\star) > 0$. Let us also consider T_n a positive sequence such that $\lim_{n\to\infty} T_n = 0$ and the probability densities given by $p_n(\boldsymbol{x}|\theta) = [p(\boldsymbol{x}|\theta)]^{\frac{1}{T_n}}$. Then the probability measures $\pi_n = \int p_n d\mu$ converge in total variation to the uniform distribution on $\Omega_\star$. Furthermore, the distributions π_n satisfy*

$$\sum_{n=1}^{\infty} \| \pi_n - \pi_{n+1} \| < \infty.$$

Proof. The proof is given by ([180] Lemma 6 pp. 292-293, [292] Thm. 2 pp. 1870). □

The Markov chain that simulates p_n is not homogeneous. An additional tool is needed, in order to prove the convergence of algorithms that simulate such chains.

Let $P(\cdot,\cdot)$ be a transition kernel on $(\Omega, \mathcal{F})$. The *Dobrushin's contraction coefficient* is

$$c(P) = \sup_{\boldsymbol{x},\boldsymbol{y}\in\Omega} \| P(\boldsymbol{x},\cdot) - P(\boldsymbol{y},\cdot) \| .$$

In the following, P_n^δ denotes the iterated application δ times of the transition kernel P_n. The kernel P_n is defined by updates of a Metropolis-Hastings algorithm with adapted births (5.1) and uniform deaths (4.17). This kernel admits π_n as unique invariant distribution.

Theorem 6.1. *Let $(X_n)_{n\geq 0}$ be a non-homogeneous Markov process on $(\Omega, \mathcal{F})$ defined by the transition kernels $P_n^\delta(\boldsymbol{x}, F)$ and consider the following conditions*

$$\sum_{n=1}^{\infty} \| \pi_n - \pi_{n+1} \| < \infty \tag{6.4}$$

and

$$\lim_{n\to\infty} c(P_{n_0,n}) = 0 \tag{6.5}$$

respected for all $n_0 \geq 0$ and the transition probabilities $P_{n_0,n}(\boldsymbol{x}, F) = \mathbb{P}(X_n \in F | X_{n_0} = \boldsymbol{x})$. Then the limit $\lim_{n\to\infty} \pi_n = \pi_\infty$ exists and $\lim_{n\to\infty} \pi_0 P_{0,n} = \pi_\infty$ in total variation, for any initial distribution π_0.

Proof. By construction, π_n is the invariant law of P_n^δ. The condition (6.4) follows from Lemma 6.1. If condition (6.5) is verified, then the statements ([180] Thm. 2 pp. 285-286, [292] Thm. 3 pp. 1871) complete the demonstration. □

The Dobrushin condition (6.5) can be obtained by means of the following result. The proof is of pedagogical interest, because it can be used to determine a cooling schedule.

Lemma 6.2. *Let $P_n^\delta(\boldsymbol{x}, F)$ be the transition kernel defined above. If the step δ is a finite integer, such that $n(\boldsymbol{x}) \leq \delta$ for all $\boldsymbol{x} \in \Omega$, then the Dobrushin coefficient satisfies the condition* (6.5).

Proof. By using the properties of the Dobrushin coefficient and according to ([180] Lemma 3 pp. 287), it is sufficient to check that

$$c(P_n) \leq 1 - \frac{1}{n}.$$

The Kolmogorov-Chapman equations are used to calculate the probability $P_n^\delta(\boldsymbol{x}, \Xi_0)$ of going from any configuration $\boldsymbol{x} \in \Omega$ to the empty configuration Ξ_0 in δ steps. We obtain

$$P_n^\delta(\boldsymbol{x}, \Xi_0) \geq P_n^{n(\boldsymbol{x})}(\boldsymbol{x}, \Xi_0) P_n^{\delta - n(\boldsymbol{x})}(\emptyset, \Xi_0) \geq \left(\frac{p_b}{\nu(W)\Lambda^{1/T_n}} \right)^\delta = \Delta_n^\delta, \tag{6.6}$$

the Δ_n term being calculated from a bound on the probability of accepting a death (4.14).

By developing the inequality (6.6) we have

$$P_n^\delta(\boldsymbol{x}, A) \geq (\Delta_n)^\delta \mathbf{1}\{A = \Xi_0\} \tag{6.7}$$

and also

$$P_n^\delta(\boldsymbol{x}, A) \leq 1 - (\Delta_n)^\delta \mathbf{1}\{A \subseteq \Omega \setminus \Xi_0\}. \tag{6.8}$$

The equations (6.7) and (6.8) are grouped together in order to obtain

$$|P_n^\delta(\boldsymbol{x}, A) - P_n^\delta(\boldsymbol{y}, A)| \leq 1 - (\Delta_n)^\delta[\mathbf{1}\{A = \Xi_0\} + \mathbf{1}\{A \subseteq \Omega \setminus \Xi_0\}]. \tag{6.9}$$

If $A = \Xi_0$ or $A \subseteq \Omega \setminus \Xi_0$, the inequality (6.9) becomes

$$|P_n^\delta(\boldsymbol{x}, A) - P_n^\delta(\boldsymbol{y}, A)| \leq 1 - (\Delta_n)^\delta. \tag{6.10}$$

Now, if $A = \Xi_0 \cup A_1$ with $A_1 \subseteq \Omega \setminus \Xi_0$

$$\begin{aligned} P_n^\delta(\boldsymbol{x}, A) &= \\ &= P_n^\delta(\boldsymbol{x}, \Xi_0) + P_n^\delta(\boldsymbol{x}, A_1) \\ &= P_n^\delta(\boldsymbol{x}, \Xi_0) + 1 - P_n^\delta(\boldsymbol{x}, \Xi_0) - P_n^\delta(\boldsymbol{x}, \Omega \setminus \Xi_0 \setminus A_1) \\ &= 1 - P_n^\delta(\boldsymbol{x}, B) \end{aligned}$$

where $B = (\Omega \setminus \Xi_0 \setminus A_1) \subseteq (\Omega \setminus \Xi_0)$. It follows that

$$|P_n^\delta(\boldsymbol{x}, A) - P_n^\delta(\boldsymbol{y}, A)| = |P_n^\delta(\boldsymbol{y}, B) - P_n^\delta(\boldsymbol{x}, B)| \leq 1 - (\Delta_n)^\delta. \tag{6.11}$$

The equations (6.10) and (6.11) give for the Dobrushin coefficient

$$c(P_n) \leq 1 - (\Delta_n)^\delta.$$

A cooling schedule can be found if

$$(\Delta_n)^\delta = \left(\frac{p_b}{\nu(W)\Lambda^{1/T_n}}\right)^\delta \geq \frac{1}{n}$$

which gives, under weak assumptions

$$T_n \geq \frac{\delta \log \Lambda}{\log(n) + \delta \log \frac{p_b}{\nu(W)}}, \tag{6.12}$$

whenever $n \to \infty$. □

Taking all the necessary precautions, the above results justify the construction of a simulated annealing algorithm capable of detecting hidden patterns in the data. The patterns obtained represent approximate solutions to the optimisation problem (6.2).

Algorithm 6.1. Simulated Annealing for marked point processes. *Data: the probability density p of the locally stable marked point process of interest, the adapted MH transition kernel P, the initial temperature T_0, the final temperature T, the maximum number of objects in a configuration δ, the counter k. Output: $\boldsymbol{x}$ a pattern approaching the maximum of p.*

1. *Choose an initial condition $\boldsymbol{x}_0$ and set $k = 0$*

2. While $(T_k > T)$ *do*

$$\begin{aligned}
&\{\\
&\boldsymbol{x}_{k+1} = \textit{Update}\,(\boldsymbol{x}_k, p, P, T_k, \delta)\\
&T_{k+1} = T_0/[\log(k+1)+1]\\
&k = k+1\\
&\}
\end{aligned}$$

3. Return $\boldsymbol{x}_k$.

Algorithm 6.1 takes as input the initial temperature and the maximum number of objects in a configuration. In theory, these parameters can be calculated from p, the point process model to be used. This is not always a trivial task. Furthermore, the values obtained can cause the algorithm to run for an extremely long time. In practice, the use of a sub-optimal cooling schedule is considered acceptable if it respects the main principle of simulated annealing in its general lines: high initial values for the temperature with respect to the energy function, together with a rather slow cooling scheme. The *Update* function is implemented using a Metropolis-Hastings dynamics, as described by Algorithm 4.2 with the birth and death proposal previously specified. Clearly, the most important component of this algorithm is the probability density $p(\boldsymbol{x}|\theta)$, which represents the model of the pattern of interest.

Example 6.1 (Galactic filaments detection: the first steps). To illustrate the application of the simulated annealing algorithm, this example presents the very first results for the detection of galactic filaments using marked point processes. All the mathematical details, the analysis of the results and a comparison with other methods can be found in [294].

The methodology for detecting galactic filaments using marked point processes was first tested on two-dimensional simulated cosmological data and then applied to real observations [294, 295]. The procedure for obtaining the two-dimensional data is described in [201]. A random field representing the spatial distribution of dark matter evolves with time and exhibits structures. Figure 6.1 shows the dark matter density obtained from such a field. Regions of low density are called voids. Filamentary structures are regions of high density that bound empty "cells". The field of galaxies is obtained by a Cox process: conditional on the dark matter density, the galaxies are distributed mainly in the filamentary regions, with very few in the empty regions. Three such data sets are shown in the first row of Figure 6.2. The (B) data set was obtained using the dark matter density in Figure 6.1. Each data set contains a different number of galaxies, forming a network with its own morphology.

As explained in Section 3.1, the key assumption is that the filamentary pattern can be approximated by a collection of thin cylinders. These cylinders appear

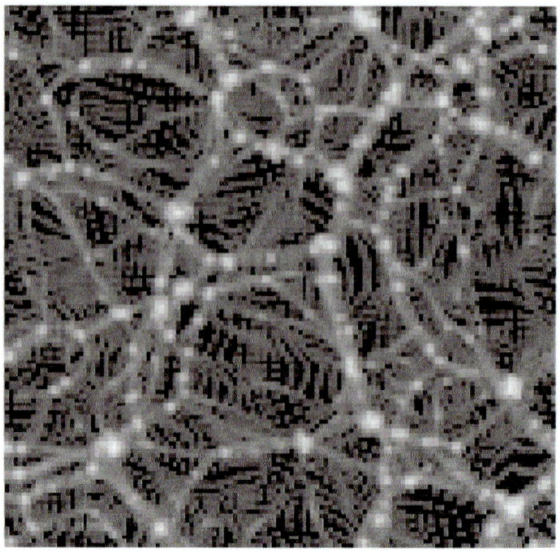

FIGURE 6.1
Dark matter density (logarithmic scale) for the data set B.

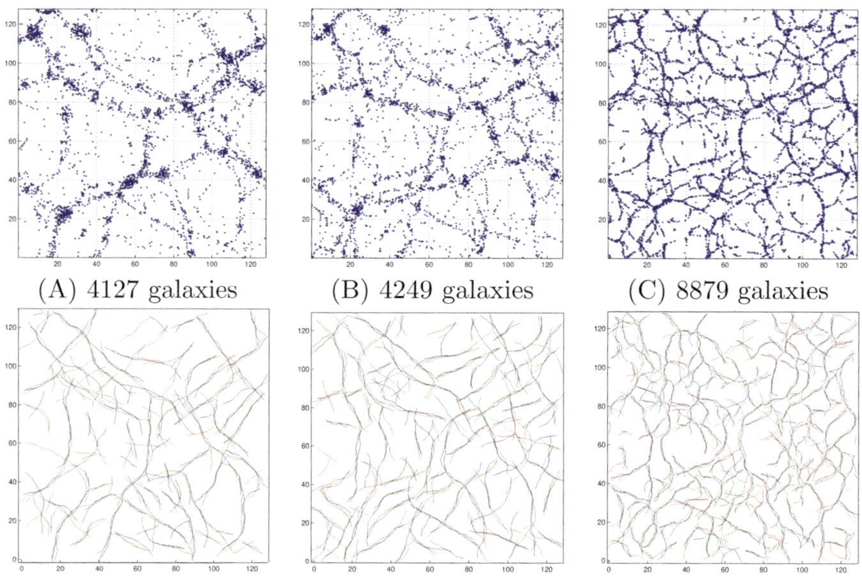

FIGURE 6.2
First row: three two-dimensional simulated catalogues together with their corresponding number of galaxies. Second row: the detection results for the considered parameter set for each catalogue. (Credit: Stoica et al., A & A, 434, 423–432, 2005, reproduced with permission ©ESO.)

TABLE 6.1
Interaction parameters.

Parameters	Catalogue		
	A	B	C
$\log \gamma_2$	10	10	10
$\log \gamma_0$	–8	–7	–7
$\log \gamma_1$	–2	–2	–1
$\log \gamma_o$	–3	–3	–3
$u_{\max}$	–25	–25	–25

as thin rectangles in two dimensions. The symmetry axes of these objects play a very important role in the interactions designed to form filaments. Therefore, for simplicity, these thin cylinders and rectangles can be called segments. The segments are connected and aligned. The probability density of the Candy model was used to control the distribution of such segments. Thus, according to (6.2), the configurations of segments that maximise this probability density are estimates of the hidden filamentary pattern in the data.

Simulated annealing was implemented according to an adapted Metropolis-Hastings dynamics[192, 294]. The results were obtained by annealing according to a sub-optimal polynomial cooling schedule

$$T_{k+1} = cT_k, k \geq 0,$$

with $T_0 = 10$ and $c = 0.9995$. The algorithm ran for 10^7 iterations, cooling the temperature every 10^3 iterations.

The parameters of the Candy model must be carefully set to obtain a good representation of the filaments in the data. The marks or segment parameters have been chosen so that the model can detect filaments in each dataset. For catalogues (A) and (B) the length of the segments was chosen between $l_{\min} = 3$ and $l_{\max} = 5$, for a thickness varying between $w_{\min} = 1$ and $w_{\max} = 2$. For the catalogue (C) smaller segments were used $l_{\min} = 2, l_{\max} = 3, w_{\min} = 0.95, w_{\max} = 1.05$. For all three data sets, the connectivity region around the segments was defined by circles with radius $r_c = 0.5$. The curvature defining the relative orientation of two adjacent segments was set to $\tau = 0.5$ radians for the (A) and (B) catalogues, and to 0.75 radians for the (C) catalogue.

Regarding the parameters of the data and interaction energies, described by (3.5) and (3.8) respectively, several values were tested. Table 6.1 shows the values with which our results were obtained. By this choice we have favoured segments connected at both ends, penalising free segments strongly, while segments connected at only one end are penalised less. The relative alignment of two segments was also favoured. Any segment whose data potential does not pass the local tests is not forbidden, but is assigned a very low probability.

This mechanism allows the network to be extended into regions where the data is of lower quality, but where the general configuration of the network suggests such an extension.

The results obtained are shown in the second row of Figure 6.2. According to the analysis in [294], the best result for all catalogues was obtained with the first parameter set. All parameter sets correctly detect the filaments found by visual inspection of the data. Clusters of galaxies do not destroy the filament structure. Difficulties arise in regions with a lower concentration of galaxies. However, the structure of the network always appears to be natural. from the point of view of the cosmologists, who also compared this approach with the techniques that were in use at the time.

This work was the starting point of a collaboration with my cosmologist friends that continues to this day. Many questions were raised, some of which are discussed in the continuation of this book. This work was also the first source of motivation for the construction of the Bisous model.

6.2.2 Level sets based estimator for robust pattern detection

At convergence, the output of the presented simulated annealing algorithm represents a realisation of the uniform distribution over the sets of pattern configurations that maximises the probability density of interest. This solution is not guaranteed to be unique. Furthermore, it is often the case that the prior knowledge related to the model parameters is built using quite empirical strategies. These facts introduce uncertainties regarding the quality of the proposed solution. There is a clear need to reduce the impact of these uncertainties.

One way to deal with this problem is to consider the expectation of the detected pattern. Extrapolating the classical concept of random variable expectation to general random patterns is not a trivial task. This is due to the fact that patterns may have qualitative and quantitative features that cannot always be expressed numerically in a simple way [220]. The authors in [134] present an approach for the construction of expectation estimators for patterns together with their convergence properties. In the following, the main ideas of this paper are further explained and detailed. All proofs of the results presented in this section can be found in [134].

Let us consider the more general setting where a marked point process is seen as a random set [54, 207, 220]. In the literature there are several ways to define the mean of a random set, such as the Aumann and the Vorob'ev expectation [14, 219, 301]. The Aumann expectation is interesting for averaging convex random sets. This is because this expectation is often convex, even when the random set is neither convex nor connected. The Vorob'ev expectation is applicable to compact random sets. However, it is not recommended for random sets of null measure [14]. In the following, the Vorob'ev expectation is preferred because the patterns encountered in the applications considered are not convex and do not have zero volume.

Level sets: a tool for averaging random sets

Let $(\mathcal{S}, \mathcal{A}, \mathbb{P})$ be a probability space and $(W = [0,1]^d, \mathcal{B}_W, \nu)$ be a measured space where $\mathcal{B}_W$ is the Borelian tribe on W and ν is the Lebesgue measure. We also consider $\mathcal{C}$ to be the class of compact sets in W.

A *compact random set* Γ in W is a measurable application of $\mathcal{S}$ in $\mathcal{C}$, in the sense that

$$\forall C \in \mathcal{C}, \quad \{\omega : \Gamma(\omega) \cap C \neq \emptyset\} \in \mathcal{A}.$$

The *coverage function* or the *contact probability* or the *volumic fraction* is given by

$$p(w) = \mathbb{P}(w \in \Gamma).$$

For $\alpha \in [0,1]$, the corresponding *level set* is

$$Q_\alpha = \{w \in W : p(w) > \alpha\}. \tag{6.13}$$

The set Q_α is deterministic, and sometimes noted simply $\{p > \alpha\}$.

The *Vorob'ev expectation* is the Borel set $\mathbb{E}_V\Gamma$ satisfying

$$\nu\left(\mathbb{E}_V\Gamma\right) = \mathbb{E}\left[\nu\left(\Gamma\right)\right] \tag{6.14}$$

as well as

$$\{p > \alpha^*\} \subset \mathbb{E}_V\Gamma \subset \{p \geq \alpha^*\},$$

where

$$\alpha^* = \inf\{\alpha \in [0,1] : \nu\left(Q_\alpha\right) \leq \mathbb{E}\left[\nu\left(\Gamma\right)\right]\}.$$

The Vorob'ev expectation is a level set α^* of volume equal to the average volume of the random set Γ. Figure 6.3 shows the evolution of the volume $F(\alpha) = \nu(Q_\alpha)$ of the set Q_α as a function of the level α. This is a càdlàg function with possible plateaus. Plateaus in $p(w)$ imply discontinuities in $\nu(Q_\alpha)$, while plateaus in $\nu(Q_\alpha)$ imply discontinuities in $p(w)$.

The uniqueness of $\mathbb{E}_V(\Gamma)$ is guaranteed if F is continuous in α^*. In this case, we can choose

$$\mathbb{E}_V(\Gamma) = \{p \geq \alpha^*\}$$

which is a compact set since $p(w)$ is upper semicontinuous [220]. It is possible to show that $\mathbb{E}_V$ minimises

$$B \rightarrow \mathbb{E}\left[\nu\left(B \triangle \Gamma\right)\right]$$

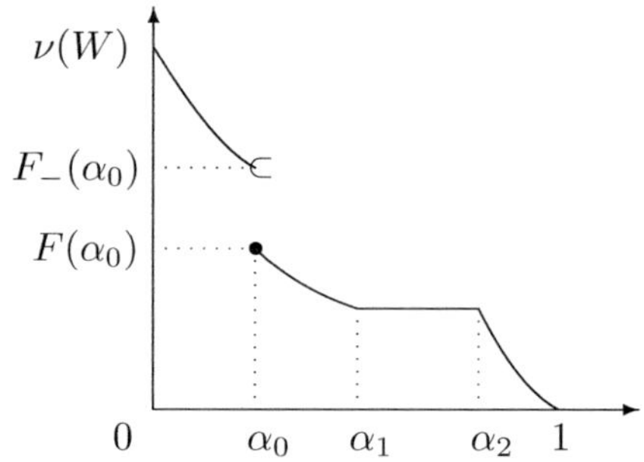

FIGURE 6.3
Behaviour of the function $F(\alpha) = \nu(Q_\alpha)$.

under the constraint $\nu(B) = \mathbb{E}[\nu(\Gamma)]$, $\triangle$ being the symmetrical difference ([220], Thm. 2.2.6, pp. 284).

However, the direct use of Vorob'ev's expectation raises a number of difficulties. First, the contact probability can only be calculated analytically for a limited number of models. Second, the level sets must be computed at each point in W. To overcome these drawbacks and construct estimators, Monte Carlo approximations and discretisation must be considered.

Estimation of level sets

Let $\Gamma_1, \Gamma_2, \ldots, \Gamma_n$ be n i.i.d. realisations of Γ. The approximation of $p(w)$ is of course written as

$$p_n(w) = \frac{1}{n} \sum_{i=1}^{n} \mathbf{1}_{\{w \in \Gamma_i\}}.$$

Then the *plug-in estimator* of a set of levels α is defined by

$$Q_{n,\alpha} = \{p_n > \alpha\}. \tag{6.15}$$

The convergence of such an estimator has already been discussed in the literature[64, 65, 215, 216, 218, 256, 314]. Several distances can be considered for the study of this convergence. We are particularly interested in the pseudo-distance given by

$$d(A, B) = \nu(A \triangle B) = \nu((A \setminus B) \cup (B \setminus A))$$

for all $A, B \in \mathcal{B}_W$. The L^1 convergence has been proved by [65] without requiring $p(w)$ to be continuous. Convergence results have also been obtained with the Hausdorff distance, subject to additional regularity and continuity conditions [215, 216, 65]. The case where $p(w)$ is a density that can be estimated by a kernel method has also been treated in the literature. Convergence rates are calculated by [218, 256, 314] and asymptotic normality is established by [204]. An adaptive estimation taking into account the regularity of $p(w)$ can also be performed [278].

In the following the approach proposed by [65] is generalised.

Let $r > 0$ be the step defining the grid $W_r = W \cap r\mathbb{Z}^d$. We seek to approximate $Q_{n,\alpha}$ by a quantity $Q^r_{n,\alpha}$. This quantity must be close to $Q_{n,\alpha}$ when r tends towards 0 and it must be easy to calculate.

For any Borel set B in W and any $r \in 2^{-\mathbb{N}}$, its *grid approximation* B^r is defined by

$$B^r = \bigsqcup_{w \in B \cap r\mathbb{Z}^d} [w, w + r)^d.$$

Let $N_r(\partial B)$ be the number of cells of W_r required to cover the boundary of B.

$$N_r(\partial B) = \text{Card}\{w \in r\mathbb{Z}^d : [w, w + r)^d \cap \partial B \neq \emptyset\},$$

and let $\overline{\dim_{\text{box}}}(\partial B)$ be the "upper box counting dimension"

$$\overline{\dim_{\text{box}}}(\partial B) = \limsup_{r \to 0} \frac{\log N_r(\partial B)}{-\log r}.$$

Proposition 6.1. *If* $\overline{\dim_{\text{box}}}(\partial B) < d$, *then for any* $\varepsilon > 0$, *there exists* r_ε *such that*

$$0 < r < r_\varepsilon \Longrightarrow d(B^r, B) \leq r^{d - \overline{\dim_{\text{box}}}(\partial B) - \varepsilon}.$$

The following result extends the one obtained by ([65], Thm. 3, pp. 14). By adding a discretisation scheme, the plug-in estimators of the level sets can be used in practice.

Proposition 6.2. *Assume there exists* $\kappa > 0$ *such that* $\overline{\dim_{\text{box}}}(\partial \Gamma) \leq d - \kappa$ *almost surely. Then, for any* α *such that* $\nu(\{p = \alpha\}) = 0$,
(i) with probability 1

$$\lim_{\substack{r \to 0 \\ n \to \infty}} \nu\left(Q^r_{n,\alpha} \triangle Q_\alpha\right) = 0,$$

(ii) for any $\varepsilon > 0$,

$$\mathbb{E}\left[\nu\left(Q^r_{n,\alpha} \triangle Q_\alpha\right)\right] \leq r^\kappa + 2\exp^{-2n\varepsilon^2} + F(\alpha - \varepsilon) - F(\alpha + \varepsilon).$$

The level α of the estimator $Q^r_{n,\alpha}$ is a fixed value. The level in the Vorob'ev expectation is a quantity that needs to be estimated. We do this in two steps. First, we use an approximation α^*_n in the expression of the Vorob'ev expectation. This quantity is known in the literature as the Kovyazin expectation [161]. Second, the discrete version of this quantity is proposed as an estimator of the Vorob'ev expectation.

The *Kovyazin's expectation* is the Borel set K_n verifying

$$\nu(K_n) = \frac{1}{n}\sum_{i=1}^{n} \nu(\Gamma_i)$$

and such that

$$\{p_n > \alpha_n^*\} \subset K_n \subset \{p_n \geq \alpha_n^*\},$$

where

$$\alpha_n^* = \inf\{\alpha \in [0,1] : \nu(\{p_n > \alpha\}) \leq \nu(K_n\}.) \tag{6.16}$$

Theorem 6.2. *If $\nu(\{p = \alpha^*\}) = 0$, then:*

$$\lim_{n\to\infty} \nu(K_n \triangle \mathbb{E}_V \Gamma) = 0.$$

almost surely.

The discrete counterpart of the Kovyazin expectation is obtained by considering the Borel set $K_{n,r}$ such that

$$\{p_n > \alpha_{n,r}^*\}^r \subset K_{n,r} \subset \{p_n \geq \alpha_{n,r}^*\}^r,$$

where

$$\alpha_{n,r}^* = \inf\{\alpha \in [0,1] : \nu(\{p_n > \alpha\}^r) \leq \nu(K_n)\}.$$

The set $K_{n,r}$ is the practicable estimator we propose. Its construction follows the following procedure. First, we approximate the mean volume of the random set Γ by $\nu(K_n)$. This approximation determines the number of grid cells to be selected. These cells are then grouped together to form the set $K_{n,r}$, according to their estimated probability of overlap p_n.

The following result shows that the estimator thus constructed is consistent, and that it improves existing estimators, even if the exact calculation of $\nu(K_n)$ is not easy for random sets of complex structure.

Theorem 6.3. *Assume that for $\kappa > 0$, we have $\overline{\dim}_{\text{box}}(\partial\Gamma) \leq d - \kappa$ almost surely, and that $\nu(\{p = \alpha^*\}) = \nu(\{p = \beta^*\}) = 0$ with*

$$\alpha^* = \inf\{\alpha \in [0,1] : \nu(Q_\alpha) \leq \mathbb{E}[\nu(\Gamma)]\}$$

and

$$\beta^* = \sup\{\alpha \in [0,1] : (Q_\alpha) \geq \mathbb{E}[\nu(\Gamma)]\}.$$

Then, almost surely:

$$\lim_{\substack{r\to 0\\ n\to\infty}} \nu(K_{n,r} \triangle \mathbb{E}_V \Gamma) = 0.$$

The proof of the result needs the following two lemmas.

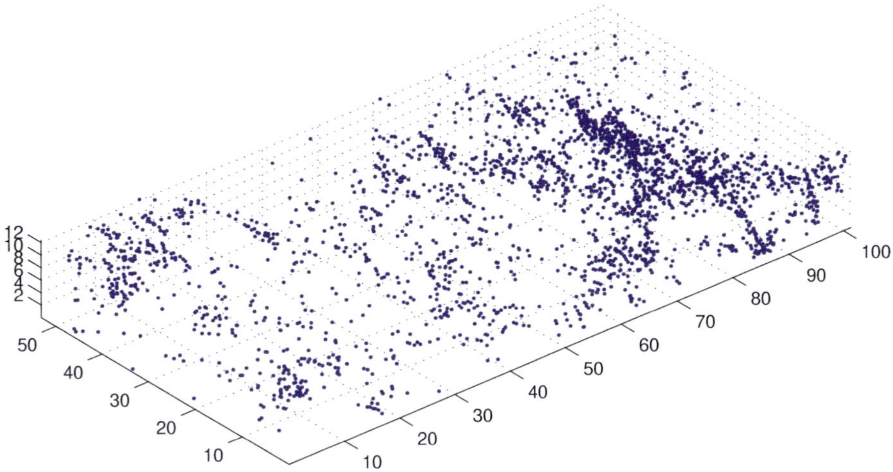

FIGURE 6.4
The NGP150 galaxy catalogue.

Lemma 6.3. *Assume that* $\mathbb{P}(\overline{\dim}_{\text{box}}(\partial\Gamma) \leq d - \kappa) = 1$ *for some* $\kappa > 0$. *Then, with probability one*

$$\alpha^* \leq \liminf_{\substack{n\to\infty \\ r\to 0}} \alpha^*_{n,r} \leq \limsup_{\substack{n\to\infty \\ r\to 0}} \alpha^*_{n,r} \leq \beta^*.$$

The following inequalities also hold

$$\alpha^* \leq \liminf_{n\to\infty} \alpha^*_n \leq \limsup_{n\to\infty} \alpha^*_n \leq \beta^*.$$

Lemma 6.4. *Assume that* $\mathbb{P}(\overline{\dim}_{\text{box}}(\partial X) \leq d - \kappa) = 1$ *for some* $\kappa > 0$. *Then, with probability one*

$$\limsup_{\substack{n\to\infty \\ r\to 0}} (K_{n,r}, K_n) \leq 2\Big[\lim_{\substack{\alpha\to\alpha^* \\ \alpha<\alpha^*}} F(\alpha) - F(\beta^*)\Big].$$

Proof. The proof is given by ([134], Thm. 2.5, pp. 50). We write

$$\nu\left(K_{n,r}\triangle\mathbb{E}_V\Gamma\right) \leq \nu\left(K_{n,r}\triangle K_n\right) + \nu\left(K_n\triangle\mathbb{E}_V\Gamma\right)$$

and we use Theorem 6.2 and the lemmas 6.3 and 6.4 to conclude. □

Example 6.2 (Average shape of the cosmic filaments network). The Vorob'ev expectation can be applied to the study of galactic filaments. The data set considered here is given by the positions of the galaxies in the *NGP*150 catalogue (see Fig. 6.4). This catalogue contains 2499 galaxies in the range $W = [0, 53.1] \times [0, 101.5] \times [0, 12.4]$. The unit of length is the mega-parsec $h^{-1}Mpc$.

The contact probability can be written as

$$p(w) = \int_{\Omega}\int_{\Theta} \mathbf{1}\{w \in \boldsymbol{x}\} p(\boldsymbol{x}|\theta)p(\theta)d\mu(\boldsymbol{x})d\theta, \tag{6.17}$$

and represents the probability that the point $w \in W$ is touched by the filament network modelled by randomly connected and aligned cylinders. The cylinder configurations are governed by $p(\boldsymbol{x}|\theta)$ given by (3.5) and (3.8). The uncertainties of the different model parameters are taken into account by the prior law $p(\theta)$.

The volume of the level sets induced by (6.17) is

$$F(\alpha) = \int_{W} \mathbf{1}\{p(w) > \alpha\}\nu(dw).$$

The formula (6.17) is approximated by its Monte Carlo counterpart. This is done by sampling from $p(\boldsymbol{x}|\theta)$ using a tailored Algorithm 4.2 as explained in Section 5.1. The prior $p(\theta)$ is chosen so that its sampling is easy to perform [262, 267].

The expression of the coverage function $p(w)$ is not available in analytic closed form. This makes it difficult to verify the continuity of $F(\alpha)$. Therefore, the conditions of Theorem 6.3 are assumed to be satisfied. The Monte Carlo approximation of $F(\alpha)$ is shown in Figure 6.5. Visual inspection of the plot does not indicate the presence of discontinuities. For the samples obtained, the empirical values $\nu(K_n) = 1158$ and $\alpha_n^* = 0.36$ were obtained. These calculations were made using $n = 1000$ samples and $r = 1h^{-1}Mpc$ for the grid size.

The Vorob'ev expectation estimator is shown in Figure 6.6. This tool provides the average shape of the filament pattern. Its interest here is that it completes the pattern detection methodology. It provides a more robust estimate of the hidden structure with respect to the data and parameter uncertainties, while preserving the local morphological information provided by the model used. However, the definition and construction of higher order moments of a structure is still an open problem.

6.3 Pattern characterisation through parameter estimation

Let W be a bounded window in which the pattern $\boldsymbol{x}$ is observed. The pattern is considered to be a realisation of a statistical model $(\Omega, \mathcal{F}, \mathbb{P}_\theta, \theta \in \Theta)$, where

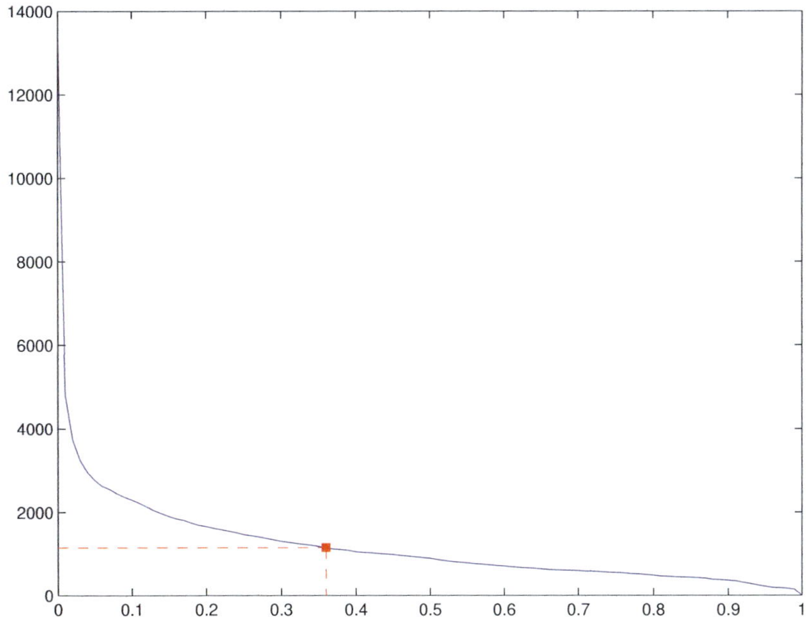

FIGURE 6.5
Monte Carlo approximation of the function $F(\alpha)$ for the $NGP150$ catalogue.

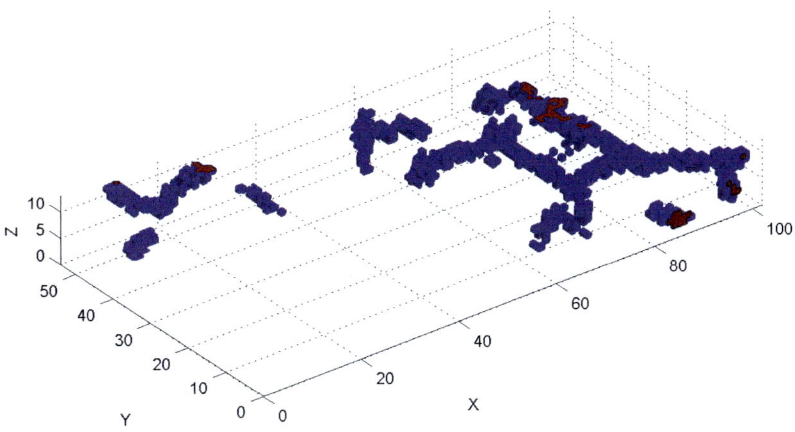

FIGURE 6.6
The Vorob'ev expectation estimator for the galaxy catalogue $NGP150$.

Θ is the parameter space. It is further assumed that there exists μ a positive $\sigma-$ finite measure such that $\mathbb{P}_\theta$ admits a probability density $p(\boldsymbol{x}|\theta)$ with respect to it.

For example, if $\boldsymbol{x}$ is an observed pattern of marked points, then $p(\boldsymbol{x}|\theta)$ is the probability density of the model assumed to produce the observed pattern, where μ is given by the unit intensity Poisson point process in W with independent marks (2.20).

Let us assume $p(\boldsymbol{x}|\theta)$ can be written as

$$p(\boldsymbol{x}|\theta) = \frac{\exp\langle t(\boldsymbol{x}), \theta\rangle}{Z(\theta)} \tag{6.18}$$

where $t(\boldsymbol{x}) = (t_1(\boldsymbol{x}), \ldots, t_d(\boldsymbol{x}))$ is the vector of sufficient statistics, $\theta = (\theta_1, \ldots, \theta_d) \in \Theta$ the vector of parameters and $\Theta \subset \mathbb{R}^d$ the compact space of these parameters. The dimension d corresponds to the number of parameters to be estimated.

The statistics $t(\boldsymbol{x})$ contain all available information about the parameters. This means that from an observed pattern $\boldsymbol{x}$ it is sufficient to calculate the vector of statistics $t(\boldsymbol{x})$ to estimate the parameters θ of the model (6.18). The morphological and statistical characteristics of a pattern are thus synthesised into a real-valued vector. In general, Gibbs point processes given by (1.1) can be written in the form (6.18). Point processes described by a probability density (6.18) are *exponential family models.*

Within this context, the reformulated $Q2$ question is what are the θ parameters of the marked point process that is capable of reproducing patterns similar to the observation $\boldsymbol{x}$. In the following, it is also assumed the pattern $\boldsymbol{x}$ is observed entirely. When this is not the case, the necessary explanations are given.

6.3.1 Maximum pseudo-likelihood

The concept of pseudo-likelihood was introduced by Julian Besag to estimate the parameters of discrete random fields [32, 33], then of point processes with pairwise interaction [34, 35]. It was then extended to broader classes of point processes [259, 144, 114, 271]. For more details and further reading, the following references are suggested [12, 101, 229, 181, 186, 318].

The construction of the *pseudo-likelihood* for point processes is based on the conditional intensity (2.24). Intuitively, the conditional intensity $\lambda(\eta; \boldsymbol{x})d\nu(\eta)$ can be interpreted as the conditional probability that the object η belongs to the process $\mathbb{X}$ knowing that the rest of the process coincides with the configuration $\boldsymbol{x}$.

For a marked point process defined according to (6.18), the conditional intensity is written

$$\lambda_\theta(\eta; \boldsymbol{x}) = \frac{p(\boldsymbol{x} \cup \{\eta\}|\theta)}{p(\boldsymbol{x}|\theta)} = \exp\langle t(\boldsymbol{x} \cup \{\eta\}) - t(\boldsymbol{x}), \theta\rangle, \tag{6.19}$$

when $\eta \notin \boldsymbol{x}$. For any $x_i \in \boldsymbol{x}$ we have

$$\lambda_\theta(x_i; \boldsymbol{x}) = \frac{p(\boldsymbol{x}|\theta)}{p(\boldsymbol{x}\backslash\{x_i\}|\theta)} = \exp\langle t(\boldsymbol{x}) - t(\boldsymbol{x}\backslash\{x_i\}), \theta\rangle.$$

Note that this conditional intensity is a priori discontinuous at the points $x_i \in \boldsymbol{x}$, and that it does not depend on the normalising function $Z(\theta)$.

The pseudo-likelihood for marked point processes is constructed by analogy with the discrete case. Let $(W_i)_{i=1,\ldots,m(\delta)}$ be a family of pairwise disjoint sets which form a partition of W such that $\delta = \max_{1 \leq i \leq m} \nu(W_i)$ and $m(\delta)\delta^2 \underset{\delta \to 0}{\to} 0$. Then, the pseudo-likelihood is obtained by computing:

$$PL_{W \times M}(\theta; \boldsymbol{x}) = \lim_{\delta \to 0} \prod_{i=1}^{m(\delta)} p(\boldsymbol{x}_{W_i}|\boldsymbol{x}_{W \backslash W_i}, \theta).$$

Proposition 6.3. *Let W be a bounded domain in which the marked point pattern $\boldsymbol{x} = \{x_i = (w_i, m_i), 1 \leq i \leq n\}$ is observed. The observation $\boldsymbol{x}$ is assumed to be the realisation of a marked point process of probability density* (6.18) *and conditional intensity* (6.19)*. Then, its pseudo-likelihood is*

$$\begin{aligned} PL_{W \times M}(\theta; \boldsymbol{x}) &= \\ &= \left[\prod_{x_i \in \boldsymbol{x}} \lambda_\theta(x_i; \boldsymbol{x})\right] \exp\left[-\int_{W \times M} \lambda_\theta((w, m); \boldsymbol{x})\nu(dw)\nu_M(dm)\right]. \end{aligned} \tag{6.20}$$

The pseudo-likelihood of point processes given by probability densities belonging to the exponential family, is a concave function [144]. The maximum pseudo-likelihood estimator $\hat{\theta}$ is obtained by maximising (6.20). Consistency and asymptotic normality of pseudo-likelihood estimators have been investigated by [143, 203, 38, 57].

All the ingredients are present to build a gradient descent type optimisation scheme to maximise the pseudo-likelihood. The delicate point of such a method remains the calculation of the integral in (6.20). Several integration schemes have been proposed by [12, 29, 20, 206]. The **R** library **spatstat**, built by A. J. Baddeley, R. Turner and contributors from the scientific community, provides procedures for computing the pseudo likelihood estimator for several point process models [18, 12].

The pseudo-likelihood estimator can be obtained as a solution of a particular set-up of the *Takacs-Fiksel parameter estimation method.* This method uses the equality

$$\mathbb{E}\left[\sum_{(w,m)\in X} h((w,m), X\setminus\{(w,m)\})\right] = \tag{6.21}$$
$$= \mathbb{E}\left[\int_{W\times M} h((w,m), X)\lambda((w,m); X)d\nu(w)d_n u_M(m)\right],$$

to estimate the parameters of the model under consideration. The principle of the Takacs-Fiksel method is to choose appropriate functions h and estimates of the right and left terms of (6.21) in order to build equations with solutions the model parameters we are looking for [56, 63, 85, 95, 181, 271, 303, 304]. The equality (6.21) is known in the literature as the *Georgii-Nguyen-Zessin formula.* It can be derived by introducing and developing the Palm distributions within the Campbell-Mecke formula (2.4) as it can be seen in [54, 181, 229]. Assuming that the conditional intensity is derivable with respect to the model parameters it is possible to choose for the h functions

$$h(\eta, \boldsymbol{x}) = \frac{\partial}{\partial\theta}\lambda_\theta(\eta; \boldsymbol{x})$$

for any $\eta \in W \times M$. Then, the equations (6.21) become

$$\mathbb{E}\left[\frac{\partial}{\partial\theta}\log PL_{W\times M}(\theta; X)\right] = 0,$$

showing that in this case the Takacs-Fiksel and pseudo-likelihood estimators coincides [54, 85, 181, 229, 271, 272].

6.3.2 Monte Carlo maximum likelihood

In the parameter estimation setup presented at the beginning of this section, the *log-likelihood function* can be written as

$$l(\theta) = \log p(\boldsymbol{x}|\theta) = \rangle t(\boldsymbol{x}), \theta\rangle - \log Z(\theta).$$

The normalisation constant $Z(\theta)$ does not always have an analytically closed form. In this context, statistical inference can be performed by using Monte Carlo simulations of the model under consideration to approximate the log-likelihood function [85, 104, 107, 108, 121, 181, 192, 229, 243, 244, 245, 246, 248, 285, 286, 318].

To reduce the number of simulations needed for such an approximation, it is convenient to consider the log-likelihood written with respect to a reference point in the parameter space. To do this, we consider the log-likelihood given by

$$l(\theta) = \langle t(\boldsymbol{x}), \theta - \psi \rangle - \log \frac{Z(\theta)}{Z(\psi)}, \tag{6.22}$$

where $\psi \in \Theta$ is a known fixed value.

The calculation of the ratio of the normalisation constants in the previous formula

$$\begin{aligned}
\frac{Z(\theta)}{Z(\psi)} &= \frac{1}{Z(\psi)} \int_\Omega p(\boldsymbol{x}|\theta) d\mu(\boldsymbol{x}) \\
&= \frac{1}{Z(\psi)} \int_\Omega p(\boldsymbol{x}|\theta) \frac{p(\boldsymbol{x}|\psi)}{p(\boldsymbol{x}|\psi)} d\mu(\boldsymbol{x}) \\
&= \int_\Omega \frac{p(\boldsymbol{x}|\theta)}{p(\boldsymbol{x}|\psi)} \frac{p(\boldsymbol{x}|\psi)}{Z(\psi)} d\mu(\boldsymbol{x}) \\
&= \mathbb{E}\left[\frac{p(\mathbb{X}|\theta)}{p(\mathbb{X}|\psi)}\right]
\end{aligned}$$

gives

$$\frac{Z(\theta)}{Z(\psi)} = \mathbb{E}\left[\exp\langle t(\mathbb{X}), \theta - \psi \rangle\right], \tag{6.23}$$

which again cannot be expressed in analytic closed form.

However, the ratio of normalising constants can be approximated by Monte Carlo importance sampling

$$\frac{Z(\theta)}{Z(\psi)} \approx \frac{1}{n} \sum_{i=1}^{n} \exp\langle t(\mathbb{X}_i), \theta - \psi \rangle, \tag{6.24}$$

where $\mathbb{X}_1, \mathbb{X}_2, \ldots, \mathbb{X}_n$ are samples drawn from the distribution $p(\boldsymbol{x}|\psi)$. This approach reduces the number of simulations required. This is due to the fact that the evaluation of the log-likelihood function does not require simulations of the model at each point $\theta \in \Theta$.

Plugging (6.24) into the original log-likelihood (6.22) gives its Monte Carlo counterpart

$$l_n(\theta) = \langle t(\boldsymbol{x}), \theta - \psi \rangle - \log \left(\frac{1}{n} \sum_{i=1}^{n} \exp\langle t(\mathbb{X}_i), \theta - \psi \rangle \right).$$

If the simulated Markov chain generating the samples $\mathbb{X}_1, \mathbb{X}_2, \ldots, \mathbb{X}_n$ is irreducible, then $\frac{1}{n} \sum_{i=1}^{n} \exp\langle t(\mathbb{X}_i), \theta - \psi \rangle$ almost surely converges to $Z(\theta)/Z(\psi)$. This implies that $l_n(\theta)$ almost surely converges to $l(\theta)$ as well. Let $\widehat{\theta}$ and $\widehat{\theta}_n$ be the maxima of $l(\theta)$ and $l_n(\theta)$, respectively. Under the hypothesis that $p(\boldsymbol{x}|\theta)$

belongs to the exponential family whenever $\theta \in \Theta$, then $\widehat{\theta}_n \to \widehat{\theta}$ almost surely [110, 108, 229, 230].

The likelihood function $l(\theta)$ of exponential family models is convex [230]. In principle, the computation of the maximum of a convex function can be done by a local optimisation method. The almost sure convergence of $l_n(\theta)$ to $l(\theta)$ encourages the calculation of the *gradient of the approximated log-likelihood function.*

$$\nabla l_n(\theta) = t(\boldsymbol{x}) - \mathbb{E}_{n,\theta,\psi} t(\mathbb{X}) \tag{6.25}$$

with

$$\mathbb{E}_{n,\theta,\psi} t(\mathbb{X}) = \frac{\sum_{i=1}^{n} t(\mathbb{X}_i) \exp\langle t(\mathbb{X}_i), \theta - \psi\rangle}{\sum_{i=1}^{n} \exp\langle t(\mathbb{X}_i), \theta - \psi\rangle}$$

the *Monte Carlo approximation by importance sampling* of $\mathbb{E}_\theta t(\mathbb{X})$.

Similarly, the *Hessian matrix of the approximated log-likelihood* is

$$-\nabla^2 l_n(\theta) = \mathbb{V}ar_{n,\theta,\psi}[t(\mathbb{X})] \tag{6.26}$$

where

$$\mathbb{V}ar_{n,\theta,\psi}[t(\mathbb{X})] = \mathbb{E}_{n,\theta,\psi} t(\mathbb{X}) t(\mathbb{X})^t - \mathbb{E}_{n,\theta,\psi} t(\mathbb{X}) \mathbb{E}_{n,\theta,\psi} t(\mathbb{X})^t.$$

The estimates of the gradient and the Hessian presented earlier are consistent. It is not very difficult to check that the gradient and the Hessian of the log-likelihood have the same expressions as their estimators when the Monte Carlo expectations are replaced by the exact expectations in (6.25) and (6.26), respectively.

We now have all the ingredients to build iterative local optimisation methods to find $\widehat{\theta}_n$. Using the Newton-Raphson Monte Carlo method, we get

$$\theta_{k+1} = \theta_k - [\nabla^2 l_n(\theta_k)]^{-1} \nabla l_n(\theta_k) \tag{6.27}$$

for $k = 1, 2, \ldots$, where $l_n(\cdot)$ is calculated from n samples of $p(\boldsymbol{x}|\theta_0)$. Calculations of the gradient and the inverse of the Hessian can be expensive and sometimes numerically unstable. A possible alternative to the (6.27) is the *stochastic gradient method* [231, 321, 322, 323, 318]:

$$\theta_{k+1} = \theta_k + \epsilon_k [t(\boldsymbol{x}) - t(\mathbb{X}_k)] \tag{6.28}$$

where $\epsilon_k > 0$ is a decreasing sequence and $\mathbb{X}_k$ is a sample drawn according to $p(\boldsymbol{x}|\theta_k)$. This method is a priori easier to implement, but the choice of an optimal sequence ϵ_k remains a delicate problem [108].

An alternative to these two strategies is to use the optimal step gradient method [55, 80, 250, 282, 192]. In this case the parameters follow the iterative scheme:

$$\theta_{k+1} = \theta_k + \rho(\theta_k)\nabla l_n(\theta_k),$$

the descent step $\rho(\theta_k)$ being chosen in an optimal way

$$l_n(\theta_k + \rho(\theta_k)\nabla l_n(\theta_k)) = \max_{\rho \in \mathbb{R}} l_n(\theta_k + \rho\nabla l_n(\theta_k)).$$

At each iteration, $\rho(\cdot)$ "directs" the parameters to the values that maximise the likelihood function.

As k tends to infinity, and under regularity conditions, all these methods approach the *Maximum Likelihood Estimator (MLE)*. All three methods are sensitive to the initial values. For models of exponential families, where the likelihood function is convex, this choice is theoretically of little importance. However, an initial condition that is too far from the solution can lead to numerical instability. Periodic resampling of the model removes this difficulty, but at the cost of computational time. A quite common strategy is to use the pseudo-likelihood estimator as the initial condition for the methods that compute the MLE. The **R** library **spatstat** implements this strategy for a variety of models. Studies and comparisons of different methods performing parameter estimation and computing the MLE can be found in [85, 205].

Let us now suppose that a value $\widehat{\theta}_n$ has been obtained by maximising an approximation of the likelihood function $l_n(\theta)$. This function was constructed from n samples $\mathbb{X}_1, \mathbb{X}_2, \ldots, \mathbb{X}_n$ drawn according to $p(\boldsymbol{x}|\theta_0)$. The difference between $\widehat{\theta}_n$ and $\widehat{\theta}$, the true but unknown maximum likelihood estimator, is characterised by a central limit theorem [107, 108].

Thus, when $n \to \infty$, the random variable $\sqrt{n}(\widehat{\theta}_n - \widehat{\theta})$ converges in law to a zero mean normal variable of variance $I(\widehat{\theta})^{-1}\Gamma I(\widehat{\theta})^{-1}$:

$$\sqrt{n}(\widehat{\theta}_n - \widehat{\theta}) \to \mathcal{N}(0, I(\widehat{\theta})^{-1}\Gamma I(\widehat{\theta})^{-1}).$$

The matrix

$$I(\widehat{\theta}) = \mathbb{V}ar_{\widehat{\theta}}[t(\mathbb{X})] = -\nabla^2 l(\widehat{\theta})$$

represents the *Fisher information* of $\widehat{\theta}$, while Γ is the asymptotic covariance matrix of the normalised Monte Carlo gradient $\sqrt{n}\nabla l_n(\widehat{\theta})$.

The matrix $I(\widehat{\theta})$ is approximated by $-\nabla^2 l_n(\widehat{\theta}_n)$. Concerning Γ, the following estimator is considered

$$\frac{C_n}{\left\{\frac{1}{n}\sum_{i=1}^{n}\exp\left[t(\mathbb{X}_i)^t(\widehat{\theta}_n - \psi_0)\right]\right\}^2}$$

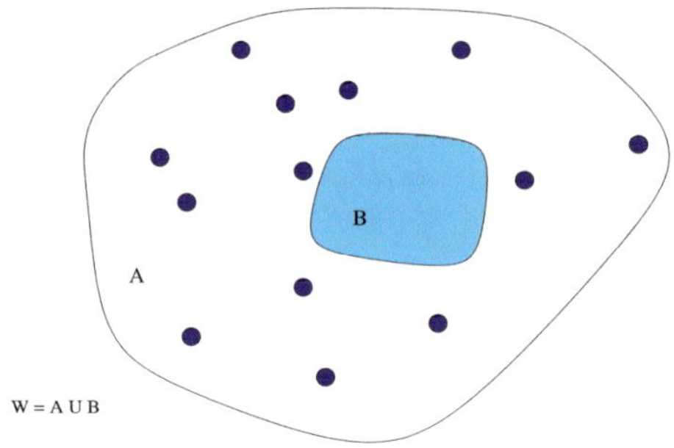

FIGURE 6.7
Observation scheme in the case of incomplete data: the process takes place in the region $W = A \cup B$, but it is observed only in the region A. The realisation of the process is only partially observed.

where C_n is the empirical covariance matrix of $(t(\boldsymbol{x}) - t(\mathbb{X})) \exp\left[t(\mathbb{X})^t(\widehat{\theta}_n - \psi_0)\right]$, based on the samples $\mathbb{X}_1, \mathbb{X}_2, \ldots, \mathbb{X}_n$ obtained from $p(\boldsymbol{x}|\psi_0)$.

Under assumptions concerning the asymptotics of the maximum likelihood, the matrix $-\nabla^2 l_n(\widehat{\theta}_n)$ can be considered as the inverse of the asymptotic variance of the maximum likelihood estimator [107, 108, 230]. Thus, by taking the square root of the diagonal elements of the inverse of this matrix, an estimate of the error $\widehat{\theta} - \theta_0$ is obtained. This error represents the difference between the maximum likelihood estimator and the true values of the model parameters.

6.3.3 EM algorithm: parameter estimation for incomplete observed data

It is not always possible to observe a pattern in its complete form. Figure 6.7 illustrates such an observation scheme. A point process takes place in $W = A \cup B$. Only the pattern of points in the region A can be observed, while there are probably "hidden" or "missing" points in the region B. Since the point configuration in $A \cup B$ is only partially observed, the data are considered incomplete. The representation of regions A and B as two nested regions is not important. The only thing required of these sets is that they are $\nu-$measurable. This is equivalent to saying that well-defined structures can coexist within these two regions.

The problem is to estimate the model parameters of a process that takes place in $W = A \cup B$, while its observation is only available in A. In this case, one solution to estimating the model parameters is to use *Expectation-Maximisation (EM) algorithms*. The idea of such an algorithm is to "complete" the data by simulating it in the regions where observations are not available. This type of algorithm allows the estimation of model parameters for the whole process. It thus gives the possibility to perform statistical inference both in the observable and in the hidden regions [75, 74, 103, 107, 196, 191, 285, 286].

Let $p(\boldsymbol{x}, \boldsymbol{y}|\theta)$ be the joint probability density of all objects, expressed as

$$p(\boldsymbol{x}, \boldsymbol{y}|\theta) = \frac{\exp\langle t(\boldsymbol{x}, \boldsymbol{y}), \theta\rangle}{Z(\theta)}$$

where $\boldsymbol{x}$ is the configuration of the observed objects (in A) and $\boldsymbol{y}$ that of the hidden objects (in B). The normalisation constant is written here

$$Z(\theta) = \int \exp\langle t(\boldsymbol{x}, \boldsymbol{y}), \theta\rangle \mu_A(d\boldsymbol{x})\mu_B(d\boldsymbol{y})$$

where μ_A and μ_B denote the restrictions at A and B of the Poissonian reference measure μ on $A \cup B$.

In this context, the likelihood is the marginal law

$$L(\theta) = p(\boldsymbol{x}|\theta) = \int p(\boldsymbol{x}, \boldsymbol{y}|\theta)\mu_B(d\boldsymbol{y}).$$

It can also be written as

$$L(\theta) = \frac{1}{Z(\theta)} \int \exp\langle t(\boldsymbol{x}, \boldsymbol{y}), \theta\rangle \mu_B(d\boldsymbol{y}) = \frac{Z(\theta|\boldsymbol{x})}{Z(\theta)}, \tag{6.29}$$

which involves the normalisation constant $Z(\theta|\boldsymbol{x})$ of the conditional probability density $p(\boldsymbol{y}|\boldsymbol{x}, \theta)$.

Let us consider the likelihood ratio with respect to a known parameter ψ, then take its logarithm. We obtain

$$l(\theta) = \log \frac{p(\boldsymbol{x}|\theta)}{p(\boldsymbol{x}|\psi)} = \log \frac{Z(\theta|\boldsymbol{x})}{Z(\psi|\boldsymbol{x})} - \log \frac{Z(\theta)}{Z(\psi)}.$$

As before, the components of the log-likelihood are expressed in terms of mathematical expectations (6.23). Hence

$$\frac{Z(\theta)}{Z(\psi)} = \mathbb{E}\left[\exp\langle t(\mathbb{X}, \mathbb{Y}), \theta - \psi\rangle\right] \tag{6.30}$$

and

$$\frac{Z(\theta|\boldsymbol{x})}{Z(\psi|\boldsymbol{x})} = \mathbb{E}\left[\exp\langle t(\mathbb{X}, \mathbb{Y}), \theta - \psi\rangle | \mathbb{X} = \boldsymbol{x}\right]. \tag{6.31}$$

The expectations (6.30) and (6.31) are not analytically tractable. But they can be approximated by their Monte Carlo counterparts, which leads to the following expression for the likelihood:

$$l_n(\theta) = \log\left(\frac{1}{n}\sum_{i=1}^{n}\exp\langle t(\boldsymbol{x}, \mathbb{Y}_i^*), \theta - \psi\rangle\right) - \log\left(\frac{1}{n}\sum_{i=1}^{n}\exp\langle t(\mathbb{X}_i, \mathbb{Y}_i), \theta - \psi\rangle\right), \tag{6.32}$$

$(\mathbb{X}_i, \mathbb{Y}_i), i = 1, 2, \ldots$ denoting samples of the joint distribution and $(\mathbb{Y}_i^*)$ samples of the conditional distribution on the observed objects $\boldsymbol{x}$.

The convergence results obtained for the approximation of the likelihood for complete data remain valid for the incomplete data case [107, 108]. Using importance sampling, the gradient of the loglikelihood approximation (6.32) is

$$\nabla l_n(\theta) = \mathbb{E}_{n,\theta,\psi}[t(\mathbb{X}, \mathbb{Y})|\mathbb{X} = \boldsymbol{x}] - \mathbb{E}_{n,\theta,\psi}[t(\mathbb{X}, \mathbb{Y})],$$

with the empirical expectations computed using samples simulated according to distributions depending of the parameter ψ.

All the ingredients are now in place to implement an iterative method of local optimisation [323, 74, 120]. For exponential family models, theoretical results ensure that the likelihood increases with each iteration and convergence to the local optimum closest to the initial configuration is established [74], Thm. 1, pp. 101,). The same authors also demonstrate central limit theorem results. However, it is very difficult to say anything about the convexity of the likelihood (6.29).

6.4 Pattern characterisation through posterior sampling

Let us go back to the case of complete data, in the context given in Sections 6.3.1 and 6.3.2. In addition, suppose that knowledge about the model parameters is available in the form of a *prior distribution* $p(\theta)$ over Θ.

The *posterior distribution* is constructed using the Bayes rule:

$$p(\theta|\boldsymbol{x}) \propto p(\boldsymbol{x}|\theta)p(\theta) \tag{6.33}$$

where $\boldsymbol{x}$ is the observed pattern and $p(\boldsymbol{x}|\theta)$ is given by (6.18). In this context, the posterior (6.33) can be developed further

$$p(\theta|\boldsymbol{x}) = \frac{\exp\langle t(\boldsymbol{x}), \theta\rangle p(\theta)}{Z(\boldsymbol{x})Z(\theta)}$$

where $Z(\boldsymbol{x})$ is its normalising constant. More generally, all methods presented in this section apply to models specified by probability densities that are continuously differentiable functions, i.e. $p(\boldsymbol{x}|\cdot) \in \mathcal{C}^1(\Theta)$ [297, 287].

Sampling posterior distributions (6.33) is of interest because it allows the application of a wider range of statistical inference procedures to the model parameters.

To this purpose, let us first consider a classical Metropolis-Hastings (MH) algorithm:

Algorithm 6.2. General MH for posterior sampling. *Data: $\boldsymbol{x}$ the observed pattern, $p(\boldsymbol{x}|\theta)$ the assumed model, $p(\theta)$ the model prior, $q(\cdot \to \cdot)$ the proposal density and θ the initial state of the system.*

1. *Propose a new value ψ according to $q(\theta \to \psi)$.*
2. *Accept the value ψ with probability*

$$\alpha(\theta \to \psi) = \min\left\{1, \frac{p(\psi|\boldsymbol{x})}{p(\theta|\boldsymbol{x})} \frac{q(\psi \to \theta)}{q(\theta \to \psi)}\right\}. \tag{6.34}$$

   The acceptance is done via the procedure

   (a) Draw $u \sim \mathcal{U}[0, 1]$.

   (b) If $u \leq \alpha(\theta \to \psi)$ then $\theta \leftarrow \psi$, else θ remains unchanged.

3. *Return θ.*
4. *If more samples are needed, go to 1.*

The previous algorithm simulates a Markov chain with transition kernel

$$\begin{aligned} P(\theta, A) &= \int_A \alpha(\theta \to \psi) q(\theta \to \psi) \mathbf{1}\{\psi \in A\} d\psi \\ &+ \mathbf{1}\{\theta \in A\} \left[1 - \int_A \alpha(\theta \to \psi) q(\theta \to \psi) d\psi\right] \end{aligned}$$

and unique equilibrium distribution

$$\pi(A) = \int_A p(\theta|\boldsymbol{y}) d\theta,$$

with $A \subseteq \Theta$. Furthermore, the simulated chain is uniformly ergodic, that is, there is a positive constant M and a positive constant $\rho < 1$ such that

$$\sup_{\theta \in \Theta} \| P^n(\theta, \cdot) - \pi(\cdot) \| \leq M\rho^n, \quad n \in \mathbb{N}.$$

In order to guarantee these properties rather mild assumptions regarding the proposal density $q(\theta \to \psi)$ are needed [312].

Despite these attractive characteristics, this algorithm cannot be used in practice in a straightforward manner: the computation of $p(\psi|\boldsymbol{x})/p(\theta|\boldsymbol{x})$, the posteriors ratio in (6.34) requires the evaluation of the normalising constants ratio, $Z(\psi)/Z(\theta)$. Next, several strategies able to deal with this problem are presented.

6.4.1 Auxiliary variable MH sampler

The authors [224] build an MH algorithm to sample from the joint distribution

$$\begin{aligned} p(\theta, \boldsymbol{y}|\boldsymbol{x}) &= a(\boldsymbol{y}|\theta, \boldsymbol{x})p(\theta|\boldsymbol{x}) \\ &= a(\boldsymbol{y}|\theta, \boldsymbol{x})\frac{\exp\langle t(\boldsymbol{x}), \theta\rangle p(\theta)}{Z(\boldsymbol{x})Z(\theta)}. \end{aligned} \tag{6.35}$$

where $a(\boldsymbol{y}|\theta, \boldsymbol{x})$ is the conditional probability density of the auxiliary variable $\mathbb{Y}$ defined over Ω the same configuration space as for $\mathbb{X}$.

The idea is to use a suitable proposal $q((\theta, \boldsymbol{y}) \to (\theta', \boldsymbol{y}'))$ that leads to the simplification of the normalising constants. This is achieved by using a proposal made of two components

$$q((\theta, \boldsymbol{y}) \to (\theta', \boldsymbol{y}')) = q_1(\theta'|\theta, \boldsymbol{x})q_2(\boldsymbol{y}'|\theta', \theta, \boldsymbol{y}).$$

The first component may not depend on $\boldsymbol{y}$ and it may be a very simple probability law. Here we consider the symmetric distribution

$$q_1(\theta'|\theta) = q_1(|\theta' - \theta|). \tag{6.36}$$

The second component q_2 may be chosen independently of $(\theta, \boldsymbol{y})$ as follows:

$$q_2(\boldsymbol{y}'|\theta', \theta, \boldsymbol{y}) = q_2(\boldsymbol{y}'|\theta') = \frac{\exp\langle t(\boldsymbol{y}'), \theta'\rangle}{Z(\theta')}. \tag{6.37}$$

The ratio $R_D((\theta, \boldsymbol{y}) \to (\theta', \boldsymbol{y}'))$ of the induced MH dynamics is

$$\begin{aligned} &R_D((\theta, \boldsymbol{y}) \to (\theta', \boldsymbol{y}')) = \\ &= \frac{p(\theta', \boldsymbol{y}'|\boldsymbol{x})}{p(\theta, \boldsymbol{y}|\boldsymbol{x})} \times \frac{q((\theta', \boldsymbol{y}') \to (\theta, \boldsymbol{y}))}{q((\theta, \boldsymbol{y}) \to (\theta', \boldsymbol{y}'))}. \end{aligned} \tag{6.38}$$

Introducing the joint density (6.35), the proposals (6.36) and (6.37) into the ratio (6.38) gives

$$\frac{a(\boldsymbol{y}'|\theta', \boldsymbol{x})p(\theta')\exp\langle t(\boldsymbol{x}), \theta'\rangle\exp\langle t(\boldsymbol{y}), \theta\rangle}{a(\boldsymbol{y}|\theta, \boldsymbol{x})p(\theta)\exp\langle t(\boldsymbol{x}), \theta\rangle\exp\langle t(\boldsymbol{y}'), \theta'\rangle}$$

which does not depend on the normalising constants Z.

Under these considerations, the *auxiliary variable MH sampler* for posteriors works as follows:

Algorithm 6.3. Auxiliary variable MH for posterior sampling. *Data:* $\boldsymbol{x}$ *the observed pattern,* $p(\boldsymbol{x}|\theta)$ *the assumed model,* $p(\theta)$ *the prior model,* $a(\boldsymbol{y}|\theta, \boldsymbol{x})$ *the auxiliary variable probability density,* $q(\cdot \to \cdot)$ *the previously specified proposal and* $(\theta, \boldsymbol{y})$ *the initial state of the system.*

1. *Propose a new candidate* θ' *using the density* $q_1(\theta'|\theta)$.
2. *Propose a new candidate* $\boldsymbol{y}'$ *using the proposal density* $q_2(\boldsymbol{y}'|\theta') = \frac{\exp\langle t(\boldsymbol{y}', \theta'\rangle}{Z(\theta')}$.
3. *Compute the MH ratio* $R_D((\theta, \boldsymbol{y}) \to (\theta', \boldsymbol{y}'))$ *using* (6.38).
4. *The new state* $(\theta', \boldsymbol{y}')$ *is accepted with probability*

$$\alpha = \min\left\{1, R_D((\theta, \boldsymbol{y}) \to (\theta', \boldsymbol{y}'))\right\},$$

   using the procedure

   (a) *Draw* $u \sim \mathcal{U}_{[}0, 1]$.

   (b) *If* $u \le \alpha$ *then* $(\theta, \boldsymbol{y}) \leftarrow (\theta', \boldsymbol{y}')$, *else* $(\theta, \boldsymbol{y})$ *remains unchanged.*
5. *Return* $(\theta, \boldsymbol{y})$.
6. *If more samples are needed, go to* 1.

If exact samples are drawn from $q_2(\boldsymbol{y}'|\theta')$, the distribution of the outputs of Algorithm 6.3 converges to the unique equilibrium distribution $p(\theta, \boldsymbol{y}|\boldsymbol{x})$. The requirement of exact simulation can lead to some difficulties from a practical point of view. First, exact simulation algorithms are only available for a rather limited class of models. Second, the convergence of existing exact simulation algorithms depends on the model parameters [195]. Furthermore, as the authors explain in [224], there is no clear strategy for proposing a conditional probability density for the auxiliary variable that avoids poor mixing. One possible strategy, which may be fair enough, is to set

$$a(\boldsymbol{y}|\theta, \boldsymbol{x}) = \frac{\exp\langle t(\boldsymbol{y}), \widehat{\theta}(\boldsymbol{x})\rangle}{Z(\widehat{\theta}(\boldsymbol{x}))}$$

where $\widehat{\theta}(\boldsymbol{x})$ is the pseudo-likelihood estimate of the model parameters based on the observation of $\boldsymbol{y}$. Nevertheless, it is important to acknowledge the importance of the work done by [224], since they provided the first solution to this problem, while inspiring important research developments in this direction.

6.4.2 Exchange MH algorithm

The *exchange MH algorithm* for sampling posterior densities with analytically intractable normalisation constants was proposed by [233]. Its convergence properties have been analysed by [177, 178]. Applications to posterior distributions of exponential random graph models are presented by [48].

Consider the augmented probability density

$$p(\theta, \boldsymbol{y}, \psi|\boldsymbol{x}) \propto p(\theta|\boldsymbol{x})p(\psi|\theta)p(\boldsymbol{y}|\psi) \tag{6.39}$$

where the law $p(\theta|\boldsymbol{x}) = p(\boldsymbol{x}|\theta)p(\theta)$ is the posterior distribution of interest (6.33), the first auxiliary variable $\boldsymbol{y}$ is controlled by the distribution $p(\boldsymbol{y}|\psi)$ from the same family as $p(\boldsymbol{x}|\theta)$ and the second auxiliary variable $\psi \in \Theta$ is given by the probability density $p(\psi|\theta)$.

The idea of the exchange algorithm is to introduce an additional auxiliary variable and an update mechanism based on an exchange procedure. Its description is given below.

Algorithm 6.4. Exchange MH for posterior sampling. *Data: $\boldsymbol{x}$ the observed pattern, $p(\boldsymbol{x}|\theta)$ the assumed model, $p(\theta)$ the model prior, $p(\psi|\theta)$ the second auxiliary variable probability density and $(\theta, \boldsymbol{y}, \psi)$ the initial state of the system.*

1. *Generate ψ using the density $p(\psi|\theta)$.*
2. *Generate $\boldsymbol{y}'$ using the density $p(\boldsymbol{y}'|\psi) = \frac{\exp\langle t(\boldsymbol{y}'), \psi\rangle}{Z(\psi)}$.*
3. *Propose to exchange θ with ψ, that is $\theta' \leftarrow \psi$ and $\psi' \leftarrow \theta$, using the following mechanism*

   (a) *Compute the exchange ratio*

$$\begin{aligned}
R_E &= \frac{p(\theta', \boldsymbol{y}', \psi'|\boldsymbol{x})}{p(\theta, \boldsymbol{y}', \psi|\boldsymbol{x})} \\
&= \frac{p(\boldsymbol{x}|\theta')p(\theta')p(\psi'|\theta')p(\boldsymbol{y}'|\psi')}{p(\boldsymbol{x}|\theta)p(\theta)p(\psi|\theta)p(\boldsymbol{y}'|\psi)} = \frac{p(\boldsymbol{x}|\psi)p(\psi)p(\theta|\psi)p(\boldsymbol{y}'|\theta)}{p(\boldsymbol{x}|\theta)p(\theta)p(\psi|\theta)p(\boldsymbol{y}'|\psi)} \\
&= \exp\langle t(\boldsymbol{x}) - t(\boldsymbol{y}'), \psi - \theta\rangle \times \frac{p(\psi)p(\theta|\psi)}{p(\theta)p(\psi|\theta)} \times \frac{Z(\theta)Z(\psi)}{Z(\psi)Z(\theta)} \\
&= \exp\langle t(\boldsymbol{x}) - t(\boldsymbol{y}'), \psi - \theta\rangle \times \frac{p(\psi)p(\theta|\psi)}{p(\theta)p(\psi|\theta)}
\end{aligned} \tag{6.40}$$

   (b) *The new state $(\theta', \boldsymbol{y}', \psi')$ is chosen following via the procedure*

      i. *Compute*

$$\alpha = \min\{1, R_E\},$$

   ii. *Draw* $u \sim \mathcal{U}[0,1]$.
   iii. *If* $u \leq \alpha$ *then* $(\theta', \boldsymbol{y}', \psi') = (\psi, \boldsymbol{y}', \theta)$, *else return* $(\theta', \boldsymbol{y}', \psi') = (\theta, \boldsymbol{y}', \psi)$ *where* ψ *and* $\boldsymbol{y}'$ *are given by steps* 1 *and* 2, *respectively.*

4. *Return* $(\theta', \boldsymbol{y}', \psi')$

5. *If more samples are needed, go to* 1.

The hypotheses regarding the distribution $p(\psi|\theta)$ in Algorithm 6.4 are mild. In general, a symmetric probability density centred around θ is the choice made in practice. The convergence of the algorithm towards (6.39) is guaranteed provided that exact samples are drawn for the auxiliary variable $\boldsymbol{y}$ [233, 177]. In this sense, the auxiliary variable and exchange algorithms are equivalent, but the exchange algorithm avoids the poor mixing problem. Strategies to avoid exact sampling of the auxiliary variables are proposed by [178]. The development of the convergence results presented relies on the work of [97, 124]. The proposed strategies require discretisation of the parameter space and storage of as many pre-simulated auxiliary patterns as the cardinality of the discretised parameter set. From a practical point of view, this can be a rather serious limitation when considering complex models.

6.4.3 Approximate Bayesian Computation

Approximate Bayesian Computation (ABC) is the generic name given to numerical methods and algorithms that allow statistical inference based on approximate sampling of the posterior distribution $p(\theta|\boldsymbol{x})$ [4, 26, 118, 200]. The algorithm below illustrates the general principle of these methods.

Algorithm 6.5. General ABC for posterior sampling. *Data:* $\boldsymbol{x}$ *the observed pattern,* $p(\boldsymbol{x}|\theta)$ *the assumed model,* $p(\theta)$ *the model prior,* t *the statistics vector,* ϵ *the tolerance threshold and* d *the distance function.*

1. *Generate* θ *according to* $p(\theta)$.

2. *Generate* $\boldsymbol{y}$ *according to the probability density* $p(\boldsymbol{y}|\theta)$ *and compute the statistics vector* $t(\boldsymbol{y})$.

3. *If*

$$d(t(\boldsymbol{x}), t(\boldsymbol{y})) \leq \epsilon$$

   then return θ, *else go to* 1.

4. *If more samples are needed, go to* 1.

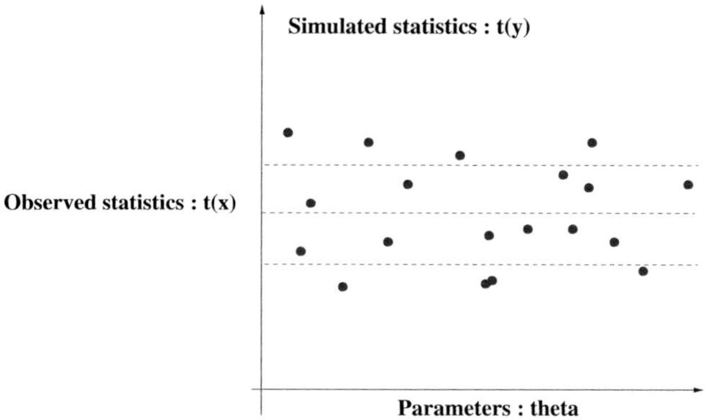

FIGURE 6.8
Graphical representation of the outputs of the General ABC algorithm.

Figure 6.8 is a picture of possible results obtained by Algorithm 6.5. The outputs of the algorithm are distributed according to $p(\theta|d(t(\boldsymbol{x})-t(\boldsymbol{y})) \leq \epsilon)$. The choice of the distance d and the tolerance parameter ϵ affects the quality of the result obtained. Sometimes not only the normalisation constant but the whole expression of $p(\boldsymbol{x}|\theta)$ is analytically intractable or difficult to manipulate, as for Cox or determinantal point processes. In this case, the choice of the number and the type of statistics to construct the vector $t(\cdot)$ is crucial [277, 316]. For exponential family models, $t(\cdot)$ is given by the vector of sufficient statistics.

The Algorithm 6.8 is easy to implement and does not suffer from poor mixing. Convergence properties of ABC algorithms have been analysed by [41, 36]. Kernel estimation can be performed on the outputs of Algorithm 6.5 to approximate the posterior distribution. The distance condition in step 3 can be replaced by requiring that the considered parameter value produces statistics within the $k-$nearest neighbours of the given observation, where k is a prefixed integer. These strategies allow the derivation of asymptotic quantities, bias and variances, related to the posterior estimation. To be efficient from a numerical point of view, the algorithm should not reject too many proposed θ values. As the theoretical studies suggest, and as can be seen from Figure 6.8, all of this requires that the prior $p(\theta)$ is close enough to the unknown posterior. For this purpose, numerical strategies that preselect the interesting θ values can be used [41, 277, 316].

6.4.4 Posterior approximation via the ABC Shadow algorithm

The *ABC Shadow* algorithm is an iterative sampling method that approximates the posterior distribution. The algorithm evolves as close as desired to the posterior distribution of interest. It is directly inspired by the auxiliary variable algorithm proposed by [224]. The ABC Shadow algorithm does not require exact simulation and does not suffer from poor mixing. The idea of the algorithm is to construct two Markov chains. The first, the *ideal chain*, has as its unique equilibrium distribution the posterior distribution of interest. This chain cannot be implemented in practice. The second Markov chain is the *shadow chain*. It is easy to implement in practice and it closely follows the ideal chain. The fact that the two chains evolve as closed as desired allows the shadow chain to approximate the posterior distribution.

This principle is well-known in stochastic approximation. For example, the stochastic gradient principle is to build an algorithm that uses random updates and is able to closely follow a theoretically deterministic ascending algorithm to the desired solution. The need to build a stochastic algorithm is due to the fact that the deterministic algorithm cannot be implemented in practice [25, 198, 231, 318].

In the following, the ABC Shadow algorithm is first introduced. Then its theoretical construction and properties will be discussed.

Algorithm 6.6. ABC Shadow for posterior sampling. *Data:* $\boldsymbol{x}$ *the observed pattern,* $p(\boldsymbol{x}|\theta)$ *the assumed model,* $p(\theta)$ *the model prior,* Δ *the control parameter,* m *an integer value and* $(\theta_0, \boldsymbol{y}_0)$ *the initial state of the system.*

1. *Generate* $\boldsymbol{y}$ *according to* $p(\boldsymbol{y}|\theta_0)$.
2. *For* $k = 1$ *to* m *do*
   (a) *Generate a new candidate* ψ *following the uniform density over* $b(\theta_{k-1}, \Delta/2)$*, the ball centred in* θ_{k-1} *and of radius* $\Delta/2$.
   (b) *The new state* $\theta_k = \psi$ *is accepted with probability*

   $$\alpha = \min\{1, R_S(\theta_{k-1} \to \psi)\}$$

   with R_S *the ratio given by*

   $$R_S(\theta \to \psi) = \exp\langle t(\boldsymbol{x}) - t(\boldsymbol{y}), \psi - \theta\rangle \times \frac{p(\psi)}{p(\theta)} \qquad (6.41)$$

3. *Return* θ_m
4. *If another sample is needed, go to step* 1 *with* $\theta_0 = \theta_m$.

Examination of the acceptance ratios (6.41) and (6.39) shows that the ABC Shadow algorithm is equivalent to the exchange algorithm using a uniform proposal for the parameter variable. The difference is that step 1 of the ABC

Shadow algorithm does not require an exact simulation to guarantee its theoretical control. The auxiliary variable $\boldsymbol{y}$ is used as a parameter for the proposal densities. This strategy makes the algorithm ready to use, while not requiring discretisation of the parameter space and storage of pre-simulated patterns [178]. In the following, theoretical details and results related to the construction of the ABC Shadow algorithm are given [297].

Let us think at a possible proposal density for the ideal Markov chain simulated by Algorithm 6.2. For this purpose, let us consider a fixed $\Delta > 0$, a parameter value $\nu \in \Theta$, a realisation $\boldsymbol{y}$ of the model $p(\cdot|\nu)$ given by (6.18) and set for the proposal the following density

$$\begin{aligned} q(\theta \to \psi) &= q_{\Delta}(\theta \to \psi|\boldsymbol{y}) \\ &= \frac{h(\boldsymbol{y}|\psi)/c(\psi)}{I(\theta, \Delta, \boldsymbol{y})} \mathbf{1}_{b(\theta,\Delta/2)}\{\psi\}, \end{aligned} \tag{6.42}$$

with $h(\boldsymbol{y}|\psi) = \exp\langle t(\boldsymbol{y}), \psi\rangle$. Here $\mathbf{1}_{b(\theta,\Delta/2)}\{\cdot\}$ is the indicator function over $b(\theta, \Delta/2)$, which is the ball of centre θ and radius $\Delta/2$. The quantity $I(\theta, \Delta, \boldsymbol{x})$ is the given by the integral

$$I(\theta, \Delta, \boldsymbol{y}) = \int_{b(\theta,\Delta/2)} h(\boldsymbol{y}|\phi)/Z(\phi)\, d\phi.$$

The resulting acceptance probability for the ideal chain is

$$\begin{aligned} &\alpha_I(\theta \to \psi) = \\ &= \min\left\{1, \frac{p(\psi|\boldsymbol{x})}{p(\theta|\boldsymbol{x})} \times \frac{h(\boldsymbol{y}|\theta)Z(\psi)I(\psi, \Delta, \boldsymbol{y})\mathbf{1}_{b(\psi,\Delta/2)}\{\theta\}}{h(\boldsymbol{y}|\psi)Z(\theta)I(\theta, \Delta, \boldsymbol{y})\mathbf{1}_{b(\theta,\Delta/2)}\{\psi\}}\right\} \\ &= \min\left\{1, \frac{h(\boldsymbol{x}|\psi)p(\psi)}{h(\boldsymbol{x}|\theta)p(\theta)} \times \frac{h(\boldsymbol{y}|\theta)I(\psi, \Delta, \boldsymbol{y})\mathbf{1}_{b(\psi,\Delta/2)}\{\theta\}}{h(\boldsymbol{y}|\psi)I(\theta, \Delta, \boldsymbol{y})\mathbf{1}_{b(\theta,\Delta/2)}\{\psi\}}\right\}. \end{aligned} \tag{6.43}$$

The proposal density (6.42) guarantees the convergence of the chain towards its unique equilibrium regime given by the posterior distribution. Still, this chain cannot be used in practice, since the computation of normalising constant ratio is replaced by the computation of

$$\frac{I(\psi, \Delta, \boldsymbol{y})}{I(\theta, \Delta, \boldsymbol{y})}. \tag{6.44}$$

The idea at the basis of the construction of the shadow chain is the approximation of (6.44) in the acceptance ratio of the ideal chain.

First, let V_{Δ} denote the volume of the ball $b(\theta, \Delta/2)$, and let

$$U_{\Delta}(\theta \to \psi) = \frac{1}{V_{\Delta}} \mathbf{1}_{b(\theta,\Delta/2)}\{\psi\}$$

be the uniform probability density over $b(\theta, \Delta/2)$.

The following result leads to the construction of the shadow chain. The construction is done in two steps. First, the uniform law $U_\Delta(\theta \to \psi)$ is used for proposing new values instead of $q_\Delta(\theta \to \psi)$. Next, the proposals ratio in the acceptance probability of the ideal chain is approximated by a quantity that is easy to calculate.

Theorem 6.4. *Let $\boldsymbol{y}$ be a point in Ω such that the function $p(\boldsymbol{y}|\phi)$ is strictly positive and continuous with respect to ϕ, then:*

1. *The probability distributions given by the proposal densities $q_\Delta(\theta \to \cdot)$ and $U_\Delta(\theta \to \cdot)$ are uniformly as close as desired in θ as Δ approaches 0. That is, for any fixed $\theta \in \Theta$ and $A \in \mathcal{T}_\Theta$, we have*

$$\lim_{\Delta \to 0_+} \int_A |q_\Delta(\theta \to \psi) - U_\Delta(\theta \to \psi)| d\psi = 0.$$

2. *For any fixed $\theta \in \Theta$, the quotient functions $\frac{q_\Delta(\theta \to \cdot)}{q_\Delta(\cdot \to \theta)}$ and $\frac{\frac{h(\boldsymbol{y}|\cdot)}{Z(\cdot)} \mathbf{1}_{b(\theta,\Delta/2)}(\cdot)}{\frac{h(\boldsymbol{y}|\theta)}{Z(\theta)} \mathbf{1}_{b(\cdot,\Delta/2)}(\theta)}$ are uniformly as close as desired in θ as Δ approaches 0. That is, for any fixed $\theta \in \Theta$, we have*

$$\lim_{\Delta \to 0_+} \sup_{\psi \in \Theta} \left| \frac{q_\Delta(\theta \to \psi|\boldsymbol{x})}{q_\Delta(\psi \to \theta|\boldsymbol{x})} - \frac{\frac{h(\boldsymbol{x}|\psi)}{Z(\psi)} \mathbf{1}_{b(\theta,\Delta/2)}(\psi)}{\frac{h(\boldsymbol{x}|\theta)}{Z(\theta)} \mathbf{1}_{b(\psi,\Delta/2)}(\theta)} \right| = 0$$

uniformly in $\theta \in \Theta$.

Moreover, if $p(\boldsymbol{y}|\cdot) \in \mathcal{C}^1(\Theta)$, then rates of convergence in (i) and (ii) can be provided.

The shadow chain appears as a natural consequence of the previous result. Assuming the chain in an initial state θ, a new value ψ is proposed uniformly in the ball $b(\theta, \Delta/2)$ and accepted with probability

$$\begin{aligned} &\alpha_S(\theta \to \psi) = \\ &= \min \left\{ 1, \frac{p(\psi|\boldsymbol{x})}{p(\theta|\boldsymbol{x})} \times \frac{h(\boldsymbol{y}|\theta) Z(\psi) \mathbf{1}_{b(\psi,\Delta/2)}\{\theta\}}{h(\boldsymbol{y}|\psi) Z(\theta) \mathbf{1}_{b(\theta,\Delta/2)}\{\psi\}} \right\} \\ &= \min \left\{ 1, \frac{h(\boldsymbol{x}|\psi) p(\psi)}{h(\boldsymbol{x}|\theta) p(\theta)} \times \frac{h(\boldsymbol{y}|\theta) \mathbf{1}_{b(\psi,\Delta/2)}\{\theta\}}{h(\boldsymbol{y}|\psi) \mathbf{1}_{b(\theta,\Delta/2)}\{\psi\}} \right\}, \end{aligned} \tag{6.45}$$

otherwise the chain remains in θ.

Also a direct implication of Theorem 6.4 is the following corollary.

Corollary 6.1. *The acceptance probabilities of the ideal and shadow chains given by (6.43) and (6.45) respectively, are uniformly as closed as desired when Δ approaches 0.*

The next result shows that it is possible to choose $\Delta > 0$ such that the ideal and shadow chains evolve as closed as desired during a fixed number of steps m.

Proposition 6.4. *Let P_I and P_S be the transition kernels for the ideal and the shadow Markov chains using a general $\Delta > 0$ and a configuration $\boldsymbol{y} \in \Omega$ as in Theorem 6.4. Then for every $\epsilon > 0$ and $m \in \mathbb{N}$, there exists $\Delta_0 = \Delta_0(\epsilon, m) > 0$ such that for every $\Delta \leq \Delta_0$ we have*

$$|P_I^m(\theta, A) - P_S^m(\theta, A)| < \epsilon$$

uniformly in $\theta \in \Theta$ and $A \subset \Theta$. If $p(\boldsymbol{y}|\theta) \in \mathcal{C}^1(\Theta)$, then a description of $\Delta_0(\epsilon, m)$ can be provided.

Proofs of the previous results are given in [297]. A practical interpretation of the Proposition 6.4 is that for chosen ϵ and m, then taking Δ such that

$$\Delta \leq K(\boldsymbol{y}, p, \Theta)\frac{\epsilon}{m} \tag{6.46}$$

ensures that the ideal and the shadow chain evolve during m steps within the desired ϵ confidence tube. The constant K depends on the auxiliary pattern, the posterior and the parameter space.

If more than one value of θ is needed, Algorithm 6.6 should be restarted with the current value of θ as initial condition and a new auxiliary variable $\boldsymbol{y}$. The question arises whether there exists a sequence $(m_k, \Delta_k)_{k\geq 1}$ which guarantees the convergence of the algorithm. In the following we present some elements given by [136] as a possible starting point for answering this question.

For any integer $k \geq 1$, let us consider the sequence $(m_j, \Delta_j, \boldsymbol{y}_j)$ $1 \leq j \leq k$ of integers, positive reals and auxiliary variables given by any parameters, respectively. The product of k ideal kernels is given by

$$\mathcal{P}_{I,k} = P_{I,\Delta_1}^{m_1} \cdots P_{I,\Delta_k}^{m_k}$$

with $P_{I,\Delta_j}^{m_j}$ the m_j applications of the ideal transition kernel using Δ_j and $\boldsymbol{y}_j$. The product of k shadow kernels, $\mathcal{P}_{S,k}$ is defined analogously.

The ideal chain is uniformly ergodic. Hence, there exists $\rho \in]0, 1[$ such that

$$\| P_{I,\Delta}^m - \pi \| \leq 2\rho^m$$

for any $(m, \Delta, \boldsymbol{y})$. The ergodicity coefficient is known in the literature as the Dobrushin coefficient [122, 123, 180, 214]. Clearly, for each element $(m_j, \Delta_j, \boldsymbol{y}_j)$ there is a ρ_j.

Let us compute the distance between the product of k shadow kernels $\mathcal{P}_{S,k}$ and π the distribution given by the posterior density of interest (6.33). We have

$$\begin{aligned}
\| \mathcal{P}_{S,k} - \pi \| &= \| \mathcal{P}_{S,k} - \mathcal{P}_{S,k-1} P_{I,\Delta_k}^{m_k} + \mathcal{P}_{S,k-1} \mathcal{P}_{I,\Delta_k}^{m_k} - \pi \| \\
&\leq \| \mathcal{P}_{S,k-1} \| \| P_{S,\Delta_k}^{m_k} - P_{I,\Delta_k}^{m_k} \| + \| \mathcal{P}_{S,k-1} \mathcal{P}_{I,\Delta_k}^{m_k} - \pi \| \\
&\leq \| P_{S,\Delta_k}^{m_k} - P_{I,\Delta_k}^{m_k} \| + \| \mathcal{P}_{S,k-1} \mathcal{P}_{I,\Delta_k}^{m_k} - \pi \| \\
&\leq K m_k \Delta_k + 2\rho_k^{m_k}
\end{aligned} \tag{6.47}$$

The last line in (6.47) is obtained using $\| \mathcal{P}_{S,k-1} \| \leq 1$, Proposition 6.4, the uniform ergodicity of the ideal chain and by considering that with a high probability there exists K a superior bound of all $K(\boldsymbol{y}, p, \Theta)$. A proof and a discussion of the latter point is given in [136]. The condition (6.47) may seem contradictory since large values for m_k are needed to cancel its second term while rather small values are needed to reduce the first term. For a desired error ϵ, information about the Dobrushin coefficient can be used to control the second term, derive conditions for m_k, and then obtain an appropriate Δ_k using the first term. In the next chapter, an algorithm will be presented that is able to perform parameter estimation while controlling the Δ_k sequence.

In terms of sampling of posteriors with continuously differentiable parameters, the exchange algorithm is the first option to consider whenever exact simulation is available. However, for highly structured models, the convergence time of the existing exact simulation algorithms grows exponentially depending on the model parameters [195]. Therefore, the ABC Shadow algorithm is a serious alternative in this context. Its associated control and convergence results are equivalent to those known for classical approximation algorithms such as the stochastic gradient [318, 178, 231].

Example 6.3 (ABC Shadow and stochastic gradient algorithms for parametric inference for the Strauss process). Let W be the unit square $[0,1]^2$ and let us consider the Strauss point process specified by the probability density

$$\begin{aligned}
p(\boldsymbol{x}|\theta) &\propto \beta^{n(\boldsymbol{x})} \gamma^{s_r(\boldsymbol{x})} = \\
&= \exp\left[n(\boldsymbol{x}) \log \beta + s_r(\boldsymbol{x}) \log \gamma\right] \\
&= \exp\langle t(\boldsymbol{x}), \theta \rangle.
\end{aligned} \tag{6.48}$$

with respect to the reference measure represented by the standard Poisson point process [149, 302]. Here, $\boldsymbol{x}$ denotes the point pattern in the domain W. The sufficient statistics vector $t(\boldsymbol{x}) = (n(\boldsymbol{x}), s_r(\boldsymbol{x}))$ is given by the total number of points and the number of pairs of points that are situated within a distance less than the range parameter r. The vector of interaction parameters is $\theta = (\log \beta, \log \gamma)$.

A sample made of m point patterns is observed. It is further assumed that the observed patterns are independent realisations of a Strauss model (6.48) with parameters r and θ. The range parameter is assumed to be known. Inference is performed with respect to the interaction parameters. It is easy to see that the maximum likelihood estimate $\widehat{\theta}$ satisfies the equation

$$\frac{1}{m}\sum_{i=1}^{m} t(\boldsymbol{y}_i) - \mathbb{E}_{\widehat{\theta}} t(\mathbb{X}) = 0$$

with $\boldsymbol{y}_1, \ldots, \boldsymbol{y}_m$ the observed point patterns forming the sample. The calculations for the sum and expectation are done component by component. It is clear that the true parameters of the model are approached by $\widehat{\theta}$ as m increases. Furthermore, since Θ is a compact region in $\mathbb{R}^2$, there is an equivalence between likelihood-based inference and inference based on posterior distributions built with uniform priors $p(\theta)$ on Θ.

The ABC Shadow dynamics can be tested using this property. Within this context, the maximum a posteriori should be close to the maximum likelihood and the true model parameters. To save computational time, it can be assumed that instead of m independent identically distributed realisations, an average pattern with sufficient statistics $\frac{1}{m}\sum_{i=1}^{m} t(\boldsymbol{y}_i)$ is observed.

Similar to [297, 287] the following experiment was performed. First, 1000 independent samples were generated from the Strauss model with parameters $(\log\beta, \log\gamma) = (4.60, -0.69)$ and $r = 0.1$ using the CFTP algorithm 4.5. The resulting sufficient statistics vector was $\bar{t}(\boldsymbol{y}) = (\bar{n}(\boldsymbol{y}), \bar{s_r}(\boldsymbol{y})) = (45.30, 17.99)$. This vector was used as input for the ABC Shadow algorithm 6.6. The other parameters of the algorithm were $\Delta = (0.01, 0.01)$ and $m = (200, 200)$. The parameter space was set to $\Theta = [0, 7] \times [-7, 0]$. The ABC Shadow dynamics was run for 10^6 iterations and samples were taken every 10^2 iterations, for a total of 10^4 samples. The panel (a) of Figure 6.9 shows the obtained realisations of the approximated joint posterior distribution. Panels (b, c) display the realisations distributed according to the margins of the posterior, while panels (d, e) present their associated histograms. The approximate MAP is computed by taking the maximum of the estimated densities corresponding to each histogram. The value obtained is $(4.68, -0.73)$.

The interaction parameters were also estimated using the stochastic gradient algorithm (6.28) with $\epsilon_k = \frac{0.1}{k}$, $k \geq 1$. The previously determined sufficient statistic vector $\bar{t}(\boldsymbol{y})$ was used as input data. The algorithm was run in 10^4 steps. At each step the sampling of the point pattern was done by 100 iterations of the MH Algorithm 4.2. The panel (e) of Figure 6.9 shows the joint dynamics of the interaction parameters given by the stochastic gradient algorithm. Panels (g) and (e) show the evolution of each parameter separately. The estimated value obtained is $(4.68, -0.75)$.

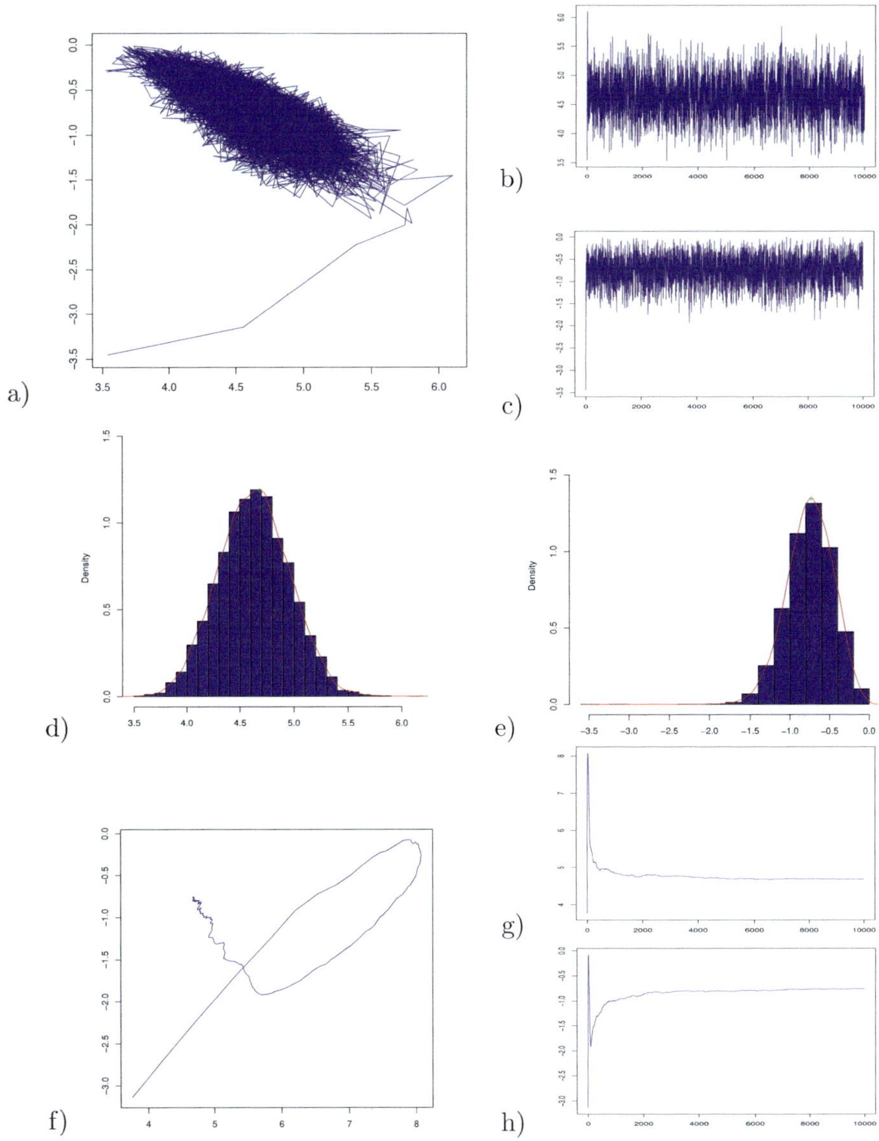

FIGURE 6.9
Parametric inference for the Strauss process. Posterior approximation using the ABC Shadow dynamics: joint distribution (a), marginal distributions (b,c) and histograms (d,e). The approximated MAP is $(4.68, -0.73)$. Estimation of parameters using the stochastic gradient algorithm: joint dynamics of the parameters (f), marginals (g,h). The approximated MLE is $(4.68, -0.75)$.

This example shows that, for the given Strauss model, estimating the interaction parameters using the ABC Shadow and the stochastic gradient algorithms produce equivalent results. For both methods, the final values obtained are close to the true model parameters. With regard to the ABC Shadow algorithm, it also allows further statistical inference based on posterior approximation, such as the implementation of statistical tests. Such examples and applications are presented in the next chapter.

6.4.5 Shadow Simulated Annealing algorithm for parameter estimation

It was explained in Section 6.3.2 that the numerical efficiency of Monte Carlo maximum likelihood estimation is only guaranteed if the initial condition is close to the desired solution. Otherwise numerical instabilities will occur. Close initial conditions can be obtained by the combined application of gradient descent methods with model resampling whenever the distance between the current and the new parameter values becomes too large. The need to resample the model dramatically increases the computational cost of these methods. Furthermore, these techniques can only be applied to convex posterior distributions.

The ABC Shadow algorithm introduced in the previous section is a versatile tool that allows posterior sampling for a large class of models. It is perfectly possible to construct an annealing process, called Shadow Simulated Annealing (SSA), that uses the ABC Shadow dynamics to maximise the posterior distribution of interest. The properties of this SSA process give rise to the construction of a new algorithm for parameter estimation. The algorithm is given below, together with a discussion of the convergence properties of the annealing process. Full technical details and proofs can be found in [287].

Algorithm 6.7. Shadow Simulated Annealing (SSA) algorithm *Data: $\mathbf{x}$ the observed pattern, $p(\mathbf{x}|\theta)$ the assumed model, $p(\theta)$ the model prior, m an integer value, $(\theta_0, \mathbf{y}_0)$ the initial state of the system, $\Delta = \Delta_0$ the control parameter, $T = T_0$ the temperature and $k_\Delta, k_T : \mathbb{R}^+ \to \mathbb{R}^+$ two positive functions.*

1. *Generate $\mathbf{y}$ according to $p(\mathbf{y}|\theta_0)$.*
2. *For $k = 1$ to m do*
   (a) *Generate a new candidate ψ following the uniform density over $b(\theta_{k-1}, \Delta/2)$, the ball centred in θ_{k-1} and of radius $\Delta/2$.*
   (b) *The new state $\theta_k = \psi$ is accepted with probability*

$$\alpha = \min\{1, R_S(\theta_{k-1} \to \psi)^{1/T}\}$$

with $R_S^{1/T}$ the ratio given by

$$R_S(\theta \to \psi)^{1/T} = \exp\left[\frac{\langle t(\boldsymbol{x}) - t(\boldsymbol{y}), \psi - \theta\rangle}{T}\right] \times \left[\frac{p(\psi)}{p(\theta)}\right]^{\frac{1}{T}} \tag{6.49}$$

3. *Return θ_m*
4. *If another sample is needed, go to step 1 with $\theta_0 = \theta_m$, $\Delta_0 = k_\Delta(\Delta)$ and $T_0 = k_T(T)$.*

The SSA algorithm has two additional elements with respect to the ABC shadow dynamics. These two elements are the sequences (T_j) and (Δ_j) with $j \geq 0$. These sequences are obtained using the functions k_T and k_Δ respectively. The function k_T is also called cooling schedule.

The key element in establishing the convergence properties of SA algorithms is the control of the Dobrushin coefficient. The main idea behind these algorithms is to iterate the transition kernel of an inhomogeneous Markov chain with a slowly varying Dobrushin coefficient. This mechanism guarantees the weak convergence of the algorithm outputs towards the uniform distribution over the subspace of configurations maximising the probability density of interest [122, 123, 180, 292]. The logarithmic cooling schedule

$$T_j = \frac{T_0}{\log(j) + 1},$$

with $j \geq 1$ and T_0 a positive constant often appears as a solution, which ensures weak convergence. Section 6.2.1 illustrates the implementation of these ideas for maximising the probability densities of point processes.

The central idea in the construction of the SSA procedure is to add to the classical cooling schedule a mechanism that controls the evolution of the parameter Δ. This parameter controls the distance between the evolution of the ideal and shadow chains, and hence the accuracy of the approximation of the posterior.

By simultaneously controlling the temperature and the distance between the posterior and its approximations, the SSA process converges to the global optimum of the posterior of interest. The authors prove in [287] that for a decreasing sequence of balls around the problem solution, it is possible to associate a sequence of

$$\Delta_j = \frac{K(\boldsymbol{y}, m, T_j)}{j}$$

in order to get as close as desired to the global optimum. Here K is a positive constant depending on the observation $\boldsymbol{y}$, the number of interactions m and the temperature T_j.

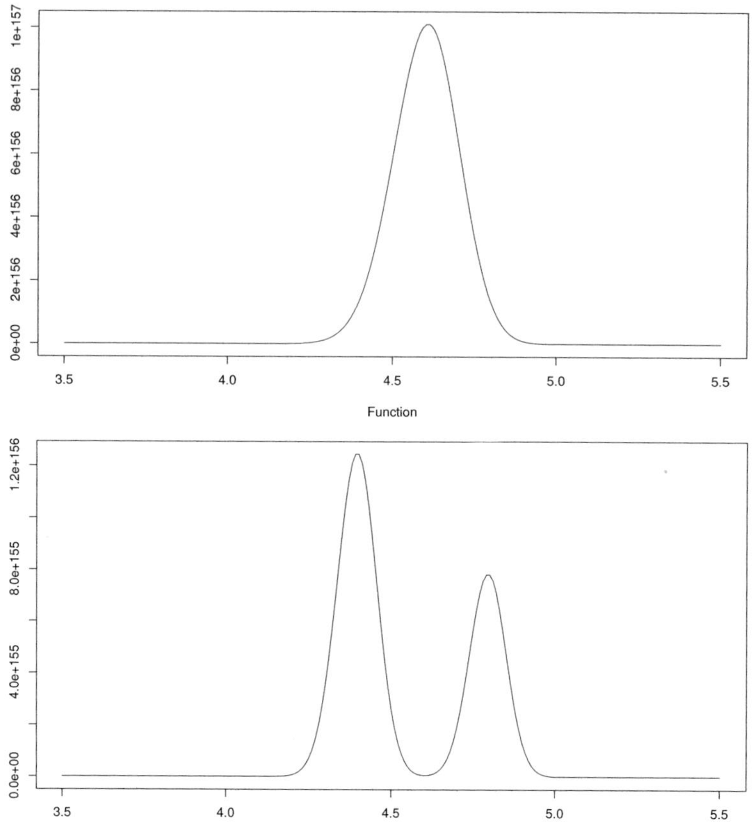

FIGURE 6.10
Log posteriors built with the probability density of a homogeneous Poisson point process and: a) a uniform prior that gives $\widehat{\theta}_{MAP} = 4.6$; b) a bi-modal prior built using a Gaussian mixture, here $\widehat{\theta}_{MAP} = 4.46$.

Example 6.4 (MAP computation of a Poisson point process likelihood with non-concave prior). Consider that $t(\boldsymbol{y}) = n(\boldsymbol{y}) = 100$ points are observed in the square $W = [0, 1]^2$. The observed data are supposed to be the realisation of a Poisson point process with probability density

$$p(\boldsymbol{x}|\theta) = \exp\left[t(\boldsymbol{x})\theta + 1 - \exp(\theta)\right] \tag{6.50}$$

with respect to standard reference measure.

Figure 6.10a shows the log of the posterior distribution built from (6.50) and a uniform prior $p(\theta)$. The log posterior is concave and the MAP estimate is

$\widehat{\theta}_{MAP} = 4.6 \approx \log(100)$ which corresponds to the maximum likelihood estimate.

Consider now, the posterior built using the following prior

$$p(\theta) = \frac{1}{\sqrt{0.045\pi}}\left\{0.6\exp\left[-\frac{(\theta-4.3)^2}{0.0225}\right] + 0.4\exp\left[-\frac{(\theta-4.9)^2}{0.0225}\right]\right\}$$

which is a mixture of two Gaussian distribution with means 4.3 and 4.9, respectively, and the same standard deviation 0.15. The log of the posterior built with the Poisson likelihood and the previous prior is shown in Figure 6.10b. As expected, the concavity is lost. Here, the application of the stochastic gradient algorithm may not provide a correct estimation of parameters. Clearly, the correctly estimated MAP is $\widehat{\theta}_{MAP} = 4.46$.

In both cases, the posterior is entirely known, hence it can be sampled using a classical Metropolis-Hastings. This allows to tune and to test the ABC Shadow and SSA algorithms. Figure 6.11a and b show the histogram of the posterior sampled using a Metropolis-Hastings and the MAP estimation done via a simulated annealing. The optimisation algorithm uses the previous Metropolis-Hastings dynamics. The obtained value for the MAP is 4.456. The results for the sampling and MAP computation using the ABC Shadow and the SSA algorithms are presented in Figure 6.11c and d, respectively. The estimated value of the MAP using the SSA algorithm is 4.457. Visually, both simulation algorithms have equivalent performances. The two maxima are well captured by both algorithms. The final numerical results of the two optimisation algorithms are really very close. The influence of the cooling schedule is clearly observable in both cases. Regarding the SSA algorithm the observed supplementary oscillations are due to the evolution of the Δ parameter.

This example recommends the use of the SSA algorithm in those situations when concavity of the considered likelihood function cannot be guaranteed.

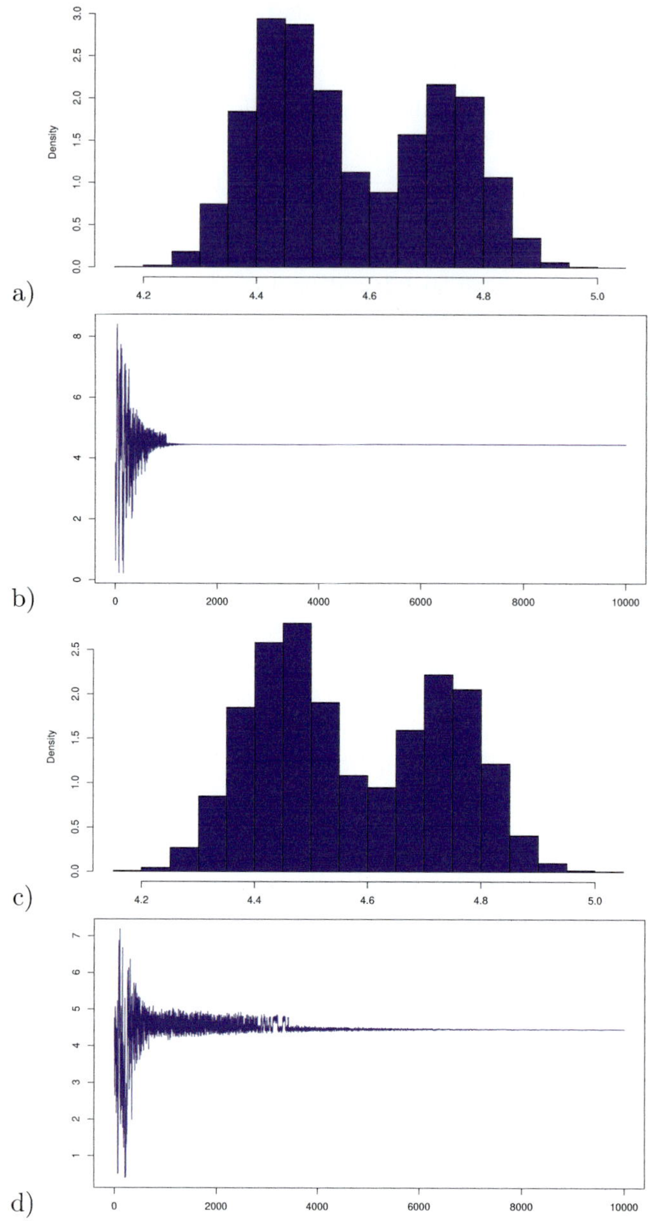

FIGURE 6.11
Sampling and maximisation of the posterior of a Poisson point process built with a non-concave prior: a) MH simulation; b) MAP computation using the simulated annealing algorithm; c) ABC Shadow simulation d) MAP computation using the SSA algorithm.

7

Applications

This chapter shows how the models, algorithms and inference procedures described previously are embedded in a coherent methodology capable of addressing precise questions arising from the analysis of data provided by different application domains. The domains considered here are astronomy, environmental sciences and network sciences.

7.1 Bayesian detection of patterns in spatial data

Pattern detection using a stochastic model requires a priori knowledge of the model parameters. Parameter estimation requires knowledge of the pattern to be detected. One way to overcome this difficulty is to set the model parameters using criteria that are constructed using the desired properties of the pattern to be detected, and then to validate the results obtained.

In the examples of filament network and seismic fault detection in two-dimensional cosmological catalogues and seismic images, respectively, the parameters were set using a two-point strategy. The first point relates to the morphology of the network. For example, filaments were formed by many segments connected at both extremities, enough segments connected at only one extremity, and very few isolated segments. Figure 7.1 illustrates such a situation. Two configurations of segments are shown. Assuming the Bisous model, the interaction energy in configuration a) is written as

$$U_a = 2\log\gamma_0 + 2\log\gamma_1 + \log\gamma_2,$$

while the energy in configuration b) is

$$U_b = 3\log\gamma_1 + 2\log\gamma_2.$$

By controlling the sign of $\triangle U = U_a - U_b$ via chosen values of the model parameters, it is possible to favour or penalise configurations containing more or less connected segments. Several such configurations can be considered in

DOI: 10.1201/9781003380382-7

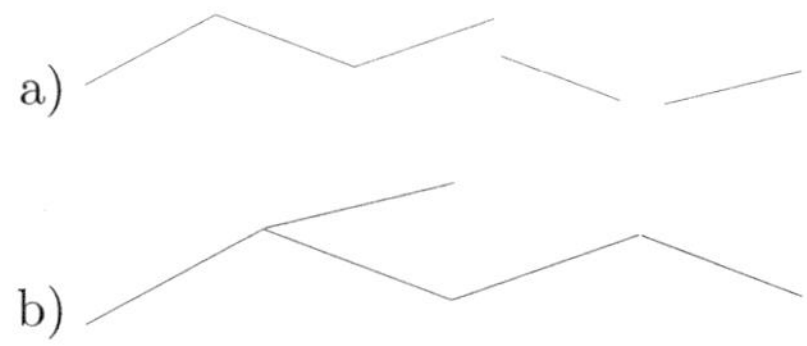

FIGURE 7.1
Two configurations of segments. a) The configuration contains 2 singly connected segments, 1 double connected segment and 2 isolated segments. b) The configuration contains 3 singly connected segment, 2 double connected segments and no isolated segments.

order to condition the model parameters to guarantee configurations with desired morphological properties. The curvature of the network has also been controlled in order to obtain more "smooth" filaments. The second point is to adjust these morphological parameters with respect to the energy parameters of the data. This avoids detecting filaments where they do not exist and compensates for poor data quality by extending the network. Several attempts were made to select parameters that were judged to be visually satisfactory [282, 289, 288, 290, 294, 305].

There are some drawbacks to this approach. One is that it is not always possible to visually check the results of the detection. This is the case with the three-dimensional cosmological catalogues of galaxies and the animal epidemiology data representing the distribution of dairy farms in France.

It is then convenient to use a Bayesian approach[295, 291, 296, 283, 285, 286]. This implies considering the joint law of pattern and parameters

$$p(\boldsymbol{x}, \theta) = p(\boldsymbol{x}|\theta)p(\theta), \tag{7.1}$$

where $p(\theta)$ is the a priori law of the parameters and $p(\boldsymbol{x}|\theta)$ is the law of the pattern given, for example, by the probability density of a marked point process.

The maximisation of (7.1) simultaneously provides the pattern and the parameters. The obtained solution can be written as

$$\begin{aligned}(\widehat{\boldsymbol{x}}, \widehat{\theta}) &= \arg\max_{(\boldsymbol{x},\theta)\in\Omega\times\Theta} p(\boldsymbol{x}|\theta)p(\theta) \\ &= \arg\max_{(\boldsymbol{x},\theta)\in\Omega\times\Theta} \frac{\exp\left[U_{\boldsymbol{d}}(\boldsymbol{x}|\theta) + U_i(\boldsymbol{x}|\theta) + U_p(\theta)\right]}{Z(\theta)Z_p},\end{aligned} \tag{7.2}$$

by setting $p(\theta) = \exp[-U_p(\theta)]/Z_p$.

This approach does not eliminate the problem of estimating the parameters, since the prior distribution has to be chosen. However, it does allow us to assign ranges of variation rather than fixed values to the parameters of the pattern, letting the algorithmic machinery do its work.

Depending on the form of $p(\theta)$, the solution (7.2) will have a more or less pronounced random character. The detected pattern has a "mean" component due to the robustness of the data, the model and the approach in general, and a "variance" component due to the uncertainties induced by the same factors. We want to strengthen the "mean" part and reduce the "variance" part.

Approaches that allow to visualise the "mean" behaviour of the pattern are the estimation of level sets (6.13) and the Vorob'ev expectation (6.14) presented in section (6.2.2). A practical level set estimator is

$$\tilde{Q}_\alpha = \{\tilde{w} \in \tilde{W} : p_n(\tilde{w}) \geq \alpha\}. \tag{7.3}$$

Here $\tilde{W}$ is the set of equal-sized cells obtained by superimposing a regular grid W_r on W. The quantity $p_n(\tilde{w})$ is the Monte Carlo counterpart of the probability $p(\tilde{w}) = \mathbb{P}(\tilde{w} \cap X) \neq \emptyset$ given by the chosen model. It represents the probability that the cell $\tilde{w}$ is touched by the pattern model. For this reason, the sets $\tilde{Q}_\alpha$ in (7.3) are also known in the literature as *visit maps* [295, 291, 296].

7.1.1 Clusters detection in animal epidemiological data

The previous pattern detection method is used here to detect clusters in animal epidemiology data [291]. The data are presented in Section 3.2. In a window W, we observe a population of points given by the locations of farms in France. To each farm it is attached its corresponding disease susceptibility score. A cluster is seen as a region where geographically close farms tend to have a rather high disease score. The cluster pattern is approximated by a configuration of discs. The disc configuration is modelled by the probability density of a marked point process $p(\boldsymbol{x}|\theta)$. The process specifies the position of the discs in the data field and their relative position through a data energy term (3.10) and an interaction energy term (3.11). The marks of the process are represented by the random radii of the discs.

The observation window is the smallest rectangle containing the position of all the farms, i.e. $W = [0.317] \times [0.318]$. The radii of the discs are chosen in the interval $[1, 10]$. In both cases the unit of length is $3km$ in the real world.

The data energy (3.10) is constructed by performing a Student's t-test on the disease scores of the farms covered by a given disc. This test checks whether the average score of these farms is greater than a predetermined threshold $d_0 = 3.75$. A disc must also cover a minimum number of farms $n_0 = 4$.

In general, discs with a large radius are less likely to be accepted than those with a small radius. This is due to the nature of the disease (mastitis). The discs covering a large number of farms are much more likely to be rejected by the Student test. Larger discs are more likely to be accepted if they are situated in an area already covered by small discs. At the same time, the presence of too many small discs increases the computational cost of the method. The simultaneous use of large and small discs reduces smoothing effects. A compromise has to be found. Since no a priori knowledge regarding the disc radius distribution was available, a non-informative uniform distribution was chosen for it.

The vector θ contains only the parameters of the interaction energy (3.11). We have defined $\theta = (\log \gamma_a, \log \gamma_o)$ on the parameter space set to $\Theta = [0, 0.5] \times [-0.1, 0]$. The parameter of the area interaction model $\log \gamma_a$ is chosen to penalise large isolated discs while encouraging the discs to superpose. The parameter of the Strauss model $\log \gamma_o$ controls this superposition. This parameter is not too strong and it allows discs to overlap. However, if the number of overlapping discs is large, each new disc is heavily penalised. Again, there is no a priori knowledge of the relative weight of each of these parameters. Consequently the prior $p(\theta)$ is built by assigning independently to each parameter the uniform distribution over its corresponding domain.

A simulated annealing algorithm was run to maximise $p(\boldsymbol{x}, \theta)$ and thus to detect the cluster pattern in a dataset collecting farms during the year 1996. Figure 7.2a shows the result obtained. The configuration of the discs shows a central cluster in the middle of the French territory and some smaller clusters around it.

The location of the farms is given by the centre of the commune to which the farms belong. One point on the map can therefore represent the position of several farms. As a result, the visual interpretation of the result is not straightforward. This argument and the nature of the proposed solution justify the calculation of contact probabilities to check the "reliability" of the clusters. To do this, the observation window was divided into cells of equal area 9km^2. For each cell the probability $p(\tilde{w})$ was calculated. The result is the visit map of the model, shown in Figure 7.2b.

The detection result and the visit map must be interpreted together. The small clusters are not very important as their contact probability varies between 0.3 and 0.6. An exception is the cluster around point $(150, 150)$, which has a contact probability greater than 0.9. From this point of view, the most important cluster is the one around the point $(175, 175)$.

The visit maps therefore reduce the "noise" of our method and make it easier to evaluate the quality of the detection. However, this technique is not a

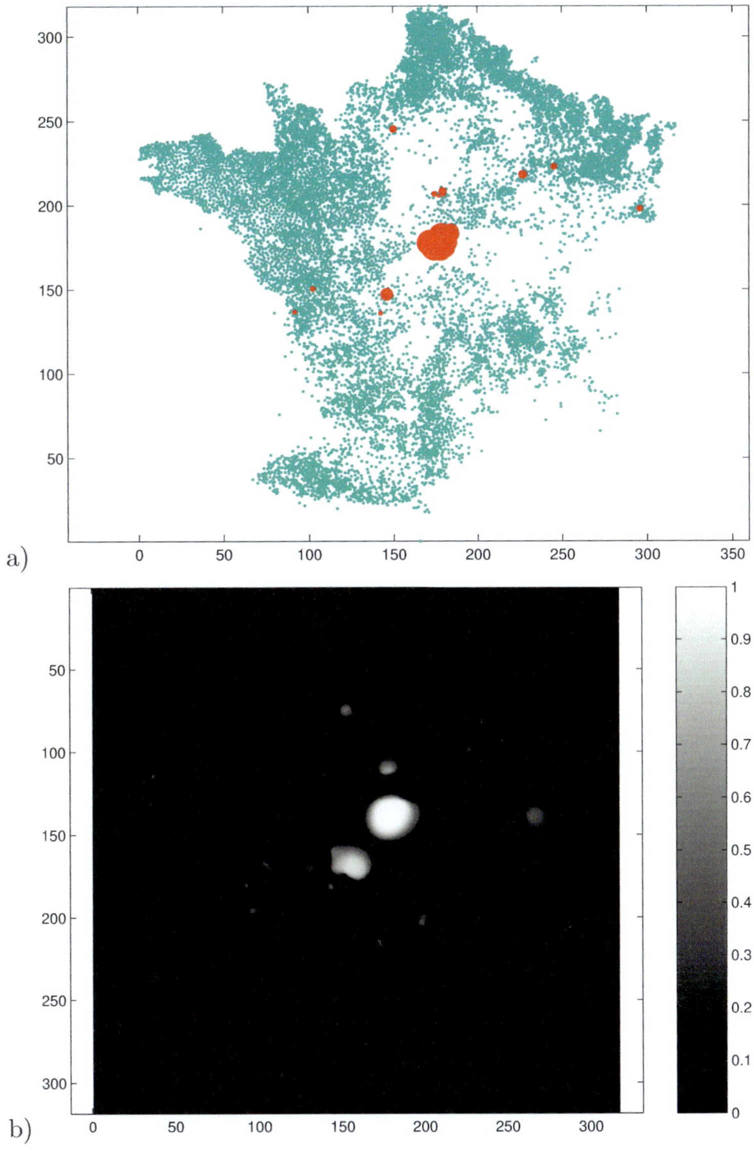

FIGURE 7.2
a) Configuration of discs (red areas) indicating the presence of clusters in the data for the year 1996: the farms are spread over almost the whole France territory (turquoise region); b) The visit map of the regions touched by the model, indicating the “strength” of each cluster.

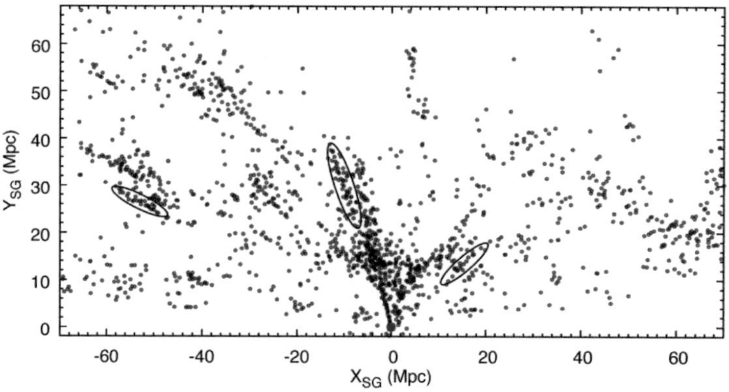

FIGURE 7.3
A galactic point field given by a slice of the 2MRS data set. The positions of the galaxies are given in supergalactic coordinates. The observer is located at the origin (blue point in the figure). (Credit: Tempel et al., A & A, 618, A81, 2018, reproduced with permission ©ESO.)

validation method in the strict sense, as we are still working under the assumption of the model. At this stage, this technique is just a very effective visualisation tool.

7.1.2 Clusters detection in cosmological data

Inspired by the previous application, the authors in [307] build a model capable of detecting clusters formed by the positions of galaxies in cosmological data. Some of their results are presented below. The data set analysed is the cosmological catalogue 2MRS. This data set has some peculiarities. Only the spherical sky coordinates are known exactly. The distance estimates of the galaxies are affected by the peculiar motions of galaxies in the Universe. Therefore, the distance estimates of galaxies in clusters are the most uncertain. As a result, clusters appear to be elongated along the line of sight. A slice of thickness 15 Mpc from this catalogue is shown in Figure 7.3. The elongation of some visually identified clusters is marked with black ellipses.

The hidden cluster pattern in the data is postulated to be the realisation of a marked point process. The marks represent the parameters of a geometric object that has been specifically designed to compensate for the elongation bias exhibited by the data. This object is shown in Figure 7.4. The object consists of two half-spheres joined by a truncated cone, which can be regarded as a cosmic potato. The symmetry axis of the object and the line of sight to the centre of the object coincide. This feature, together with the radius r and the shape parameter $t \geq 1$, fully characterises the object.

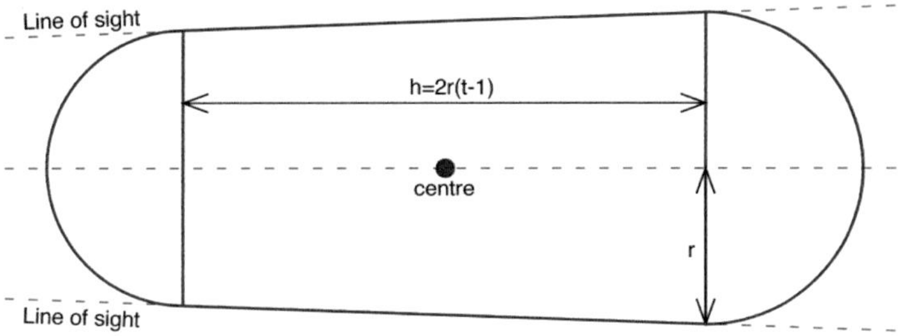

FIGURE 7.4
The shape of the object to be used in the proposed marked point process for the detection of galaxy groups and clusters in cosmological redshift surveys. (Credit: Tempel et al., A & A, 618, A81, 2018, reproduced with permission ©ESO.)

The cluster pattern model proposed by [307] is a marked point process defined by a probability density that includes both a data term and an interaction term, directly inspired by [295]. The data term determines whether a typical object contains a sufficient number of galaxies. The interaction term is a superposition of an area-interaction process and a Strauss-like process. In fact, the area-interaction process is essentially a volume interaction process. Depending on its parameters, it is possible to favour those configurations of objects that tend to reduce the volume of the set union of their corresponding objects, thereby forming clustered patterns. The Strauss-like process penalises those configurations that have an excessive number of pairs of overlapping objects. This term is used to control the strength or intensity of a cluster.

Figure 7.5 shows the results of cluster pattern detection in a region close to the Coma cluster. The thickness of the region under consideration is about 4 Mpc around SGX = 0. The probabilities that a given point within the region is touched by the cluster pattern are represented by a green colour code. The region induced by the points with probabilities greater than a given threshold is the estimated level set or visit map. This map can be considered as the average cluster pattern of interest. Similar to the calculation of the visit map, the probabilities that two given points are touched by the cluster pattern can be estimated. This allows the construction of a measure of the dissimilarity between galaxies and a hierarchical grouping of galaxies. The red points in Figure 7.5 represent galaxies in the Coma cluster. The grey points represent galaxies in groups of four or fewer members. The remaining coloured points represent galaxies belonging to groups with a minimum of five members. It can be seen that the galaxies included in groups with a relatively high number of galaxies belong to the regions that the cluster pattern visit map assigns a high probability to. Moreover, all the points belonging to the Coma cluster are touched by the visit map, which tends to form a connected region.

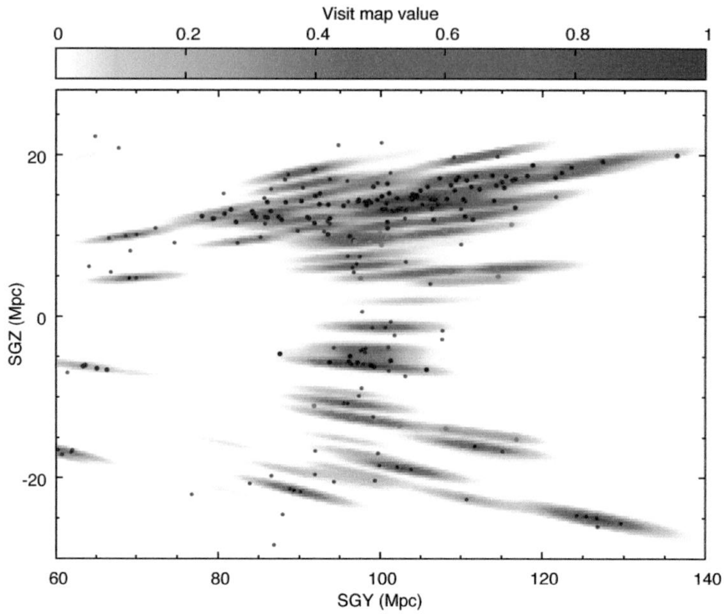

FIGURE 7.5
The distribution of galaxies in supergalactic coordinates and the visit map detected using the proposed cluster point process. (Credit: Tempel et al., A & A, 618, A81, 2018, reproduced with permission ©ESO.)

7.1.3 Bias effects minimisation in cosmological data

The clusters detected in the data set presented in the previous section appeared elongated due to uncertainties that produce the "finger-of-God" effect [201, 307]. One way of dealing with this effect is to measure the peculiar velocities of the galaxies and use these to correct the distribution of the galaxies. However, peculiar velocity measurements are affected by several types of bias, all of which are known together under the generic name of "Malmquist bias". To reduce these biases, the authors in [280] propose a Bayesian approach based on a marked point process with interactions. The main ideas of this work are presented below.

The observed data is a field of n galaxies $\boldsymbol{d} = \{(w_i, \sigma_i, z_i), 1 \leq i \leq n\}$, where w_i is the distance modulus, σ_i is its uncertainty, and z_i is the observational redshift corresponding to the ith galaxy. Based on the distance moduli, luminosity distances can be calculated. Cosmological redshifts can be obtained by solving an equation involving observational redshifts and luminosity distances. Finally, radial peculiar velocities can be estimated from the difference between the cosmological and observational redshifts.

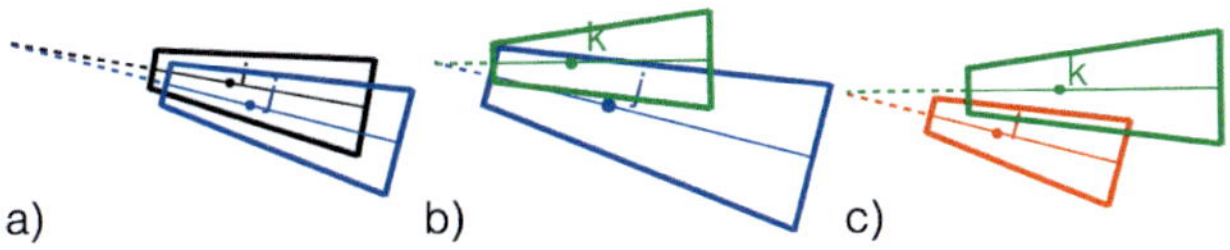

FIGURE 7.6
Schematic view of interactions within the galaxy field. In the panels a) and b) the galaxies interact, while in the panel c) the galaxies do not interact.

The aim is to obtain a corrected configuration of galaxy distance moduli $\boldsymbol{x}$ together with their new associated uncertainties $\boldsymbol{\sigma}$ that fits the observed data $\boldsymbol{d}$ and reduces the bias effects. For this purpose, an original point-field model with interaction is proposed. Its probability density has the expression

$$p(\boldsymbol{x}|\theta) \propto \exp\left[-U(\boldsymbol{x}|\boldsymbol{d}, \theta)\right],$$

with the energy function U given by

$$U(\boldsymbol{x}|\boldsymbol{d}, \theta) = U_{\boldsymbol{d}}(\boldsymbol{x}|\theta) + U_i(\boldsymbol{x}|\theta).$$

The data energy component $U_{\boldsymbol{d}}$ is the sum of the potential contributions for each galaxy. The contribution for each galaxy controls the modulus of its peculiar velocity and the difference between the current distance modulus x_i and its initial value w_i. The interaction component U_i takes into account the interactions between galaxies. To define interactions, a truncated cone S_i is placed around each galaxy i to act as a local influence region. Two galaxies interact if at least one of them is in the other's influence region. Figure 7.6 shows some examples of interactions. The interaction energy controls the pairs of interacting galaxies, while the difference in the peculiar velocity moduli and the uncertainties are used as weights in the construction of the interaction potential for each galaxy [280].

Finally, the distribution $p(\boldsymbol{x}, \theta)$ is constructed by considering a non-informative uniform distribution for the prior $p(\theta)$. Thus, the Bayesian approach given by (7.1) and (7.2) is used to propose a new galactic catalogue with reduced bias effects. The solution (7.2) is computed via a Simulated Annealing (SA) algorithm using a Metropolis-Hastings (MH) dynamics and a logarithmic cooling schedule.

First, the proposed method was tested on a mock catalogue. In this case, the true configuration of the galaxies and the perturbations used to obtain the catalogue are known. Figure 7.7 shows the results obtained in a 5 $h^{-1}Mpc$ thick slice of the XY supergalactic plane of the catalogue. The unit of length is the megaparsec Mpc. One parsec $1pc$ is the distance equal to about 3.261 light years, while $h \approx 0.677$ is a dimensionless constant called the Hubble parameter. The centres of the dots in the figure represents the positions of

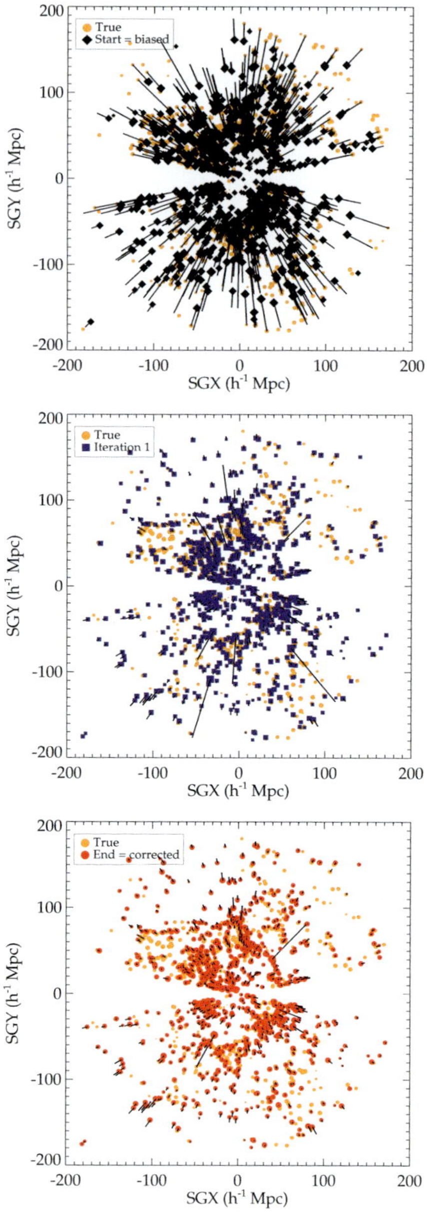

FIGURE 7.7
Results of the proposed method obtained on a mock catalogue (amber). From top to bottom: initial perturbed configuration of galaxies (black), intermediate configuration after running a few steps of the MH algorithm (purple), final configuration after running the SA algorithm (red).

the galaxies. As a rough guide, the average diameter of a galaxy is $30kpc$. The yellow dots represent the true galaxies. Here, the size of a dot is proportional to the module of its corresponding peculiar velocity. In the top panel of Figure 7.7, the black lines indicate the projected distance between the true and biased galaxies. The biased galaxies are represented by black dots. In the middle panel, the evolution of the galaxy configuration after a few MH iterations is plotted in purple. In the lower panel, the final configuration obtained with the SA algorithm is shown in red. Note that the errors in the upper panel are large, making the amber dots more difficult to distinguish. The algorithm tends to reduce errors in the positions of the data points, and thus in their peculiar velocities. Shorter purple and red lines can also be observed, as well as a better matching of the size of their associated dots.

Statistics were calculated for the three galaxy patterns obtained previously. The results are shown in Figure 7.8. It can be observed that, with the chosen parametric setting, the application of the MH and SA algorithms clearly reduces the bias effects.

After verifying and setting up the algorithms on simulated data, the method was applied to a real catalogue. Figure 7.9 shows the results obtained for two projected slices of 40 $h^{-1}Mpc$ thickness. The considered projections of the original data are shown in the top row. The red and purple dots represent galaxies with positive and negative peculiar velocities, respectively. The strong bias in the catalogue is clearly visible, as many galaxies with negative peculiar velocities are shown at large distances. The results of the SA algorithm and the average of 10 realisations of the SA algorithm with different initial conditions are shown in the middle and bottom rows, respectively. The filamentary structure is now visible in the corrected catalogues.

For all details concerning the model, the simulation algorithm and the statistical analysis of the results obtained, the reader is referred to [280]. This paper also compares the Wiener-filter reconstruction of the density and velocity fields from biased and corrected catalogues. The influence of the choice of the value of the Hubble constant is also analysed. All this confirms the good results of the proposed methodology and encourages the authors to further improve and develop this statistical tool.

7.1.4 Galactic filaments detection in cosmological data

The Bayesian pattern detection paradigm has been successfully applied to the detection and study of galactic filaments [294, 295, 283, 296, 298, 309, 310, 310]. Comparisons and syntheses of the performance of several existing methods for the detection of filaments in cosmological data are presented in [179]. As an example of application, we present here some of the very first results we have obtained on real data [295].

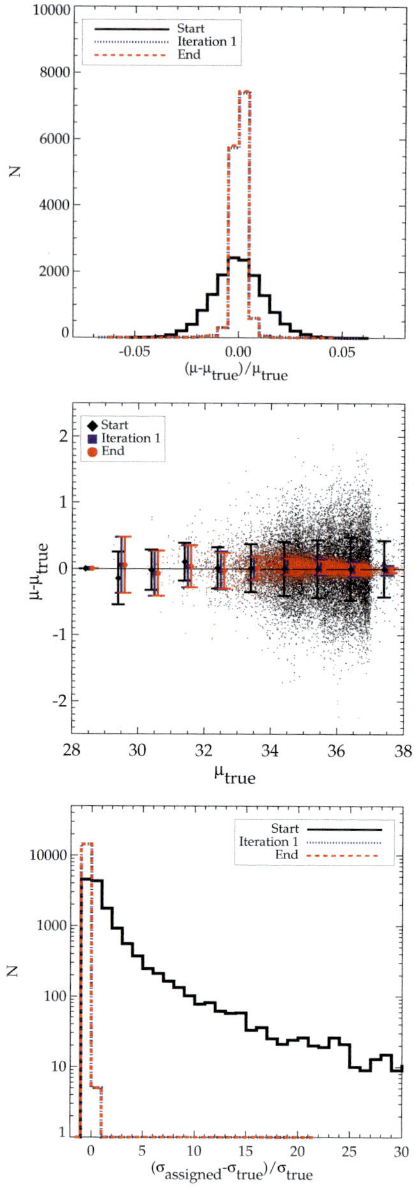

FIGURE 7.8
Statistical indicators of the following three galaxies configurations: initial perturbation (black), after few MH iterations (purple) and final SA (red). From top to bottom: histogram of the relative errors for the distance moduli, difference between true and biased distance moduli and histogram of the relative errors for the uncertainties.

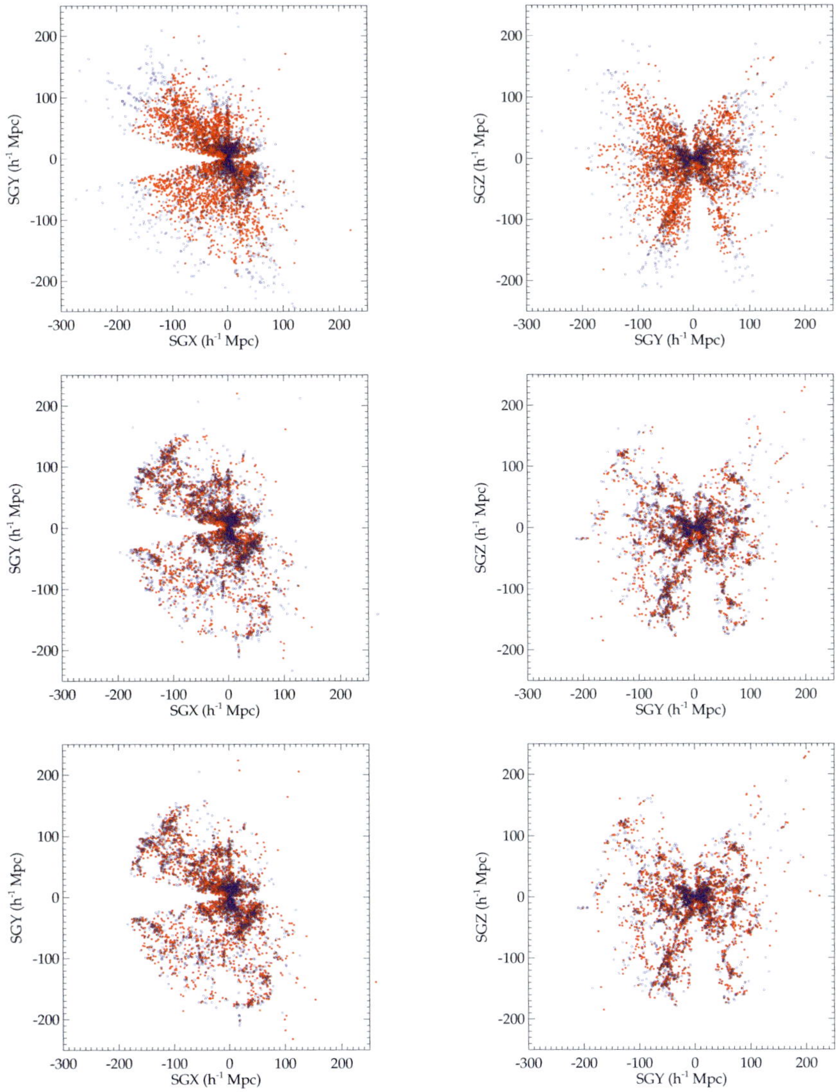

FIGURE 7.9
From top to bottom: the original two data sets, the corrected result obtained with the SA algorithm, the average of 10 realisations of the SA algorithm with different initial conditions.

Our method was tested on samples from the 2dFGRS catalogue. Such a sample is shown in Figure 7.11a. This catalogue contains 7588 galaxies in a rectangular volume W with dimensions $88.6 \times 169.1 \times 20.7\ h^{-1}$Mpc.

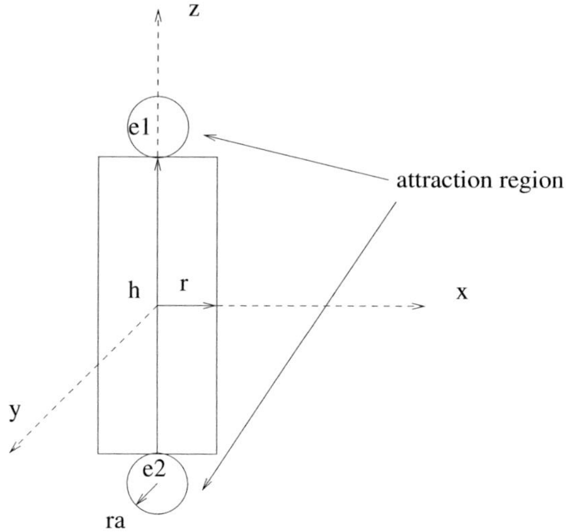

FIGURE 7.10
Cylinder with interaction regions, the generating object of the filament pattern.

As explained in Section 3.1, galactic filaments are approximated by a network of interacting segments or thin cylinders. A typical cylinder is shown in Figure 7.10. These cylinders are positioned in the galaxy field in regions where galaxies tend to be locally aligned. The dimensions of the cylinders are $r = 0.5$ and $h = 6.0$. The cylinders interact by being able to connect and align to form filaments. The radius of attraction that allows the cylinders to connect is $r_a = 0.5$. The curvature parameters that define the alignment and crossing of the cylinders are set to $\tau_{\|} = \tau_{\perp} = 0.15$. This imposes an angle of no more than $\approx 30°$ between two connected cylinders and at least $\approx 80°$ for two crossing cylinders.

All these properties are incorporated into the probability density of a marked point process $p(\boldsymbol{x}|\theta)$. The positioning of the cylinders in the galaxy field is controlled by the data energy term (3.5), while the connectivity between the cylinders is controlled by the interaction energy term (3.8). The parameter of the data term is $u_{\max} = -25$. As for the connection parameters of the interaction term, their respective definition domain is $\log \gamma_0 \in [-12.5, -7.5]$, $\log \gamma_1 \in [-5, 0]$ and $\log \gamma_2 \in [0, 5]$. If the cylinders are very close together and not orthogonal, this configuration is forbidden by setting $\gamma_h = 0$. Choosing these parameters favours the connection of the cylinders and the formation of rather long filaments. However, there is no precise knowledge of the respective weight of each of these parameters. Therefore, a uniform distribution over their respective domain is chosen for the prior $p(\theta)$ which controls the interaction parameters.

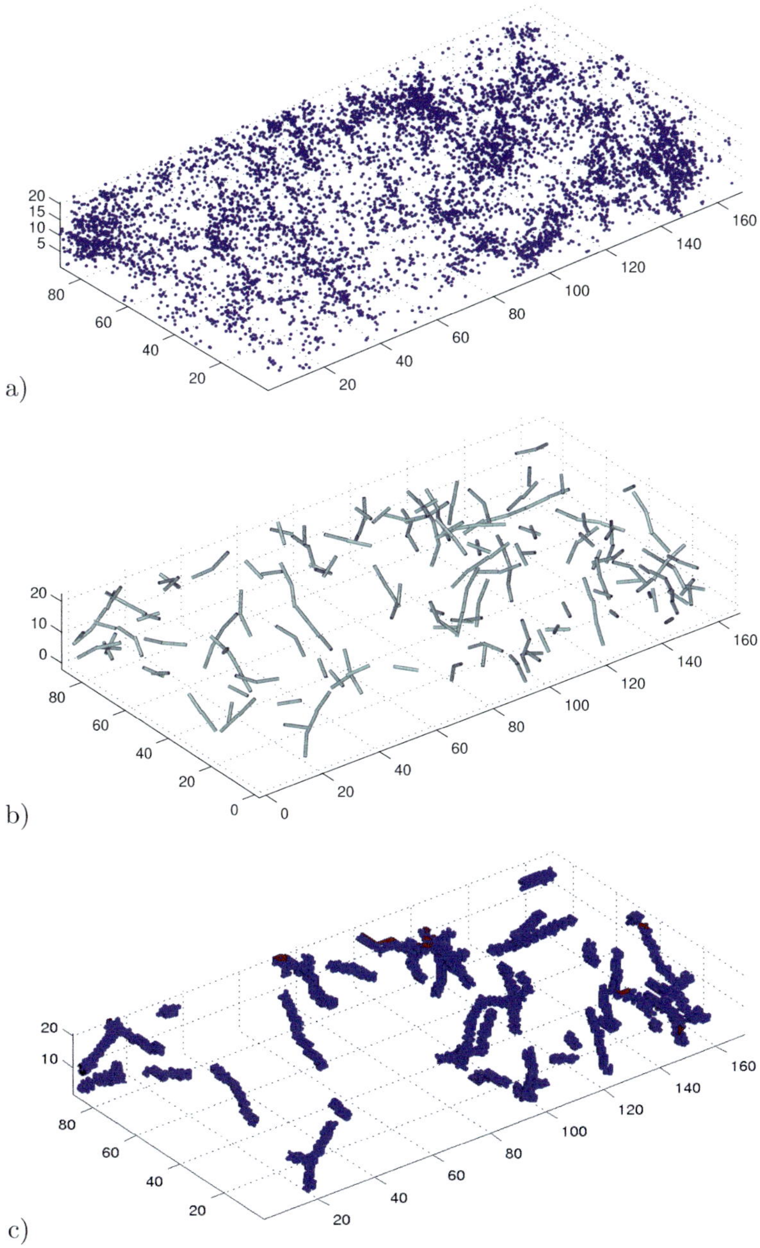

FIGURE 7.11
a) Data: the N250 catalogue; b) cylinder configuration obtained using the simulated annealing algorithm; c) visit map corresponding to the estimated level set $\tilde{Q}_{95\%}$.

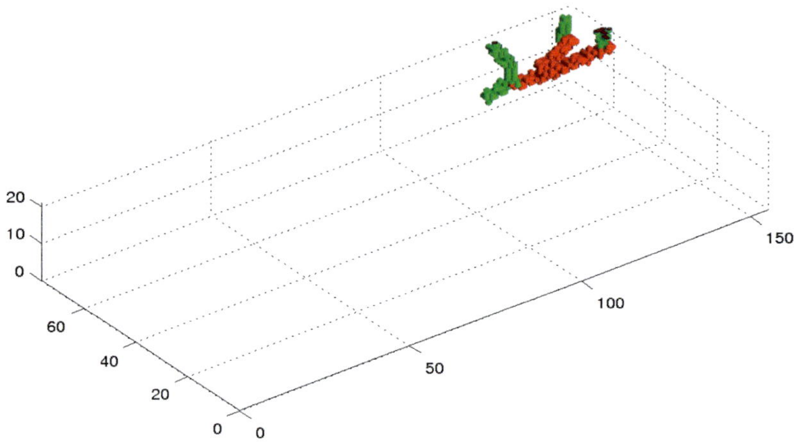

FIGURE 7.12
Simple filament from the NGP250 sample. The contact probability visit map is thresholded at 50% (green) and 95% (red).

The Simulated Annealing was used to maximise the joint distribution $p(\boldsymbol{x}, \theta)$. The algorithm was run for 250000 iterations. Samples were taken and the temperature was lowered every 250 iterations. An initial temperature of $T_0 = 1$ and a logarithmic cooling schedule were used. The final cylinder configuration is shown in Figure 7.11b. The detected cylinders are located in the regions where the galaxies appear to be most aligned.

As in the previous cases, the solution provided by Simulated Annealing is not unique and convergence is achieved only at infinity. Again, a visual diagnosis is not possible. Therefore, the computation of level sets (7.3) is used to assign an "average" shape to the galactic filament pattern. The W domain has been divided into small cubic cells of equal volume $1\,h^{-3}\mathrm{Mpc}^3$. The contact probability visit map is shown in Figure 7.11c. The filament network is clearly visible. In the computed visit map, a filament is a connected component made up of neighbouring cells with a contact probability $\geq 95\%$. Here a cell has a maximum of 26 neighbours.

Many of the filaments detected have a rather simple structure. However, many filaments have a complex morphological structure. This is a new aspect that may force cosmologists to change their initial definition of filaments as simple links or bridges between clusters of galaxies. Figures 7.12 and 7.13 show two such structures.

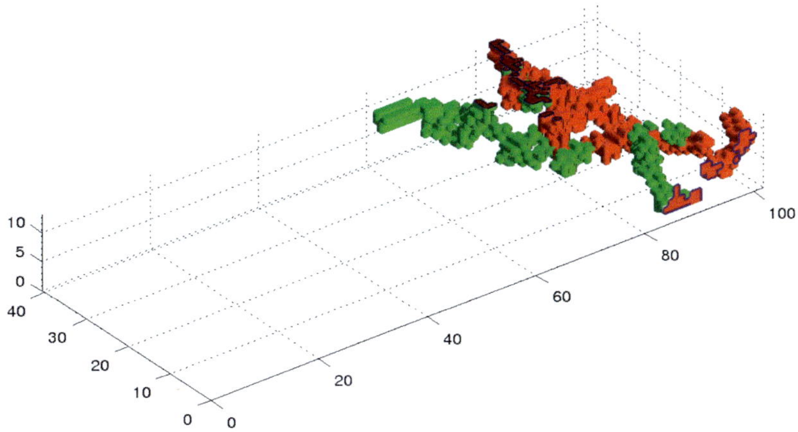

FIGURE 7.13
Complex filament from the NGP150 sample. The contact probability visit map is thresholded at 50% (green) and 95% (red).

7.2 Evaluation and analysis: does the detected pattern really exist?

The Bayesian pattern detection approach does not eliminate the need to estimate the model parameters. The contact probability visit maps are not a statistical validation method. This is still an open problem for our methodology. However, it is possible to check the stability of the method with respect to the model parameters. The mathematical framework of our methodology makes it possible to evaluate the obtained detection results via the sufficient statistics of the model. These statistics fully characterise the detected pattern over the whole observation domain. The detection of clusters in epidemiological data and of galactic filaments are analysed in the following [295, 291, 296].

7.2.1 Verification of the clusters detection in epidemiological data

Let us check whether the detected cluster structure is totally dependent on the data, or it can be considered as a random artefact of the proposed model or methodology. The data and the model used are those from Sections 3.2 and 7.1.1.

Two experiments were performed on each dataset from the years 1996 to 2000. The first consisted of running our previous detection algorithm with $T_0 = 1$

TABLE 7.1
The means of the sufficient statistics for the data sets: $\bar{n}(\boldsymbol{x})$ is the mean number of discs forming the cluster pattern, $\bar{\nu}(Z(\boldsymbol{x}))$ is the mean area of the cluster pattern and $\bar{n_o}(\boldsymbol{x})$ is the mean number of pairs of overlapping discs in the cluster pattern.

Sufficient statistics	Data sets				
	1996	1997	1998	1999	2000
$\bar{n}(\boldsymbol{x})$	74.10	50.15	33.35	20.02	24.75
$\bar{\nu}(Z(\boldsymbol{x}))$	312.46	196.66	167.13	89.53	129.59
$\bar{n_o}(\boldsymbol{x})$	555.08	247.73	131.48	16.25	25.12

and leaving the temperature unchanged throughout the dynamics. This simply amounts to simulating $p(\boldsymbol{x}, \theta)$. Keeping the temperature fixed guarantees the existence and uniqueness of a stationary equilibrium distribution. This condition is necessary to perform the desired statistical inference. The model parameters are chosen to be the same for all the data sets examined. The cluster pattern samples thus obtained were used to calculate the empirical means of the sufficient statistics of the model considered.

The results are shown in Table 7.1. The sufficient statistics of the model are descriptors of the detected cluster pattern. They allow us to quantify the presence of clusters in the data. The total number of discs making up the pattern and their area are related to the size and geometry of the clusters. The number of pairs of overlapping discs is a more subtle feature. It can be interpreted as the "concentration" or "intensity" of a cluster in the data field. The presence of clusters in the data is indicated by the magnitude of the numerical values of these three characteristics. For example, the presence of clusters in 1996 is indicated by an average number of 74.10 discs, an average surface area of the union of discs of 312.46, and a "concentration" of clusters, indicated by an average number of 555.08 pairs of overlapping discs. These are the largest values in Table 7.1. Given this interpretation, the largest cluster patterns are found in 1996, then in 1997 and 1998. The presence of clusters is lowest in the years 1999 and 2000.

The second experiment was a two-step procedure, also applied to each data set. We started by permuting the cell values for each farm, leaving their positions unchanged. This created 100 permuted data fields. This destroyed the cluster structure of the data. Cluster detection was then performed on each generated field. The model and algorithm parameters were exactly the same as in the first experiment. For each of these fields, the average of the sufficient statistics was calculated. Finally, the maximum value of the 100 averages was kept.

TABLE 7.2
The maximum of mean of the sufficient statistics over the 100 permuted fields generated for each of the data sets.

Sufficient statistics	Permuted data sets				
	1996	1997	1998	1999	2000
$\max \bar{n}(\boldsymbol{x})$	2.36	3.85	3.09	4.14	2.06
$\max \bar{\nu}(Z(\boldsymbol{x}))$	13.83	17.63	19.48	25.96	9.12
$\max \bar{n_o}(\boldsymbol{x})$	2.62	7.12	6.10	9.65	1.02

The results are shown in Table 7.2. As expected, the permuted data contains almost no clusters. All values of the sufficient statistics are smaller than those of the original data. This test discriminates between the two situations using a $p-$value less than 1%. In this way, we can ensure that the detected cluster pattern is due to the structure of the data and not to the parameters of the detection method.

7.2.2 Verification of the galactic filaments detection

The use of sufficient statistics and permutation tests naturally extends to the detection of galactic filaments [295, 296]. We present here a comparison between the observational catalogue $2dF$ and 22 mock catalogues. These simulated catalogues have been produced using physical models. This comparison between observations and simulations is of interest when one wishes to update the physical models under consideration, taking into account the observed real data.

For all data sets, the same group of parameters were used for the connected cylinder point process, for the a priori distributions and for the simulation algorithm. Full details can be found in [296]. For each catalogue, the empirical means of the sufficient statistics were recorded. The results are presented in Table 7.3. The first column of the table gives the values of $\bar{n_2}$, the mean number of 2 connected cylinders, the second column gives those of $\bar{n_1}$, the mean number of 1 connected cylinders, and the third column gives the values of $\bar{n_0}$, the mean number of 0 connected cylinders.

As for the clusters, the simulation was carried out at a fixed temperature. This induces a stationary regime that allows comparisons of the observations and simulated data via the distributions of the sufficient statistics. Figure 7.14 shows the boxplots corresponding to these distributions. The boxplots of the mock catalogues are indexed from 1 to 22. The position and spread of the distributions indicated by the boxplots allow simultaneous visual comparisons of the catalogues considered.

The distributions of the n_0 statistic are shown in the graph at the top left of Figure 7.14. This statistic represents the number of isolated (0 connected)

TABLE 7.3
Mean values of the sufficient statistics for the real data and the mocks catalogues.

Data sets	Sufficient statistics		
	$\bar{n_2}$	$\bar{n_0}$	$\bar{n_1}$
2dF	1.94	5.30	11.66
MOCK 1	2.53	5.62	13.16
MOCK 2	0.48	6.20	7.52
MOCK 3	1.29	4.65	6.88
MOCK 4	1.55	9.33	15.45
MOCK 5	1.45	10.63	9.24
MOCK 6	0.38	6.21	8.96
MOCK 7	1.36	9.08	8.12
MOCK 8	0.18	6.91	4.27
MOCK 9	2.07	6.09	9.76
MOCK 10	1.62	4.40	11.91
MOCK 11	1.28	4.65	10.14
MOCK 12	2.65	7.97	11.25
MOCK 13	0.73	6.48	7.08
MOCK 14	0.36	7.30	16.44
MOCK 15	0.98	4.36	8.47
MOCK 16	2.75	11.04	22.88
MOCK 17	0.30	5.96	7.67
MOCK 18	2.15	5.11	10.44
MOCK 19	1.59	8.02	10.99
MOCK 20	1.27	8.79	10.50
MOCK 21	2.77	10.57	11.06
MOCK 22	1.79	8.10	17.26

cylinders. A large number of isolated cylinders indicates a rather fragmented network. It can be seen that only the mock catalogue 3 has a less fragmented network than the *2dF* data. A large number of simulated catalogues have a much more fragmented network than the data: the median of the distribution for these catalogues is clearly higher than that of the real data. However, there are some catalogues with a median and a general shape similar to that of the real data. These catalogues are $1, 10, 11, 15$ and 18. In order to test these similarities, we performed a Kolmogorov-Smirnov test. The p-values obtained for the test of the 1 and 18 catalogues were 0.96 and 0.13 respectively. For the catalogues $10, 11$ and 15, the p-values were all less than 0.002. We conclude that only two of the simulated catalogues are similar to the real data in terms of the distribution of the n_0 statistics. The majority of the simulated catalogues show networks that are much more fragmented than the real data.

The top right panel of Figure 7.14 shows the distributions of the n_2 statistic. This statistic represents the number of cylinders connected at both extremities (2−connected). A network with many such cylinders tends to be composed of

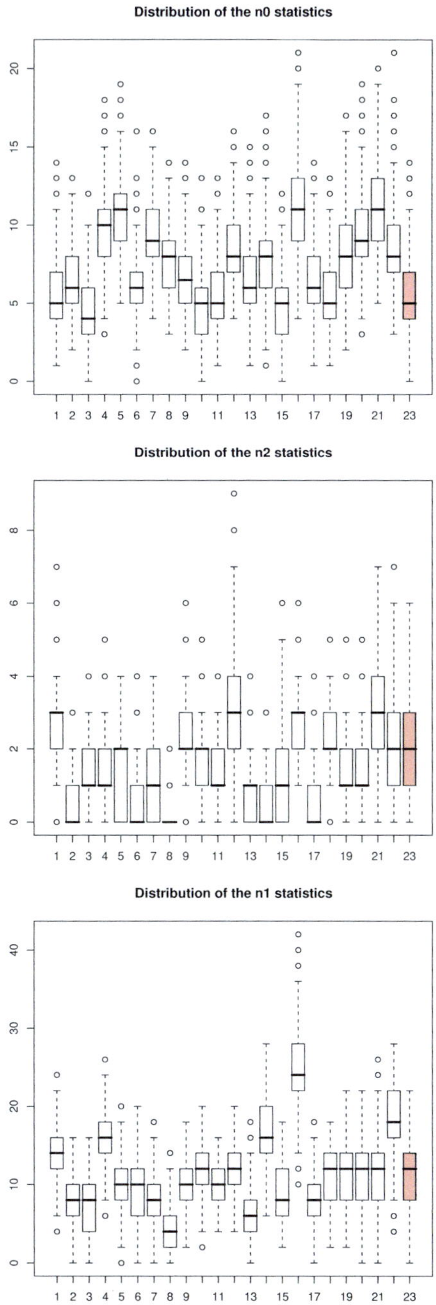

FIGURE 7.14
Comparison of the distributions of the sufficient statistics for the real data (pink) and the mock catalogues (white).

rather long filaments. Within this interpretation, the networks of catalogues 12 and 21 contain filaments that are longer than those present in the real data. The distribution of catalogue 22 is very similar to that of the data. The p-value of the associated Kolmogorov-Smirnov test is 0.09. To make a decision about the catalogues $1, 9, 16$ and 18, a one-tailed Kolmogorov-Smirnov test was performed. The alternative hypothesis was that the n_2 distribution of the real data has larger values more often than that of the mock catalogues. All p-values were very small, implying that these four catalogues have shorter filaments than those in the real data. We conclude that, with a few exceptions, the filaments in the $2dF$ catalogue tend to be longer than those in the simulated catalogues.

The distributions of the n_1 statistic are shown in the bottom panel of Figure 7.14. This statistic represents the number of cylinders connected at a single extremity ($1-$connected). A network with a high value for this statistic tends to have a high density of filaments. By looking at all three statistics together, we can get a general idea of the network topology. For example, if the values of n_1 and n_2 are both high, this indicates the possible presence of networks with filaments similar to a spaghetti plate or a tree with long branches. If the values of n_1 and n_0 are simultaneously high, this also suggests networks with filaments more similar to a macaroni plate or a bush with short branches. In line with this interpretation, catalogues $1, 4, 14, 16$ and 22 have a network that is much denser than that of catalogue $2dF$. The boxplots of catalogues $18, 19, 20$ and 21 are almost identical to those of the real data, indicating the identity of the corresponding distributions. This was confirmed by a Kolmogorov-Smirnov test. The 10 and 12 catalogues have the same median as the data, while their distributions are more concentrated and symmetrical. The Kolmogorov-Smirnov test shows that these distributions are different from the n_1 statistic in the data. We conclude that from the point of view of the n_1 statistic, four simulated catalogues have the same distribution as the data, five others clearly have a much denser network, while the other catalogues have less dense networks or contain fewer $1-$connected extremities.

In conclusion, no single statistic can be used to discriminate between simulated and observed data. However, when the three statistics are taken together, there is a clear difference between the simulations and the observations. From a topological point of view, the networks in the simulated catalogues tend to be more fragmented and contain shorter filaments than those in the data. As for the filament density of the simulated catalogues, it is in a range that includes that of the real data. The limits of this range are due to the morphological variability of the simulated catalogues.

To check that the detected filamentary network really exists, we performed a permutation test. For this, the number of galaxies in each dataset was kept the same, but the positions of the galaxies were generated uniformly in W.

TABLE 7.4
Maximum of the mean of the sufficient statistics over binomial fields generated for some mock catalogues.

Binomial data sets	Sufficient statistics		
	$\max \bar{n_2}$	$\max \bar{n_0}$	$\max \bar{n_1}$
MOCK 1	0	0.02	0
MOCK 2	0	0.015	0
MOCK 3	0	0.01	0
MOCK 5	0	0.015	0
MOCK 6	0	0.03	0
MOCK 7	0	0.02	0
MOCK 8	0	0.015	0

This gives the realisation of a binomial point process. For each catalogue, 100 realisations of a binomial process were generated. For each realisation, a filament detection was performed using the same parameters for the model and simulation dynamics as in the original data. The averages of the sufficient statistics were calculated for each field, then their maximum over the 100 realisations was taken. The results are shown in Table 7.4.

It can be observed that the method does not detect filaments in binomial fields. This method only detects filaments in data sets where something is present. The detection is due to the presence of a structure in the data and not to any random effect in the method. As with cluster detection in animal epidemiology, this verification only certifies that filaments are present in the data. It does not certify the quality of the detection.

7.3 Level sets for the study of planetary perturbations on comets dynamics

Level sets are extremely versatile tools for describing spatial patterns. This section presents a study of planetary perturbations on the dynamics of comets. Order statistics, parameter estimation of regularly varying tails distributions and Öpik theory are used to construct level sets to map the planetary perturbations on comets dynamics in our solar system [293].

The trajectory of comets that originate in the Oort Cloud and pass through our Solar System can be disturbed by the gravitational force exercised by the major or giant planets. These planets are Jupiter, Saturn, Uranus and Neptune. The effect of these perturbations on the dynamics of a comet can be negligible, but it can go so far as to eject the comet into the Universe.

Here a planetary perturbation is given by the variation $\Delta z = z_f - z_i$ of a parameter of a comet's trajectory. This parameter is the inverse of the semi-major axis of the ellipse along which the comet is moving. This variation takes place between the comet's entry into and exit from our solar system, since it is during this interval that the comet can be disturbed by a large planet. These perturbations also depend on two other parameters of the comet's trajectory: the perihelion distance and the cosine of the inclination angle, denoted by q and $\cos i$ respectively.

Several thousand perturbations were simulated using a numerical integrator [92]. These perturbations form a spatial data set of marked points. The points are situated in the domain $(q, \cos i) \in W = [0, 32AU] \times [-1, 1]$ and are marked by their corresponding planetary perturbation δz. The complexity of the integration calculation is a mathematically recognised fact. This calculation is a special case of the $n-$ body problem, where the bodies considered are the Sun, the comet and the four major planets. Assuming that these perturbations are the realisation of a stochastic process, we want to check the existence of patterns formed by their corresponding statistical descriptors. These patterns may provide useful information regarding the possibility of replacing the costly numerical integration by the simulation of a probabilistic model associated with these perturbations. The section continues with the description of the statistical analysis tools and their obtained results presented in [293].

7.3.1 Order statistics

The semi-major axes of the trajectories of the giant planets are as follows: Jupiter $a_J = 5.2AU$, Saturn $a_S = 9.6AU$, Uranus $a_U = 19.2AU$ and Neptune $a_N = 30.1AU$. The perturbations are uniformly located in W. Their associated marks Δz must show some heterogeneity: the perturbations are stronger in the regions near the orbits of the giant planets than in other regions of space.

To demonstrate this lack of stationarity by statistical analysis, a partition of $W = \cup_{i=1}^{n} W_i$ is considered. The W_i cells must be small enough to allow the marks within each cell to be considered i.i.d., but large enough to contain a sufficient number of perturbations to enable statistical inference to be performed. We chose cells with a volume of $0.1AU \times 0.1$, containing approximately 1500 perturbations.

We are interested in three questions. The first two questions concern the tails and the symmetry of the distributions of the perturbation marks. The third question is: do the perturbation marks have a particular spatial structure in the W domain under consideration?

To answer these questions, the order statistics or empirical q-quantiles were calculated for each cell. Most of these values indicate that the perturbations are distributed around the origin, while no particular pattern appears in W.

However, the situation is quite different when extreme quantiles such as $0.01, 0.05, 0.95, 0.99$ are considered: around the semi-major axes of the major planets these quantiles take on important values. To check whether these values are due to the presence of heavy-tailed distributions, the indicator $z_q - n_q$ was used. The first term is the empirical quantile of order q; the second term is the quantile of order q of the normal distribution, for which the parameters were estimated to be $z_{0.50}$ and $0.5(z_{0.84} - z_{0.16})$ respectively. For quantiles close to 1, positive values of this indicator suggest a heavy-tailed distribution type behaviour of the data. Of course, this indicator can also be used for quantiles close to 0.

Figure 7.15 shows the values of the $z_{0.99} - \widehat{n}_{0.99}$ indicator. It can be seen that the level sets have an arrow shape around the orbits of the large planets. The arrow points from right to left, and this shape is lost as the cosine of the inclination angle approaches -1. The larger the planet, the larger the levels of the sets. This is consistent with the very reasonable idea that the larger the planet, the more it can perturb a comet's orbit.

In order to verify that the appearance of this pattern is actually present in the data and is not an artefact of the methodology[1], two experiments were carried out. The first was to apply the Central Limit Theorem to the order statistics to perform a test that rejects the normality hypothesis [70, 293]. We found that normality is essentially rejected in the regions around the orbits of the large planets. The second experiment consisted of calculating the mean values of the indicators $z_q - n_q$ using a bootstrap procedure. This allowed us to confirm the results obtained for the extreme quantiles [293].

It is perfectly possible to use indicators based on the quantile difference to check the symmetry of the distribution. Looking at $z_q - |z_{1-q}|$, we would expect to get values close to 0 if the distribution is symmetric. Our analysis suggests that the perturbation distribution may or may not be symmetric, regardless of the position of the large planets. The conclusion of this analysis is that symmetry, and in particular the tails of the distribution, are elements that need to be taken into account in a probabilistic model for the perturbations.

7.3.2 Stable distributions models

Stable distributions are a very rich class of probability laws that include features such as heavy tails and symmetry. A random variable Z is *stable* or has *stable distribution* if for all $a, b > 0$ there exists $c > 0$ and $d \in \mathbb{R}$ such that

$$aZ_1 + bZ_2 \stackrel{\mathcal{L}}{=} cZ + d,$$

where Z_1 and Z_2 are independent copies of Z and "$\stackrel{\mathcal{L}}{=}$" is the equality in distribution.

[1] The only parameter in this methodology is the size of the W_i cell.

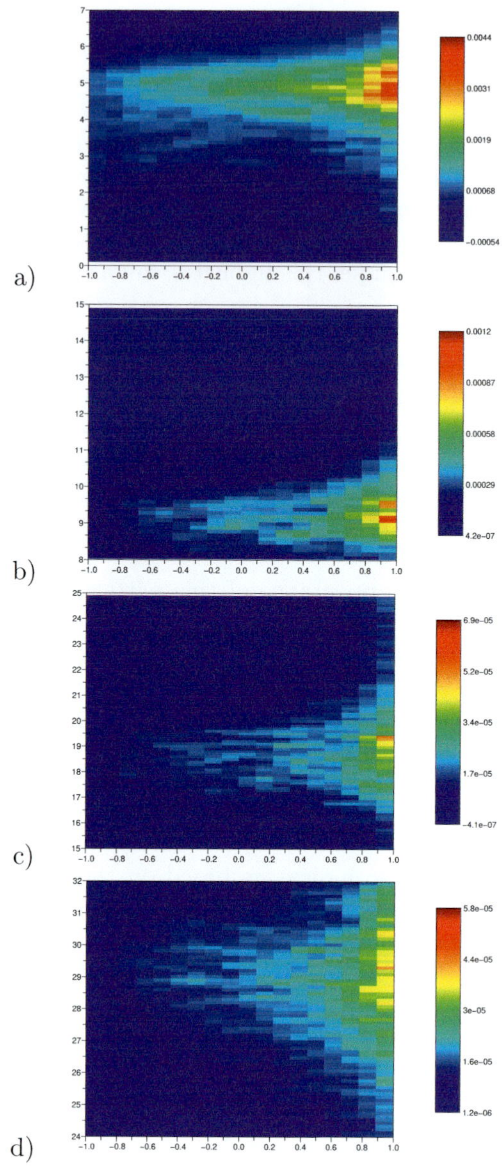

FIGURE 7.15
Values of the indicator based on the difference between the empirical quantiles and the quantiles of the normal distribution $z_{0.99} - \widehat{n}_{0.99}$ for planetary perturbations. The ranges of values for the indicator around the orbits of each major planet are: a) Jupiter ($[-0.0005, 0.0044]$), b) Saturn ($[4.2 \times 10^{-7}, 0.0012]$), c) Uranus ($[-4.1 \times 10^{-7}, 6.9 \times 10^{-5}]$) and d) Neptune ($[1.2 \times 10^{-6}, 5.8 \times 10^{-5}]$). For each curve, the y axis corresponds to the initial perihelion distance and the x axis to the cosine of the angle of inclination. We recall that the respective semi-major axis of the four giant planets are $a_J = 5.2AU$, $a_S = 9.6AU$, $a_U = 19.2AU$, $a_N = 30.1AU$.

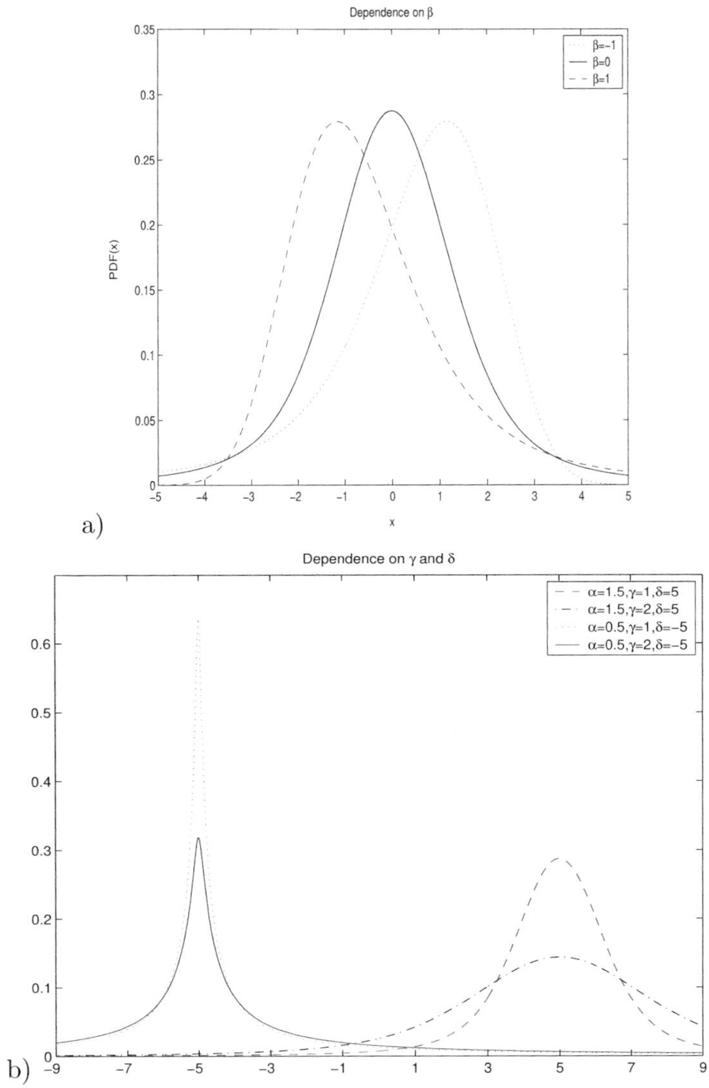

FIGURE 7.16
Parameters influence on the shape of a stable distribution: a) the parameter β, b) the parameters α, γ and δ.

A stable distribution is characterised by four parameters $\alpha \in [0, 2]$, $\beta \in [-1, 1]$, $\gamma \geq 0$ and $\delta \in \mathbb{R}$. It is denoted by $\mathcal{S}_\alpha(\beta, \gamma, \delta)$. The role of each parameter is as follows: α determines the speed with which the law converges to 0, β controls the symmetry of the law, while γ and δ are the scale and position parameters respectively. Figure 7.16 shows the influence of each parameter on the shape of the probability distribution.

The linear transform of a stable random variable is also a stable random variable. If $\alpha \in (0,2)$, then $\mathbb{E}|Z|^p < \infty$ for all $0 < p < \alpha$ and $\mathbb{E}|Z|^p = \infty$ for all $p \geq \alpha$. The distribution is Gaussian when $\alpha = 2$. The stable variable with $\alpha < 2$ has infinite variance, and its distribution tails are asymptotically equivalent to a Pareto distribution [279]. More precisely

$$\begin{cases} \lim_{z\to\infty} z^\alpha \mathbb{P}\{Z > z\} & = & \frac{(1+\beta)}{2}\sigma, \\ \lim_{z\to\infty} z^\alpha \mathbb{P}\{Z < -z\} & = & \frac{(1-\beta)}{2}\sigma. \end{cases}, \tag{7.4}$$

where $\sigma = C_\alpha \gamma^\alpha$, $C_\alpha = \frac{2}{\pi}$ if $\alpha = 1$ and $C_\alpha = \frac{1-\alpha}{2\Gamma(2-\alpha)\cos(\pi\alpha/2)}$ otherwise. The law is symmetric when $\beta = 0$. If $\alpha < 1$, the support of the law is $\mathcal{S}_\alpha(\beta,\gamma,0)$, the positive half-line if $\beta = 1$ and the negative half-line if $\beta = -1$. If $\alpha > 1$, the first moment exists and is equal to the translation parameter δ.

One of the technical difficulties in studying stable laws is that, except in a few notable cases (Gaussian, Cauchy, Lévy), the probability densities have no explicit analytical expression. This means that the estimation of the parameters of stable laws is still an open problem. Several methods exist in the literature [197, 293]. However, most of these methods only work when $\alpha \in (0,2)$. For our problem, we don't know whether the data is the outcome of a stable distribution or not. In this case, regularly varying tails can be considered.

A random variable has a distribution with *regularly varying tails distribution* with parameter $\alpha \geq 0$ if there exist numbers $p, q \geq 0, p+q = 1$ and a slowly varying function[2] L such that

$$\begin{cases} \lim_{z\to\infty} z^\alpha L(z) \mathbb{P}\{Z > z\} & = & p, \\ \lim_{z\to\infty} z^\alpha L(z) \mathbb{P}\{Z < -z\} & = & q. \end{cases} \tag{7.5}$$

It is important to notice that the conditions (7.4) can be derived from (7.5) by taking $L(z) = 1/\sigma$ and $p = (1+\beta)/2$.

Under the assumption that the sample has the asymptotic property (7.4), we used the parameter estimation algorithm proposed by [72, 73, 197]. This algorithm estimates three parameters $\widehat{\alpha}$, $\widehat{\beta}$ and $\widehat{\sigma}$. The parameter $\widehat{\delta}$ is approximated by the empirical mean of the sample when $\alpha > 1$. If this algorithm is used on data with a stable distribution, the parameter α should have values in the interval $(0,2)$. The strong point of the algorithm is that it can also be used when the data is not stable. In this case, the data is assumed to follow a law with a regularly varying tail. The weakness of the algorithm is that in this case we have no information about the central part of the distribution.

[2]A slowly varying function $L(z)$ is such that $lim_{z\to\infty} \frac{L(\lambda z)}{L(z)} = 1$ for all $\lambda > 0$.

Nevertheless, this algorithm already provides a fairly complete characterisation of a wide range of probability distributions.

Figure 7.17 shows the application of the algorithm to the estimation of the α parameter on the data of the planetary perturbations. Level sets below 2 clearly form a regional pattern around the orbit of each large planet. In fact, this pattern is less pronounced than in the case of the order statistics. This is undoubtedly due to the number of perturbations per cell, which is quite limited compared to the requirements of the algorithm. However, the two results are consistent and indicate a heavy-tailed law type behaviour for perturbations around the orbits of the giant planets.

Several tests have been carried out to support this conclusion [293]. First, using the asymptotic equivalence between Pareto type laws and the central limit theorem for order statistics, we were able to validate the estimate of the α coefficient. Secondly, for regions where $\alpha \leq 2$, a χ^2 test was performed to validate the results of the estimation of all the parameters. Finally, in order to remedy the shortcomings of the algorithm, a different law was proposed for regions where $\alpha > 2$, and a χ^2 test was constructed to validate the results.

The expression for this other law is

$$f(z) = \frac{C_{\kappa,\alpha}}{1 + \mid \kappa z - \omega \mid^{\alpha+1}}, \tag{7.6}$$

where $C_{\kappa,\alpha}$ is a normalisation constant, κ is the scale parameter, ω is the position parameter and α is the tail exponent. The estimation of the parameters of (7.6) is done in several steps. First, α is estimated using the previous algorithm. Then ω is estimated as the empirical mean of the data. Finally, the normalisation constant $C_{\kappa,\alpha}$ and the scale parameter are estimated by the method of moments.

The p-values of the χ^2 tests for the perturbations around Jupiter's orbit are shown in Figure 7.18. It can be seen that the tests confirm the parameter estimate in most cells. This supports the idea that around orbits of large planets, the perturbations behave like a heavy-tailed law. At the same time, to improve the results of these tests, we are considering another alternative symmetric distribution, as well as an increase in data to better estimate the model parameters.

7.3.3 A verification of the results through Öpik's theory

Our astronomer colleagues were able to interpret and validate our results in the light of the Öpik's theory. This theory studies close encounters between a comet and a planet orbiting the Sun. It is a three-body problem, and in this

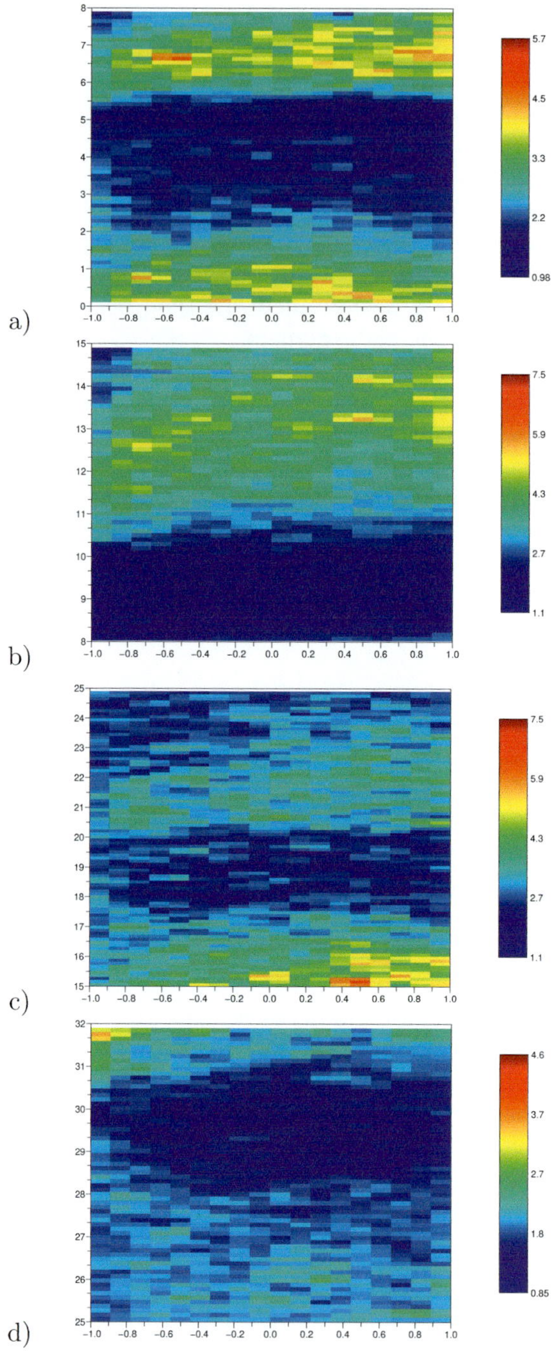

FIGURE 7.17
Result estimation of the parameter α for the perturbations marks around the orbits of the giant planets. The corresponding intervals are: a) Jupiter $[0.98, 5.7]$, b) Saturn $[1.1, 7.5]$, c) Uranus ($[1.1, 7.5]$) and d) Neptune ($[0.85, 4.6]$.

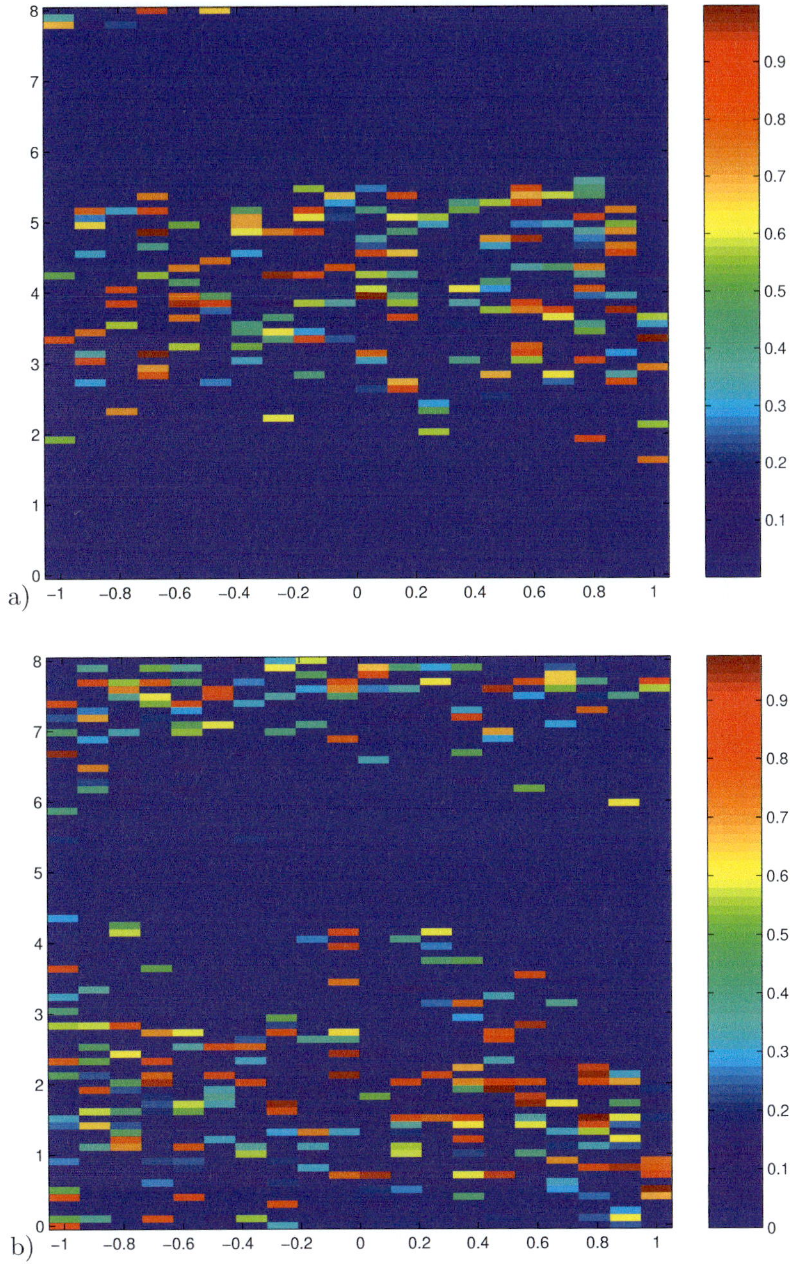

FIGURE 7.18
The p–values of the implemented χ^2-tests for the perturbations around the Jupiter orbit: a) region $\alpha \leq 2$, b) region $\alpha > 2$.

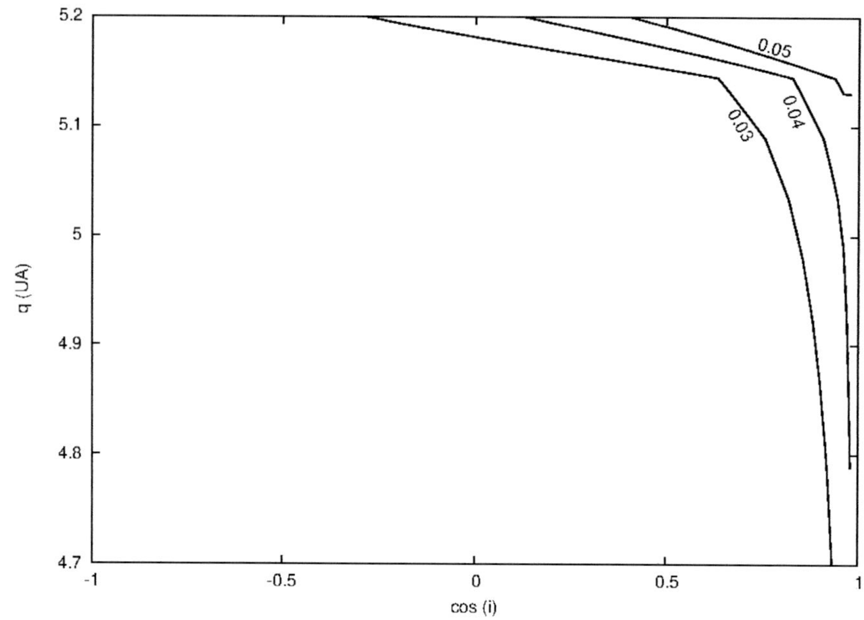

FIGURE 7.19
Level curves for the perturbations around Jupiter, calculated using Öpik's theory.

case mathematical formulae are available to calculate the planetary perturbations. Figure 7.19 shows contour plots of the perturbations around Jupiter, calculated in [293]. It can be seen that these level curves have similar characteristics to those of the patterns highlighted by the statistical analysis.

7.4 Morpho-statistical characterisation of patterns

This section is dedicated to parameter estimation. A pattern is observed and its features are registered. These features are morphological and statistical descriptors of the observed pattern. Furthermore, the pattern is assumed to be the realisation of a stochastic model. Parameters of the model are estimated to fit the observed features. The estimated model parameters allow the simulation of new patterns with characteristics that match the registered features. Therefore, the estimated parameters provide a morpho-statistical characterisation of the observed pattern.

7.4.1 Segment pattern characterisation via maximum likelihood estimation of the Candy model

Suppose a pattern of segments is being observed. This means that statistics about the number of segments, such as their connectivity and orientation, are recorded. Suppose further that the observed pattern is the realisation of the Candy model, which describes the spatial distribution of randomly connected and aligned segments [282, 192]. One wants not only to capture these features, but also to be able to create patterns that have similar statistical properties to the observed ones. Maximum likelihood estimation makes this possible. On average, the realisations of the models simulated with parameters estimated by the maximum likelihood method have the same characteristics as the observations. We present here the results obtained for the estimation of the interaction parameters of the Candy model [83, 192].

The observation domain is a compact region $W \subset \mathbb{R}^2$ of strictly positive volume $0 < \nu(W) < \infty$. A segment is an object with centre w in W and characteristics $m = (l, \theta)$, length and orientation, in the probability space $M = [l_{\min}, l_{\max}] \times [0, \pi]$. The distribution of the marks ν_M is the uniform distribution on M. Let $\boldsymbol{x} = \{(w_1, m_1), \ldots, (w_n, m_n)\}$ be a finite configuration of segments. The Candy model is defined by the probability density with respect to the Poisson measure of unit intensity (2.17)

$$\begin{aligned} p(\boldsymbol{x}) &= p(\emptyset) \prod_{i=1}^{n} \exp\left[? \frac{l_i - l_{\max}}{l_{\max}} \right] \\ &\times \gamma_0^{n_0(\boldsymbol{x})} \gamma_1^{n_1(\boldsymbol{x})} \gamma_2^{n_2(\boldsymbol{x})} \gamma_r^{n_r(\boldsymbol{x})} \gamma_o^{n_o(\boldsymbol{x})} \end{aligned} \quad (7.7)$$

where $p(\emptyset)$ is the normalisation constant and $\gamma_2, \gamma_1, \gamma_0 > 0$, $\gamma_r, \gamma_o \in (0, 1)$ are the interaction parameters of the model. The sufficient statistics of the model $n_2(\boldsymbol{x}), n_1(\boldsymbol{x}), n_0(\boldsymbol{x}), n_r(\boldsymbol{x}), n_o(\boldsymbol{x})$ are respectively the number of segments connected at both extremities, the number of segments connected at a single extremity, the number of free (unconnected) segments, the number of pairs of segments exhibiting the repulsion interaction and the number of pairs of segments not aligned. All the details about the definition of the interactions and the properties of the Candy model can be found in [192].

For this estimation problem it is assumed that the parameters l_{min}, l_{max}, r_a and τ are known. This assumption is open to debate, as the estimation of the range parameters of interactions is still an open problem. In this context, however, it is not unreasonable to imagine that these parameters are estimated using exploratory statistics or image analysis techniques [54, 142, 229].

The Candy and Bisous point processes belong to the exponential family models (6.18). The simulation dynamics of these two models are based on irreducible, Harris recurrent and geometric ergodic Markov chains. Consequently, the MCMC maximum likelihood parameter estimation framework can be applied.

Let us write the probability density (7.7) as

$$p(\boldsymbol{x}|\theta) = \frac{\exp\langle t(\boldsymbol{x}), \theta\rangle h(\boldsymbol{x})}{Z(\theta)} \tag{7.8}$$

where $t(\boldsymbol{x})$ is the vector of sufficient statistics

$$t(\boldsymbol{x}) = (n_2(\boldsymbol{x}), n_1(\boldsymbol{x}), n_0(\boldsymbol{x}), n_r(\boldsymbol{x}), n_o(\boldsymbol{x})),$$

θ is the vector of parameters

$$\theta = (\log \gamma_2, \log \gamma_1, \log \gamma_0, \log \gamma_r, \log \gamma_o)^T$$

and $h(\boldsymbol{x}) = \prod_{i=1}^{n} \exp\left[\frac{l_i - l_{\max}}{l_{\max}}\right]$.

The likelihood maximisation procedure requires the observed values of $t(\boldsymbol{x})$ and an initial value of θ_0. Since the likelihood is convex, the maximum likelihood computation will converge through a local optimisation procedure independently of the initial conditions. Due to the nature of the likelihood, the optimisation procedure is approximated by its Monte Carlo counterpart. The model is simulated using a Markov chain, which has good convergence properties. As a result, the Monte Carlo approximations are very close to the theoretical values and a central limit theorem can be used to evaluate the estimation errors. However, numerical evaluations of the log-likelihood and the gradient by importance sampling are not numerically stable. In fact, these calculations require the evaluation of expressions of the type

$$\exp\langle t(X), \theta_k - \theta_{k+1}\rangle. \tag{7.9}$$

If the distance between the new and old parameter values is too great, the numerical values of (7.9) may explode.

The solution to this problem is to resample the model [108]. By iterating this procedure we can get close enough to the maximum likelihood estimator and then compute it using a direct method or a Newton-Raphson type method. The iterative method we present is an ascending gradient method with optimal step [82, 83, 282, 192].

Algorithm 7.1. Optimal step ascending gradient iterative method for the MLE approximation. *Data: the observed sufficient statistics $t(\boldsymbol{y})$ of the pattern supposed to be a realisation of (7.8), the number of samples m, the precision parameter α, the threshold valuesτ_1 and τ_2. Output: the estimated MLE θ_{k+1} and the counter $k+1$.*

1. *Set θ_1 and $k = 1$.*
2. *Get m samples from $p(\boldsymbol{x}|\theta_k)$ and compute $\nabla l_m(\theta_k)$.*

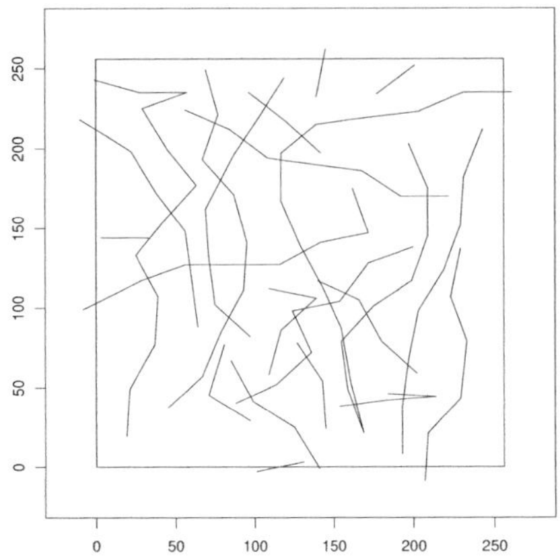

Interaction parameters	Sufficient statistics
$\log \gamma_2 = 2.5$	$n_2 = 63$
$\log \gamma_1 = -3.0$	$n_1 = 34$
$\log \gamma_0 = -8.5$	$n_0 = 4$
$\log \gamma_r = -2.5$	$n_r = 9$
$\log \gamma_o = -2.5$	$n_o = 12$

FIGURE 7.20
Observed realisation of the Candy model, the interaction parameters and the sufficient statistics.

3. *For each parameter component* $i = \{1, \ldots, 5\}$ *and gradient component* $\triangle_i$, *calculate the intervals* $I_k^i = [\theta_k^i - \alpha\triangle_i, \theta_k^i + \alpha\triangle_i]$.
4. *Maximise the log-likelihood in each interval* I_k^i *using the golden section search method to obtain a new value* θ_{k+1}.
5. *If* $\| \theta_{k+1} - \theta_k \| > \tau_1$, *then set* $k = k + 1$ *and go to Step* 2.
6. *If* $\| \nabla l_m(\theta_{k+1}) - \nabla l_m(\theta_k) \| > \tau_2$, *then set* $k = k + 1$ *and go to Step* 3.
7. *Return* θ_{k+1} *and* $k + 1$.

Figure 7.20 shows a realisation of the Candy model, its interaction parameters and the observed values of the sufficient statistics. The Algorithm 7.1 has been implemented by taking as data the observed sufficient statistics of Figure 7.20 and as initial parameters θ_1 from the first column of Table 7.5. The algorithm was run for 1000 iterations using $\alpha = 10^{-3}$, $\tau_1 = 3.0$ and $\tau_2 = 10^{-6}$. The result is the vector parameter vector θ_0 which appears in the middle column of Table 7.5. The log-likelihood $l_n(\theta)$ was calculated from $n = 2 \times 10^4$ simulations of the probability density $p(\boldsymbol{x}|\theta_0)$ of the Candy model. The log-likelihood profiles are shown in Figure 7.21. The maximum of $l_n(\theta)$ is shown in the third column of Table 7.5.

The asymptotic standard deviation of the maximum likelihood estimator with respect to the model parameters (unknown in practice) and the Monte Carlo standard error of the maximum likelihood approximation with respect to the true maximum likelihood are given in Table 7.6.

TABLE 7.5
Parameter estimation from the data presented in Figure 5.2.

Initial values	Optimal step gradient	Monte Carlo MLE
$\log \gamma_0^i = -9.5$	$\widehat{\log \gamma_0^0} = -8.37$	$\widehat{\log \gamma_0^n} = -8.32$
$\log \gamma_1^i = -4.0$	$\widehat{\log \gamma_1^0} = -2.74$	$\widehat{\log \gamma_1^n} = -2.73$
$\log \gamma_2^i = 1.5$	$\widehat{\log \gamma_2^0} = 2.46$	$\widehat{\log \gamma_2^n} = 2.47$
$\log \gamma_o^i = -3.5$	$\widehat{\log \gamma_o^0} = -2.13$	$\widehat{\log \gamma_o^n} = -2.17$
$\log \gamma_r^i = -3.5$	$\widehat{\log \gamma_r^0} = -2.42$	$\widehat{\log \gamma_r^n} = -2.42$

TABLE 7.6
Estimation errors for the MCMCML estimates of the Candy model.

Asymptotic errors	
$\|\theta - \theta^{MLE}\|$	$\|\theta^{MLE} - \theta^{MC}\|$
0.51	2e-3
0.23	3e-3
0.17	1e-3
0.30	2e-3
0.31	5e-3

7.4.2 Parametric inference for the Candy model using the ABC Shadow and the SSA algorithms

The ABC Shadow 6.6 and SSA 6.7 algorithms were also used to perform parametric inference for the Candy model. In [297, 287], the authors used the Candy model as an example of a marked Gibbs point process with multiple interactions for which the ABC Shadow and SSA can be successfully used to sample and maximise its posterior distribution.

The considered posterior distribution was built using a version of the Candy model that was used to detect cosmic filaments [294]. With respect to (7.8), the fitted probability density of the model has $\gamma_o = 1$ and $h(\boldsymbol{x}) = 1$, so only four parameters need to be inferred. If the prior $p(\theta)$ is the uniform distribution over a given compact region Θ, then the posterior becomes the likelihood restriction to Θ. Thus, this strategy can be used to verify and tune the algorithms used.

The following experiment was performed. First, the tailored Metropolis-Hastings (MH) dynamics [192, 292] was used to generate 1000 samples of a candy model on $W = [0, 3] \times [0, 1]$. The segment parameters were: length $l = 0.12$, connection range $r_c = 0.01$ and curvatures $\tau_c, \tau_r = 0.5$ radians. The interaction parameters were $\log \gamma_2 = 10, \log \gamma_0 = 3, \log \gamma_1 = 7, \log \gamma_r = -2$. On the basis of the samples obtained, the calculation of the vector of the

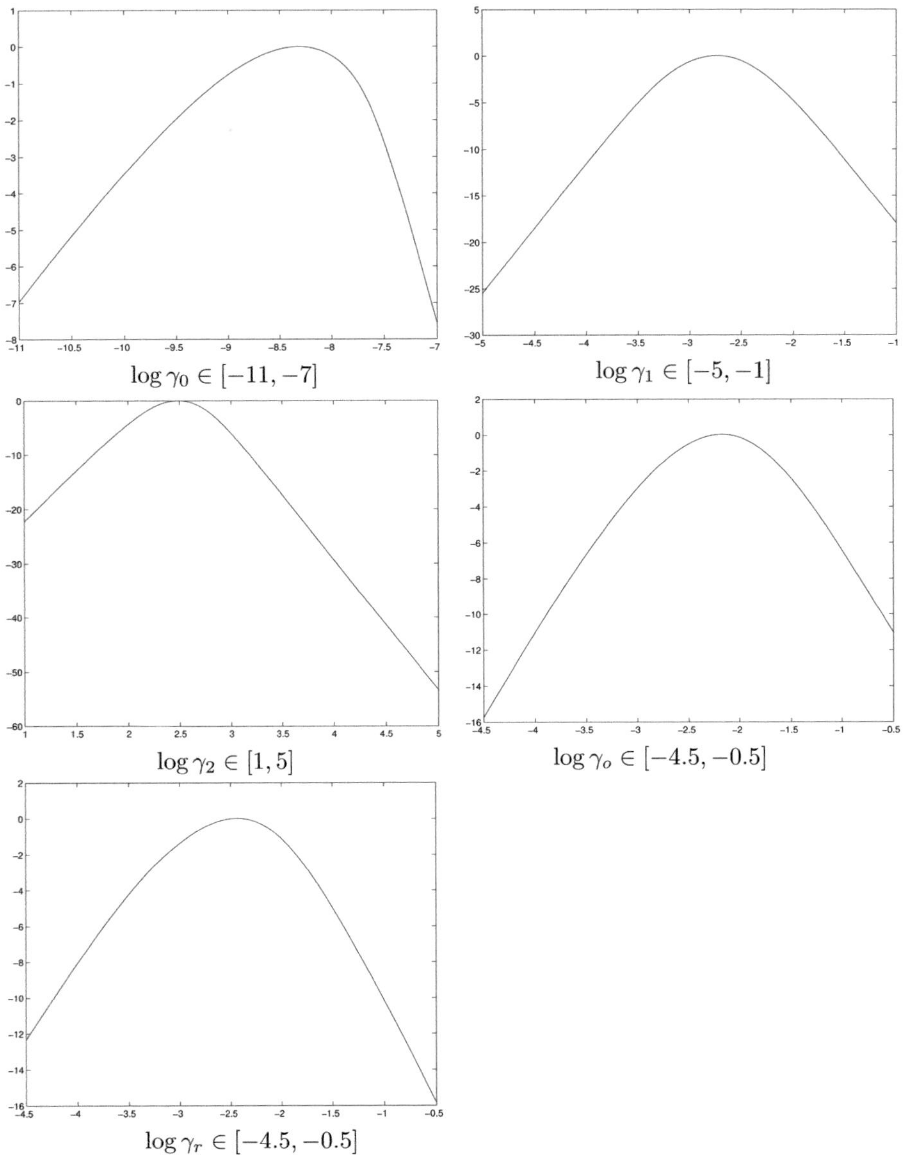

FIGURE 7.21
Monte Carlo approximation of the log-likelihood as a function of the individual components. The ranges of the values corresponding to each parameter were pre-fixed.

empirical means of the sufficient statistics gave $\bar{t}(\boldsymbol{y}) = (\bar{n}_2(\boldsymbol{y}) = 30.76, \bar{n}_0(\boldsymbol{y}) = 19.65, \bar{n}_1(\boldsymbol{y}) = 77.22, \bar{n}_r(\boldsymbol{y}) = 17.65)$. Second, the ABC shadow and SSA algorithms were launched using these statistics as input data to sample and maximise the posterior induced by these observations. The prior density $p(\theta)$ chosen was the uniform distribution over the interval $[0, 12]^3 \times [-12, 0]$. For both algorithms, the auxiliary variable was sampled each time with 2000 moves of the adapted MH dynamics algorithm. All values for each of the vectors m and Δ were set to 500 and 0.01 respectively. A total of 1000 samples were obtained after running the algorithms for 10^6 iterations and retaining samples every 10^3 iterations. For the SSA algorithm, the initial temperature was $T = 10000$ and a slow polynomial cooling schedule was implemented

$$T_n = 0.9999T_n.$$

In this case, the Δ parameters have all been set to 0.1, while for their evolution a polynomial schedule with $k_\Delta = 0.99995$ is adopted.

Figure 7.22 shows the marginals of the posterior of the Candy model obtained with the ABC Shadow algorithm. Of all the estimated parameters, the corresponding marginal posterior modes are close to the true model parameters. This is also the case for the posterior mean estimates.

The results of the MAP calculation for the Candy model using the SSA algorithm are presented in Figure 7.23. The behaviour of the SSA dynamics shows that, for high values of the temperature and important values of the control parameters Δ, the configuration space is well explored. Finally, the output of the algorithm converges to the expected values: the MAP estimates are almost identical to the true model parameters. Table 7.7 shows the asymptotic computation after [297, 287] and confirms the quality of the results obtained.

These examples show that ABC Shadow and the SSA algorithm are very interesting tools for parametric inference on marked point processes with multiple interactions. They also serve to show how these methods can be used and tuned. The next sections of the chapter present applications of these algorithms to real data from cosmology and network science.

7.4.3 Morpho-statistical characterisation of the spatial galaxy distribution through Gibbs point processes

The distribution of galaxy positions in our Universe forms an intricate network of clusters, sheets and filaments. These structures surround empty regions with very few or no galaxies, called cosmic voids [201, 202, 247, 146]. Depending on the filamentary pattern, the distribution of galaxies shows some special features. For example, the bright spiral galaxies may have a tendency to align parallel to the filaments, while the S0 elliptical galaxies appear to have their spin axes more perpendicular to the filament to which they belong [309]. Furthermore, the galaxies belonging to a filament are not evenly distributed. They tend to form clusters more or less regularly along the filament, similar to pearl necklaces [306].

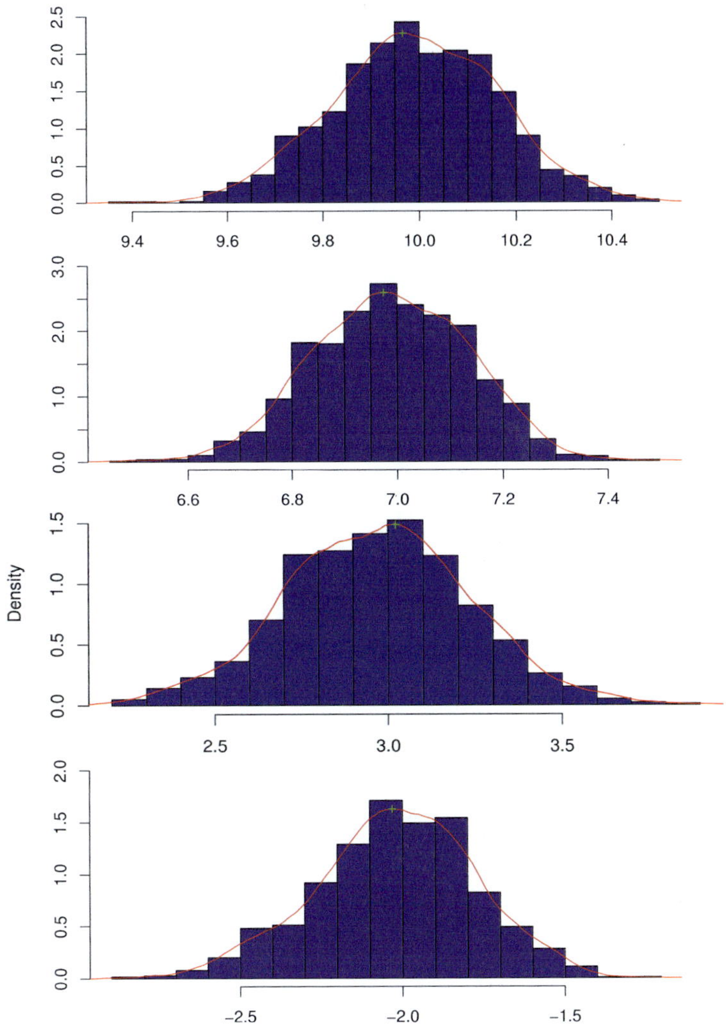

FIGURE 7.22
Marginal histogram and kernel density estimates of the Candy model posterior sampled using the ABC Shadow algorithm. The true model parameters are $\theta = (10, 7, 3, -2)$. The posterior mode and the posterior mean estimates are $(9.96, 6.97, 3.02, -2.03)$ and $(9.99, 6.98, 2.96, -2.02)$, respectively.

Marked point processes with interactions are a natural choice for the mathematical modelling framework to be adopted to tackle the study of galaxy distributions [6, 93, 201]. This idea is the natural continuation of the work done with the well-known models such as Neyman-Scott, Cox and log

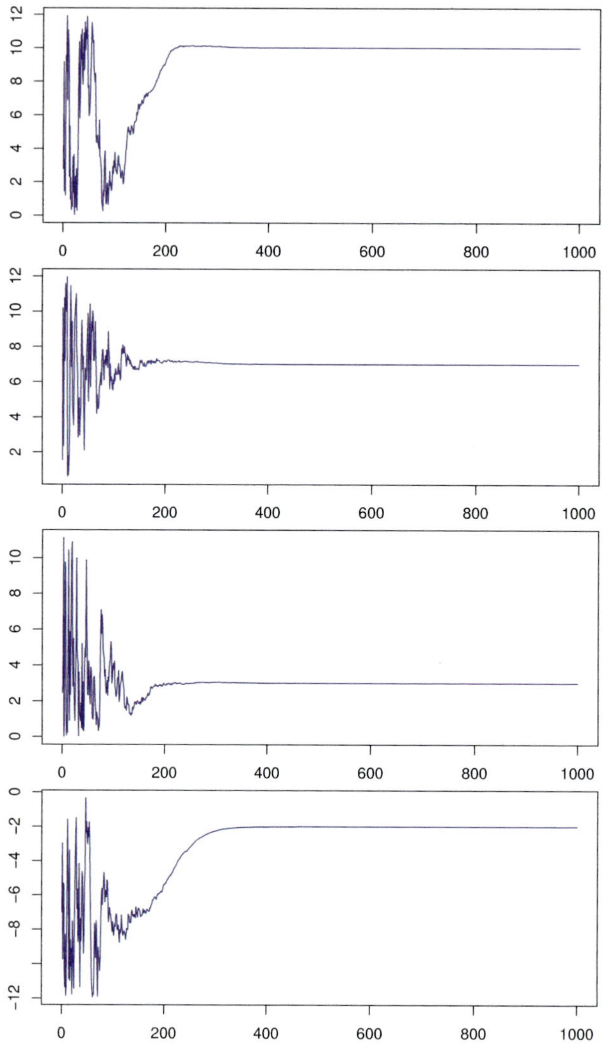

FIGURE 7.23
MAP computation of the Candy model parameters using the SSA algorithm. The true parameters of the model are $\theta = (10, 7, 3, -2)$, while the SSA estimates are $\widehat{\theta} = (10.00, 6.99, 3.00, -1.99)$.

Gaussian Cox point processes [237, 59, 227]. However, all of these processes assume independence, i.e. lack of interaction, between the points in a configuration. The use of interacting point processes is also strongly encouraged by the fact that the detection of filaments and clusters, and the restoration of galactic catalogues have been successfully carried out using methods based on interacting point processes [280, 294, 295, 296, 310, 308, 307]. This section

TABLE 7.7
Estimation errors for the SSA MAP estimates of the considered Candy model. For the computation of the MCSE, a number of 10^4 samples from the fitted model were used.

Asymptotic errors for the SSA Candy estimates							
$\sigma_{\log\gamma_2}$	$\sigma^{MC}_{\log\gamma_2}$	$\sigma_{\log\gamma_1}$	$\sigma^{MC}_{\log\gamma_1}$	$\sigma_{\log\gamma_0}$	$\sigma^{MC}_{\log\gamma_0}$	$\sigma_{\log\gamma_r}$	$\sigma^{MC}_{\log\gamma_r}$
0.16	2.6e-4	0.14	2e-4	0.25	6e-4	0.24	5.5e-4

presents the results obtained by [140, 287], which are a very first step in the very challenging and promising direction of modelling the galaxy distribution. It is important to mention the work of [175], where the authors propose to use Gibbs point processes to characterise star clusters. Thus, there is hope for the analysis and synthesis of cosmological and astronomical data under a unified mathematical framework.

The data analysed here is shown in Figure 7.24. It represents a cubic sample of size $30\,h^{-1}$ Mpc from the SDSS (Sloan Digital Sky Survey) catalogue of galaxies. The data also includes detected cosmic filaments. The data sample and filament detection were produced by [310]. The purple points represent the positions of the galaxies, while the green lines show the pattern of the main axes of the filaments. Two features of the galaxy distribution can be observed by visual inspection of the data. The first is the presence of galaxy clusters and the second is the proximity of the galaxy positions to the main axes of the filaments.

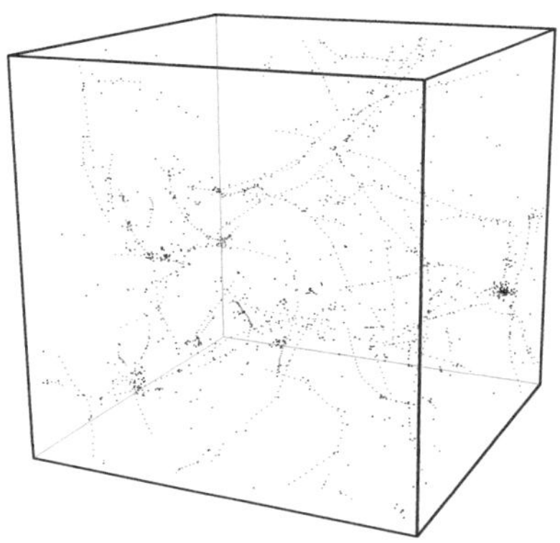

FIGURE 7.24
Data set: galaxy positions (purple points) with filaments main axes or spines (green).

The work carried out in [140, 287] fitted models to the observed data in order to characterise the aforementioned aspects and give them statistical significance. Since the area-interaction process is a point process that indicates clustering by the value of its interaction parameter, it was natural to consider a model given by the probability density function

$$p(\boldsymbol{x}|\theta) \propto \beta_1^{n(\boldsymbol{x})}\beta_2^{d_F(\boldsymbol{x})}\gamma^{-a_r(\boldsymbol{x})} \tag{7.10}$$

with respect to the standard Poisson reference. The model parameters are

$$\theta = (\log\beta_1, \log\beta_2, \log\gamma),$$

and the sufficient statistics vector is

$$t(\boldsymbol{x}) = (n(\boldsymbol{x}), d_F(\boldsymbol{x}), a_r(\boldsymbol{x})). \tag{7.11}$$

The parameter $\log\beta_1$ controls the total number of galaxies in a given configuration, represented by the statistic $n(\boldsymbol{x})$. The parameter $\log\beta_2$ controls the statistic $d_F(\boldsymbol{x})$, which measures the proximity of the galaxy centres to the filament pattern F. Its expression is

$$d_F(\boldsymbol{x}) = -\sum_{i=1}^{n(\boldsymbol{x})} d(x_i, F),$$

where $d(x_i, F)$ is the shortest distance from the galaxy position x_i to the filament pattern. With respect to the interaction parameter, if $\log\gamma > 0$ the galaxies in a configuration tend to aggregate, while for $\log\gamma < 0$ the galaxy pattern tends to be more regular or repulsive. The sufficient statistic controlled by this parameter is

$$a_r(\boldsymbol{x}) = \frac{3A(\boldsymbol{y})}{4\pi r^3},\ A(\boldsymbol{x}) = \nu[\bigcup_{i=1}^{n(\boldsymbol{x})} b(y_i, r)].$$

Since the volume $A(\boldsymbol{x})$ is divided by the volume of a sphere of radius r, the statistic $a_r(\boldsymbol{x})$ can be interpreted as the number of spheres of radius r centred at the galaxy positions needed to form a structure of volume $A(\boldsymbol{x})$. The model (7.10) allows us to see the distribution of galaxies as a point pattern, which creates a territorial influence in the vicinity of the galactic filaments.

The sufficient statistics (7.11) computed from the observed data are presented in Table 7.8. The statistics have been computed for different values of the radius r. Each such statistics vector was used to construct a particular posterior. All posteriors were constructed using the same prior $p(\theta)$, i.e. the uniform distribution over the interval $[-50, 50]^3$.

TABLE 7.8
Sufficient statistics computed for the observed galaxy pattern.

Data for the galaxy pattern							
r	0.5	1	1.5	2	2.5	3	3.5
$n(\boldsymbol{x}) = 1024,\ d_F(\boldsymbol{x}) = -\sum_{i=1}^{n(\boldsymbol{x})} d(x_i, F) = -1180.05$							
$a_r(\boldsymbol{x})$	724.29	484.01	357.16	263.10	195.08	142.86	105.30

The SSA algorithm was used to compute the MAP estimate from each posterior distribution obtained. The results are shown in Figure 7.25. The right column shows the behaviour of the estimated parameters with respect to r. As the radius r increases, the parameter $\log \beta_1$ tends to stabilise, while the parameter $\log \beta_2$ tends to increase. As for the parameter $\log \gamma$, it tends to increase before $r = 3$ and to decrease just after. This is consistent with the fact that there is a change in the clustering regime around these ranges [90]. The right column shows the evolution of the estimated parameters when $r = 2$. At high temperatures the configuration space is explored. At low temperatures the parameter distributions approach the convergence regime. The obtained estimated parameter $\widehat{\theta} = (\widehat{\log \beta_1}, \widehat{\log \beta_2}, \widehat{\log \gamma})$ is $(-0.33, 0.98, 4.57)$. Full details of the implementation of the SSA algorithm, together with the values of the asymptotic errors which allow the quality of the estimation results to be assessed, can be found in [287].

The ABC Shadow algorithm was used to test the significance of the parameters. Figure 7.26 shows the marginals of the ABC shadow approximation of the posterior given by sufficient statistics computed with $r = 2$. Based on this, a statistical Student like test can be performed to check whether the estimated parameters are significantly different from 0. For the parameter $\log \beta_1$ the obtained $p-$value was 0.16, while for the parameters $\log \beta_2$ and $\log \gamma$ the corresponding $p-$values were both less than 10^{-5}. This result indicates that the parameter $\log \beta_1$ is not significantly different from 0. It also gives statistical significance to the cosmological fact that galaxies tend to be distributed along filaments when forming clusters, and are rarely placed independently in our Universe.

The work in [140] tested and compared different clustering mechanisms. To do this, two additional clustering models were fitted to the data. The first model is called the *Geyer saturation process* [108]. Its probability density is given by

$$p(\boldsymbol{x}|\theta) \propto \beta_1^{n(\boldsymbol{x})} \beta_2^{d_F(\boldsymbol{x})} \gamma^{u_r(\boldsymbol{x})}$$

with parameterisation and statistics similar to (7.10). The difference is the statistic $u_{r,s}(\boldsymbol{x})$. This statistic depends on two positive parameters, the

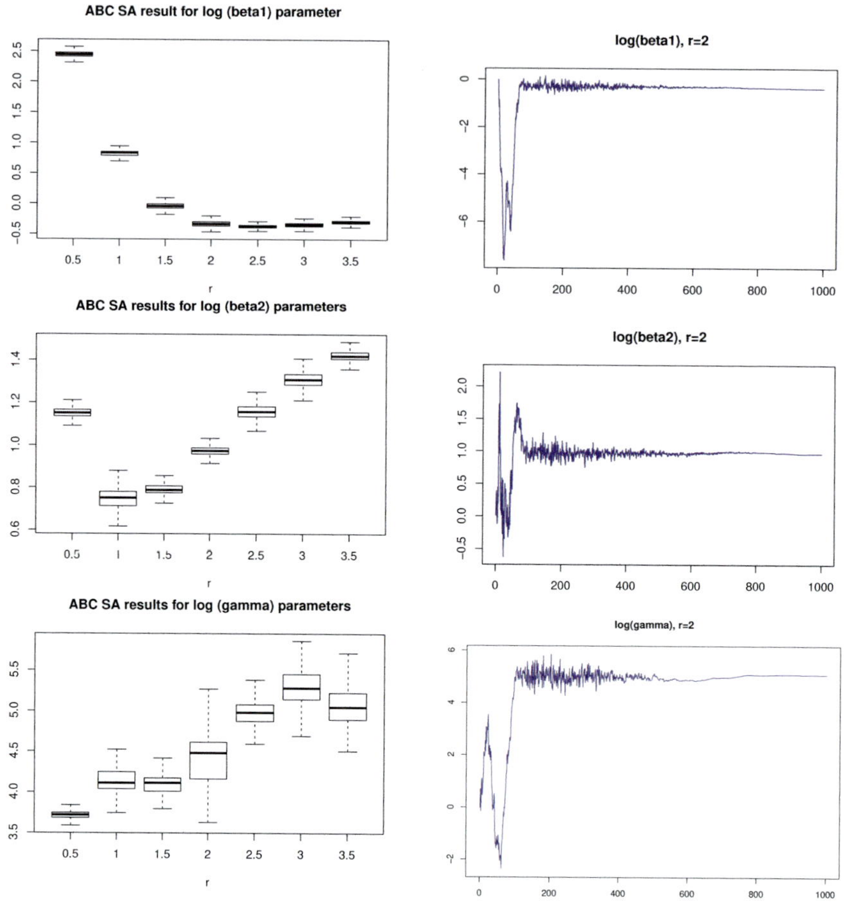

FIGURE 7.25
Left column: box plots of SSA algorithms outputs. The entry of each algorithm was the sufficient statistics vector computed from the considered SDSS sample for a given radius r. Right column: the time series of the outputs of the SSA algorithm for $r = 2$.

saturation s and the range r. Its expression is given by

$$u_{r,s}(\boldsymbol{x}) = \sum_{i=1}^{n(\boldsymbol{x})} \min\{s, m_{x_i}(\boldsymbol{x})\}$$

with

$$m_{\xi}(\boldsymbol{x}) = \sum_{i=1}^{n(\boldsymbol{x})} \mathbf{1}\{d(\xi, x_i) \leq r\}\mathbf{1}\{\xi \neq x_i\}$$

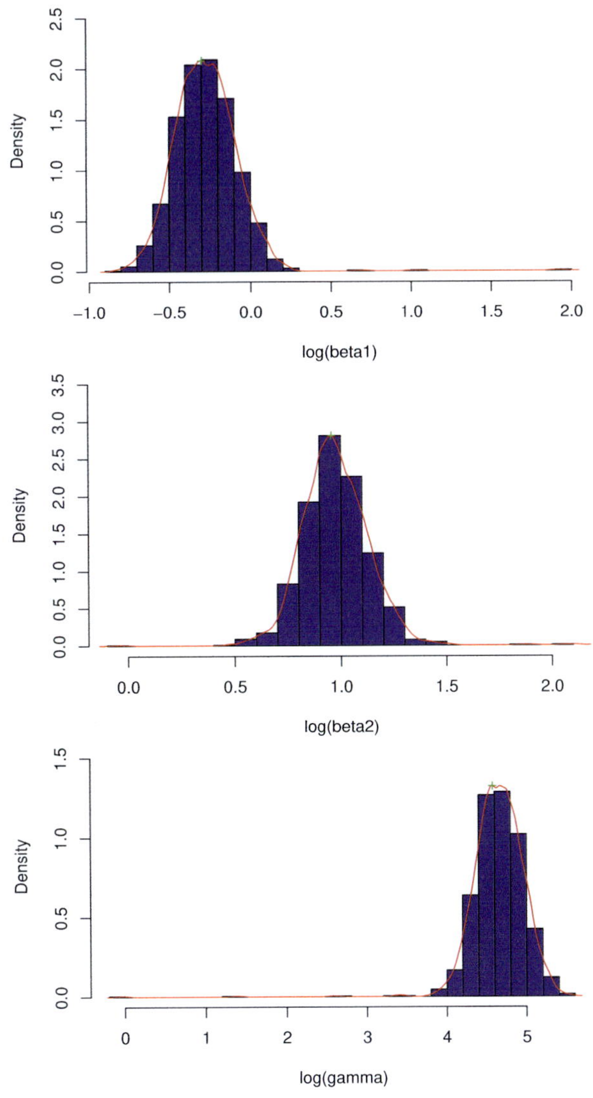

FIGURE 7.26
ABC Shadow outputs for the approximate posterior sampling of the model fitted to the SDSS sample (the range parameter is $r = 2$). The obtained MAP estimate is $\widehat{\theta} = (-0.29, 0.95, 4.57)$.

which counts the neighbours of the point ξ within a range r in the configuration $\boldsymbol{x}$. If $\gamma < 1$ the process exhibits repulsion similar to the Strauss model. For values of $\gamma > 1$ the model tends to produce clusters. The saturation parameter limits the effect of a large number of interactions, so the model is well-defined and locally stable [108]. This model understands the distribution of galaxies as a distance-based pattern.

The second model is known as the *continuum random cluster model* or the *connected component process* [15, 12, 53, 127, 221]. Its probability density is

$$p(\boldsymbol{x}|\theta) \propto \beta_1^{n(\boldsymbol{x})} \beta_2^{d_F(\boldsymbol{x})} \gamma^{-v_r(\boldsymbol{x})}$$

where $v_r(\boldsymbol{x})$ is the number of connected components in the configuration $\boldsymbol{x}$. Two points are connected if the distance between them is less than the range parameter r. A connected component is a set of points where each point is connected to at least one other point. The model is well-defined for $\gamma > 0$. The number of connected components in a configuration tends to increase when $\gamma < 1$, or to decrease when $\gamma > 1$. This model interprets the galaxy distribution as a membership-based interaction.

For different range parameters, the previous models were also fitted to the SDSS sample of galaxies using the ABC Shadow algorithm [140]. A natural question is which of these clustering mechanisms is most appropriate for the data set at hand. An answer to this question can be formulated using the *residual analysis of spatial point processes* [19, 11, 16, 12, 58, 324].

Let $\mathbb{X}$ be a locally stable marked point process on $W \times M$. For non-negative functions h and sets $A \subseteq W \times M$, the $h-$*Innovations* are defined by

$$I(A, h, \lambda) = \sum_{x_i \in \mathbb{X}_A} h(x_i, \mathbb{X} \setminus x_i) - \int_A \lambda(\eta; \mathbb{X}) h(\eta, \mathbb{X})(\nu \times \nu_M)(d\eta).$$

Assuming that the sum and integral in the definition have finite expectations, the Georgii-Nguyen-Zessin formula gives

$$\mathbb{E}I(A, h, \lambda) = 0.$$

The innovation I is a signed measure that has two components. The first component is the quantity

$$\triangle I(x_i) = h(x_i, \mathbb{X} \setminus \eta)$$

which represents the innovation increment ("error") attached to a point $\eta \in \mathbb{X}$. The second component is

$$dI(\eta) = -\lambda(\eta; \mathbb{X}) h(\eta, \mathbb{X})$$

and is interpreted as the increment attached to a background location $\eta \in W \times M$.

Suppose a parametric model for a marked point process has been fitted to an observed pattern $\boldsymbol{x}$. The conditional intensity of the process, computed with

$\widehat{\theta}$ the estimated model parameters, is denoted by $\widehat{\lambda}$. The function h may also depend on the model parameters, so wh is its estimated counterpart. In this context, the *h–residuals* are defined as

$$\begin{aligned} R(A, \widehat{h}, \widehat{\theta}) &= I(A, \widehat{h}, \widehat{\lambda}) \\ &= \sum_{x_i \in \boldsymbol{x}_A} \widehat{h}(x_i, \boldsymbol{x} \setminus x_i) - \int_A \widehat{\lambda}(\eta; \boldsymbol{x}) \widehat{h}(\eta, \boldsymbol{x})(\nu \times \nu_M)(d\eta). \end{aligned}$$

The natural application idea is to expect the residuals to be close to 0 if the model is appropriate.

There are several options for the function h [19, 12]. The raw residuals are obtained by taking $h(\eta, \boldsymbol{x}) = 1$. Their expression is

$$R(A, 1, \widehat{\theta}) = n(\boldsymbol{x} \cap A) - \int_A \widehat{\lambda}(\eta; \boldsymbol{x})(\nu \times \nu_M)(d\eta).$$

For present purposes, a smoothed version of these residuals has been used [140].

From a practical point of view, the calculation of the residuals is done by decomposing the domain $W \times M$ into subregions A_i. This decomposition affects the distribution of points in a subregion in a similar way to the χ^2 test. At the same time, if the number of dimensions of $W \times M$ is too high, the visualisation of the residuals is not straightforward. In this case, averaging the residuals over some dimensions is a strategy that allows the analysis of "marginalised" residuals. The distribution of points in a sub-region is also affected by the interaction between points. By analogy with regression-based methods, the Q-Q plot is a visualisation tool proposed by [19, 11, 12] to detect changes in the distribution of the number of points. This tool plots the quantiles of the observed residuals against the estimated expected quantiles computed under the fitted model.

Figure 7.27 shows the result of the residual analysis carried out to compare the performance of the Geyer, connected component, and area-interaction processes fitted to the SDSS sample under consideration. The results obtained are presented for the three models in this order, from top to bottom. The same range parameter $r = 1$ was considered for all models. The left column shows the averaged smoothed residuals. These results indicate that the area-interaction process is the most appropriate model among the proposed candidates. For this model, the majority of the calculated residuals are close to zero. The extreme red-coloured areas or "hot zones" indicate regions where the values of the innovation increment due to the points are rather important. The extremely purple coloured areas or "cold zones" indicate regions where the values of the innovation increment due to the background are more important. The residuals obtained by fitting an area-interaction process show only

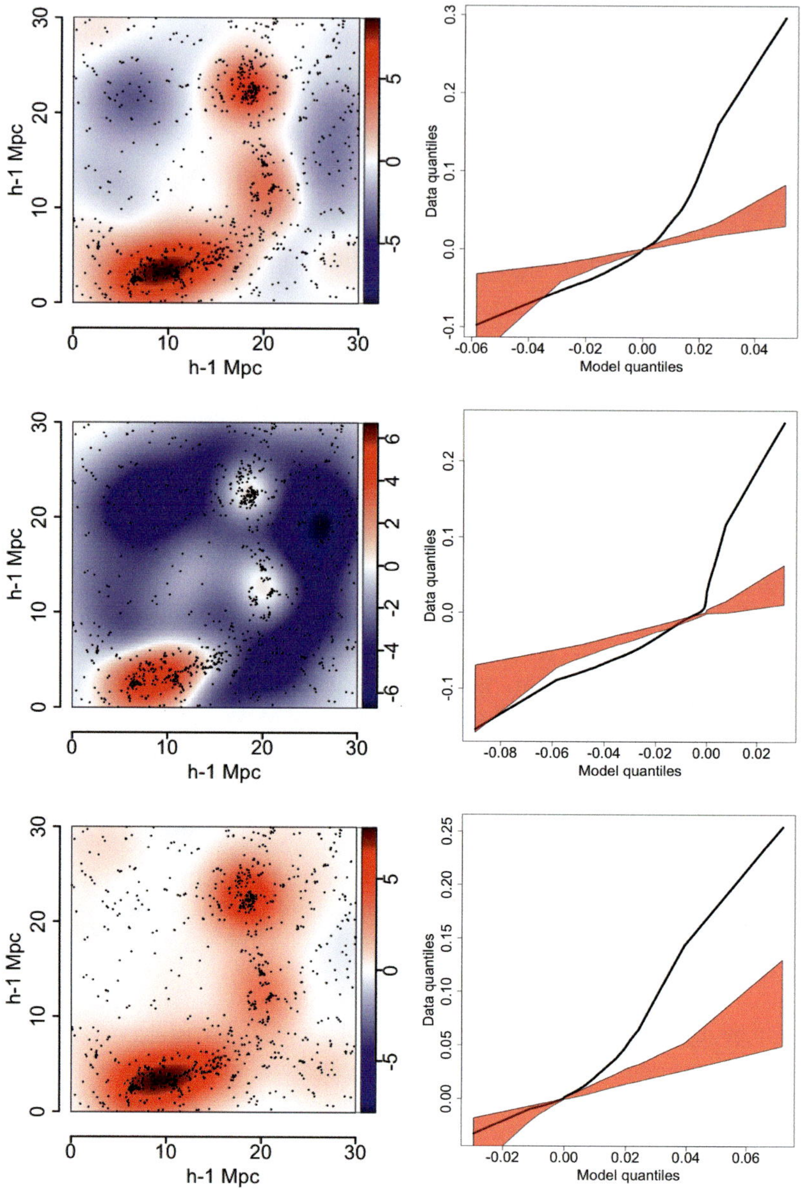

FIGURE 7.27
Left column: averaged smoothed residuals. Right column: Q-Q plots. Top to bottom: Geyer, Connected Components and Area-Interaction models. The range parameter was for all the models $r = 1\, h^{-1}$ Mpc.

two hot zones. The same hot zones are found for the Geyer model residuals. The Geyer residuals also show two cold zones. The connected components model shows one hot zone and two cold zones. The hot zone coincides with the same hot zone exhibited by the area-interaction and the Geyer residuals. The cold zones of the Geyer and connected components models clearly overlap.

The Q-Q plots shown in the right column of Figure 7.27 reinforce the previous conclusion. The black lines are obtained by plotting the residuals computed from the observed data against the expected residuals from the simulated data. The red envelopes represent 95% confidence regions obtained by plotting the residuals computed from the data against quantiles of the residuals from the simulated data. The residuals from the simulated data are computed by simulating patterns from the fitted model, estimating the model parameters on these new patterns and computing their corresponding residuals [19, 12]. These results confirm the previous analyses. The area-interaction is the model that best fits the observed data: the black curve is inside its envelope for the largest interval of values. Nevertheless, for all the models, almost all the positive values of the residual distributions are outside their confidence region.

The residual based analysis reveals the limitations of the currently chosen models and suggests the next steps that can be taken towards a possible solution to this challenging problem. First, the data show cluster regions with a local number of galaxies that is really high compared to the local density of galaxies within other regions. The introduction of an inhomogeneous term that takes this into account may improve the present result. Second, since different mechanisms of clustering or repulsion are at work, the study of embedding multiple interactions within a model should give more insights regarding the galaxy distributions [3, 111, 112]. Finally, the use of a single range parameter is simple but not entirely natural, since galaxies have different masses, luminosities, shape characteristics, etc. The very first step to be taken in this direction is to consider a marked point process, where the mark is an individual range of interaction for each galaxy.

Example 7.1 (Spatial interaction between earthworm burrows and cracks at the soil surface). During my stay at INRAe Avignon, I had many enlightening and stimulating discussions with A. Kertzschmar, J. Chadoeuf and F. Aries. At the time, I was giving seminars on my initial findings regarding the detection of cosmic filaments in two-dimensional data and their potential to motivate the construction of models capable of reproducing complex three-dimensional structures, given the recent release of three-dimensional astronomical data.

I was initially intrigued by their attempt to model the distributions of earthworm burrow positions conditional on soil surface cracks as an inhomogeneous

Poisson point process [51, 52]. Also of interest was their work on generating images of the three-dimensional burrow system [162, 49]. We were all quite impressed by the fact that similar mathematical models and descriptive tools are suitable for describing such different patterns at such different scales that have comparable morphology.

7.4.4 Morpho-statistical description of collaborative social networks

The Gibbs MRF model presented in Section 3.4 aims to describe the behaviour of teams in a research institute with respect to the way their members collaborate: inter-organisational, intra-organisational or both tendencies simultaneously. The model describes the labelling of the dual graph induced by observing the network formed by the co-authorship of publications in a given year. Its probability distribution is given by the Gibbs energy function (3.15) and is written as

$$\mathbb{P}(X = \boldsymbol{x}|\theta) = \frac{\exp\langle t(\boldsymbol{x}), \theta\rangle}{Z(\theta)}$$

with $\theta = (\theta_{11}, \theta_{12}, \theta_{22})$ the model parameters vector and $t(\boldsymbol{x}) = (t_{11}(\boldsymbol{x}), t_{12}(\boldsymbol{x}), t_{22}(\boldsymbol{x}))$ the sufficient statistics vector given by

$$\begin{aligned} t(\boldsymbol{x}) &= (t_{11}(\boldsymbol{x}), t_{12}(\boldsymbol{x}), t_{22}(\boldsymbol{x})) \qquad (7.12)\\ &= \Big(\sum_{w_i \sim w_j} \mathbf{1}\{m(w_i) = m(w_j) = 1\}, \sum_{w_i \sim w_j} \mathbf{1}\{m(w_i) + m(w_j) = 3\},\\ &\quad \sum_{w_i \sim w_j} \mathbf{1}\{m(w_i) = m(w_j) = 2\}\Big). \end{aligned}$$

Let $\boldsymbol{y}$ be an observed network. Assuming that the observation is a realisation of the model (3.15), the sufficient statistics (7.12) can be computed. Choosing a prior $p(\theta)$ for the model parameters, sampling posterior distributions

$$p(\theta|\boldsymbol{y}) \propto \frac{\exp\langle t(\boldsymbol{x}), \theta\rangle}{Z(\theta)} p(\theta)$$

allows us to search for models capable of reproducing networks with characteristics similar to the observations.

The ABC Shadow algorithm 6.6 is used for this purpose. Its application requires a sampling algorithm from (3.15). Here the Gibbs sampler is adopted [105, 318].

In order to set up the parameters of the ABC Shadow dynamics, the algorithm was used to perform inference for a known simple model that can be

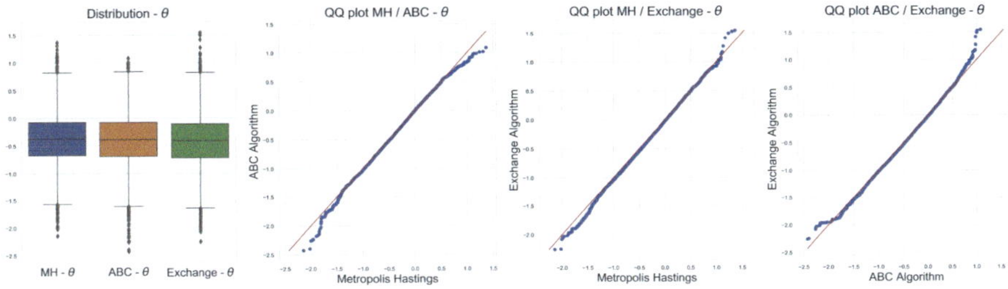

FIGURE 7.28
Posterior distribution of a binomial random variable sampled with Metropolis Hasting, ABC shadow and the exchange algorithms.

controlled from an analytical perspective. The posterior distribution of a binomial random variable $\mathcal{B}in(n, p)$ was constructed

$$\begin{aligned} p(\theta|\boldsymbol{y}) &= \binom{n}{y} p^y (1-p)^{(n-y)} p(\theta) \\ &= \exp\left[y\theta\right] \frac{\binom{n}{y}}{[1+\exp(\theta)]^n} p(\theta) \end{aligned} \tag{7.13}$$

with $\theta = \log \frac{p}{1-p}$. For the experiment considered, the parameters of the binomial variable were $n = 20$ and $p = 0.4$ (i.e. $\theta = -0.405$) and the observation was $y = 8$. The parameter n is assumed to be known and an inference is desired with respect to θ. By choosing the uniform distribution over $[-100, 100]$ as the prior for $p(\theta)$, the posterior (7.13) becomes a restriction of the likelihood to the domain under consideration. Since the prior distribution domain is large enough, the likelihood and inference based on the approximation of the posterior should give similar numerical results. The posterior (7.13) is sampled using three methods: classical Metropolis-Hastings (MH), Murray's exchange algorithm, and ABC Shadow dynamics. The MH and ABC dynamics both used a uniform proposal for θ with $\delta = 0.005$. The m parameter for the ABC Shadow was set to 100. The exchange algorithm used a Gaussian proposal with a standard deviation of $\sigma = \frac{\Delta}{2}$. The sampled posterior distributions obtained with these three algorithms are shown in Figure 7.28. The quantiles, MAP estimates and asymptotics are presented in Table 7.9. It can be seen that the performance of the three algorithms is equivalent. A verification of the performance of the ABC Shadow algorithm using as observed data samples generated with a Potts model with known parameters has also been done in [167]. In the following, the choice of ABC Shadow is justified by the fact that its control does not require an exact simulation of (3.15).

The data analysed was built on the collaborative network created by the co-authorship of papers at the LORIA research institute. In this institute,

TABLE 7.9
Statistics on the posterior of a Binomial distribution approximated with three posterior sampling algorithms.

	Q_{10}	Q_{25}	Q_{50}	Q_{75}	Q_{95}	MAP	$\hat{\sigma}_\theta$	$\hat{\sigma}_\theta^{MC}$
MH	–0.96	–0.67	–0.37	–0.07	0.35	–0.37	0.45	4.3×10^{-4}
ABC Sh.	–0.99	–0.69	–0.38	–0.07	0.34	–0.42	0.45	4.2×10^{-4}
Exchange	–1.00	–0.72	–0.41	–0.10	–0.34	–0.44	0.45	4.3×10^{-4}

researchers work on specific scientific objectives within well-defined research teams. In general, such a team consists of two categories of researchers: LORIA researchers and external collaborators. Two researchers are linked when they co-sign a publication. The induced co-authorship graph of a given research team is further pre-processed as shown in Figure 7.29. First, two types of researchers are distinguished: internal (LORIA) and external members. Next, only the edges containing at least one internal member are kept. Finally, the corresponding line graph is constructed and sufficient statistics 7.12 are computed. The assumed model is defined over a state space produced by all possible edges containing at least one internal member.

For 11 research teams, the co-authorship graphs of publications in the years 2017, 2018 and 2019 were used to construct the corresponding line graph and then to compute the sufficient statistics (7.12). Thus, for a single research team, three posterior sampling runs were performed, with each posterior obtained corresponding to a particular year of activity. The posteriors were sampled using the ABC Shadow algorithm. The results obtained are shown in Figure 7.30. The parameter θ_{11} controls the number of connections where a

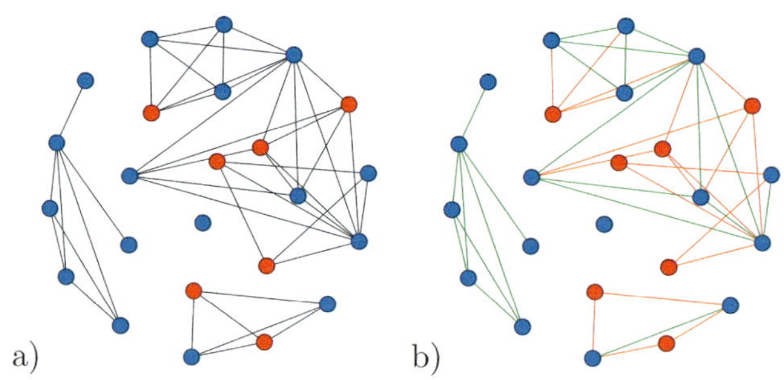

FIGURE 7.29
An example of a pre-processing performed on the collaborative graph of a team. The graph illustrates the collaborative links involving the members of the team COAST during the year 2018: a) distinguish two types of members: LORIA (blue) and external collaborators (red), b) keep edges involving at least one member of LORIA. The corresponding line graph is constructed by considering only the edges connected by at least one member of LORIA.

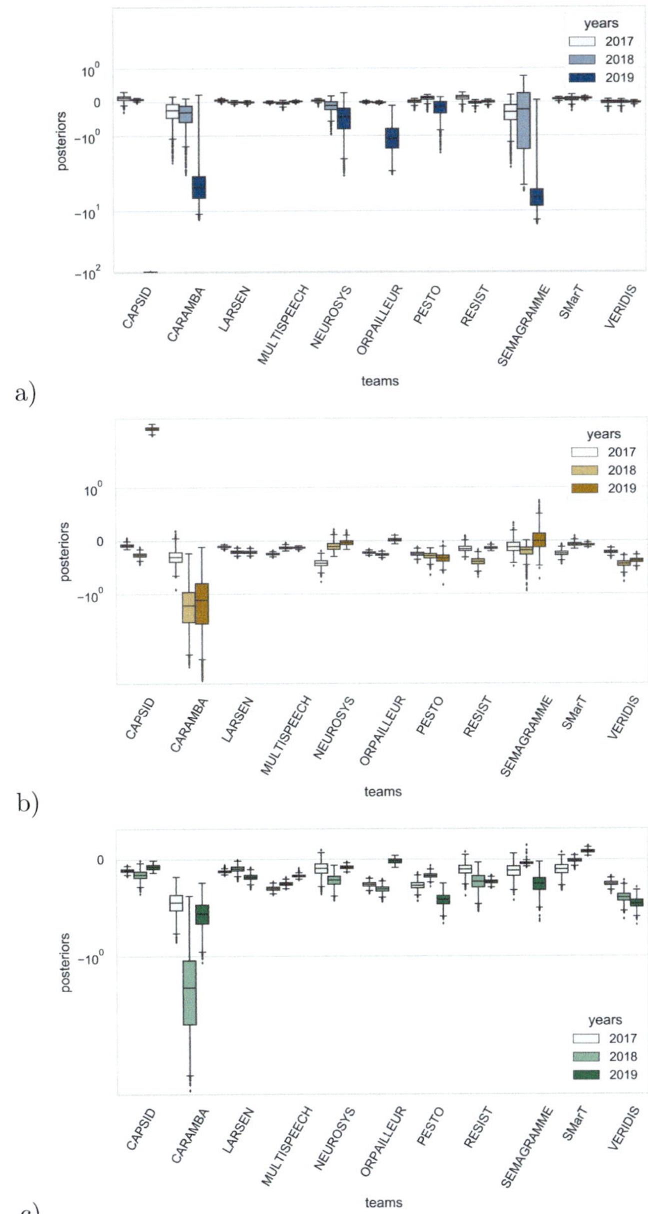

FIGURE 7.30
Box plots of the posterior distributions for the parameters estimated from the collaboration graphs: a) the θ_{11} distributions, b) the θ_{12} distributions and c) the θ_{22} distributions. For each team, three posteriors are estimated corresponding to the activity during the years 2017, 2018 and 2019, respectively.

LORIA researcher collaborates with two people from the same research team. It is an indicator of intra-organisational collaboration. For the majority of teams, the distribution of this parameter tends to be positive, i.e. they tend to form network patches, which is a natural tendency within a research team with commonly defined goals. The parameter θ_{22} controls the number of links where a LORIA researcher works with two people belonging to different organisations. This parameter is an indicator of the inter-organisational dynamics within the research team. For almost all teams, the posterior distribution of θ_{22} tends to be negative for the years considered. This may indicate a kind of preference for internal over external collaboration. Of course, these distributions do not explain the reasons for this possible preference. The parameter θ_{12} controls the number of connections where a LORIA researcher collaborates with one person from the same research team and one from another institute. This parameter may be an indicator of researchers with high connectivity impact or hubs. The very important values obtained for the CASPID team in 2019 are explained by the co-signing of papers with an important number of authors.

The work presented in [166, 167] also presents the computation of asymptotics that certify the quality of the estimation, together with a statistical test that assesses the significance of the results. An unsupervised classification of the research teams is performed with respect to each posterior parameter. The classification procedure uses the Kolmogorov-Smirnov distance between probability distributions. The problem of degeneracy is also discussed. The natural perspectives of this work are the addition of a goodness-of-fit procedure to the current methodology and also the extension of the proposed model with a temporal dimension [19, 12, 121, 101].

7.4.5 Agricultural landscape characterisation through Gibbs T–tesselations

The mosaic of fields in agriculture are natural spatial patterns that can be modelled using T–tesselations. This section presents some of the results obtained by [158, 1]. The first row of Figure 7.31 shows two examples of agricultural landscape patterns observed in France. The light green polygons represent agricultural areas, while the dark green polygons represent non-agricultural areas (mainly forests, villages and farmland). It can be observed that almost all the junctions formed by the edges of the fields are T–junctions. A pre-processing treatment is needed to transform these landscape patterns into polygonal T–tesselations, i.e. tesselations with cells that have only line segments as edges.

The T–tesselations resulting from the observed landscapes are assumed to be realisations of a Gibbs model. With respect to the reference measure (3.17),

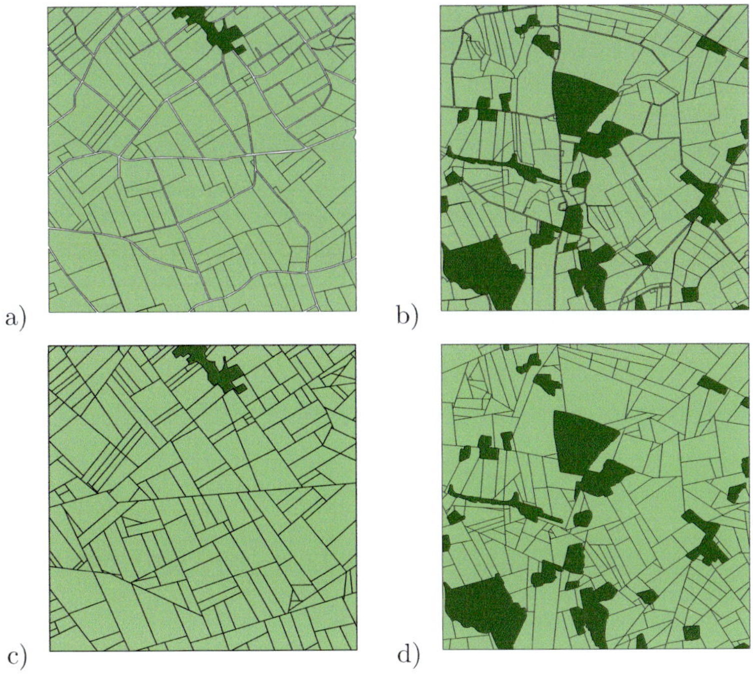

FIGURE 7.31
Two agricultural landscape patterns in France (first row) and their respective linear T–tesselations counterparts (second row): Selommes (a,c) and Kervidy (b,d).

the probability density of the model is written as

$$p(T|\theta) \propto \exp(-U(T\theta)) = \exp\langle\theta, t(T)\rangle \tag{7.14}$$

where $U(T|\theta)$ is the energy function, $t(T)$ is the sufficient statistics vector and θ is the parameter vector. The energy function is the linear combination of four sufficient statistics

$$U(T|\theta) = -\sum_{i=1}^{4} \theta_i t_i(T).$$

Each component of the sum controls a property of the tesselation morphology:

- number of cells in a tessellation: $\theta_1 t_1(T)$.
- deviation of the current tesselation from tesselations with cells of equal area: $\theta_2 t_2(T)$
- deviation of the current tesselation from rectangular tesselations: $\theta_3 t_3(T)$

- number of elongated cells: $\theta_4 t_4(T)$; a cell is considered elongated if the length/width ratio of the smallest rectangle containing the cell is greater than a given threshold.

The $t_i, 1 \leq i \leq 4$ sufficient statistics are computed from the Ttesselations in Figure 7.31 and used as data to estimate the model parameters through a Monte Carlo maximum likelihood approach. The Algorithm 5.1 used to sample (7.14) is not ergodic. To overcome this drawback, a two-step procedure is used. The first step is to obtain an initial estimate of the parameters through the maximum of the pseudo-likelihood. Then, this first estimate is used as an initial condition for the Monte Carlo MLE of the final values of the parameters.

Let T be an observed T–tesselation and $p(T|\theta)$ an identifiable, hereditary probability density. The log-pseudo-likelihood function of θ given T is defined as

$$\begin{aligned} \log PL(\theta;T) &= \sum_{M \in \mathbb{M}_{\mathbf{T}}} \log \frac{p(M^{-1}(T)|\theta)}{p(T|\theta)} - \int_{\mathbb{S}_T} \frac{p(S(T)|\theta)}{p(T|\theta)} dS \quad (7.15) \\ &+ \sum_{F \in \mathbb{F}_{\mathbf{T}}} \log \frac{p(F^{-1}(T)|\theta)}{p(T|\theta)} - \int_{\mathbb{F}_T} \frac{p(F(T)|\theta)}{p(T|\theta)} dF. \end{aligned}$$

The maximum pseudo-likelihood estimate of θ is then

$$\hat{\theta}_{\text{MPL}} = \arg\max_{\theta \in \Theta} \log PL(\theta;T). \tag{7.16}$$

The introduction of (7.14) in (7.15) leads to

$$\begin{aligned} \log PL(\theta;T) &= -\theta^T \sum_{M \in \mathbb{M}_{\mathbf{T}}} \Delta t(M,T) - \int_{\mathbb{S}_T} \exp\left(\theta^T \Delta t(S,T)\right) dS \\ &+ \theta^T \sum_{f \in \mathbb{F}_{\mathbf{T}}} \Delta t(F,T) - \sum_{f \in \mathbb{F}_T} \exp\left(\theta^T \Delta t(F,T)\right) \end{aligned} \tag{7.17}$$

where

$$\begin{aligned} \Delta t(M,T) &= t(M(T)) - t(T), \\ \Delta t(S,T) &= t(S(T)) - t(T), \\ \Delta t(F,T) &= t(F(T)) - t(T) \end{aligned}$$

represent the variations of the sufficient statistics vector induced by each considered possible transformation of the observed tesselation T.

The computation of (7.16) requires the approximation of the integral in (7.17). For this purpose, an algorithm that alternates between Newton optimisation scheme updates and refinements of the approximation of the integral over the set of splits is presented in [158, 1].

TABLE 7.10
Estimation results for the T–tesselation Gibbs model fitted to the observed landscapes.

Model parameter	Landscape	$\hat{\theta}_{MPL}$	$\hat{\theta}_n$	SE($\hat{\theta}$)	MCSE
θ_1 (number of cells)	Selommes	1.62	1.99	0.12	$1.35*10^{-6}$
	Kervidy	1.43	1.70	0.11	$1.05*10^{-6}$
θ_2 (areas)	Selommes	68.36	181.19	66.48	0.17
	Kervidy	23.73	79.19	56.34	0.20
θ_3 (angles)	Selommes	0.95	2.23	0.15	$3.28*10^{-6}$
	Kervidy	0.77	1.26	0.12	$1.59*10^{-6}$
θ_4 (elongation)	Selommes	0.48	0.28	0.17	$2.93*10^{-6}$
	Kervidy	0.31	0.07	0.15	$3.44*10^{-6}$

Table 7.10 shows the maximum pseudo-likelihood estimates $\hat{\theta}_{MPL}$ of the parameters for the Gibbs model (7.14) fitted to the two landscapes. Using these values as initial conditions, the maximum likelihood estimates $\hat{\theta}_n$ are computed. The standard error of the maximum likelihood estimate and the Monte Carlo standard error are also given.

A comparison was made between MPL (maximum pseudo-likelihood) and MCML (Monte Carlo maximum likelihood) estimates for the Selommes landscape. The results are shown in Figure 7.32. For the parameters estimated by each method, a series of 500 simulations were generated. Based on these, the distributions of the sufficient statistics are compared visually using box plots (grey – MPL, white – MCML). The values of the sufficient statistics used to estimate the model parameters are also plotted (red horizontal lines). The results clearly show a better behaviour of the MCML estimates, which allow to reproduce landscape patterns with average characteristics similar to the observations.

The model was verified using global envelope tests [235]. Figure 7.33 shows the results obtained for the summary statistics F, G and L computed from the centroids of the tesselations. The envelope tests were carried out using 500 simulations. The $p-$ values obtained for the tests based on these statistics allow us to conclude that the observations can be considered as a realisation of the fitted model.

The analysis presented in [1] considered several summary statistics to validate the fitted model. Among them, we mention the distribution of the direction of the vector connecting the centroid of a cell with its nearest neighbour [12]. This statistic also allows the detection of anisotropies. Furthermore,

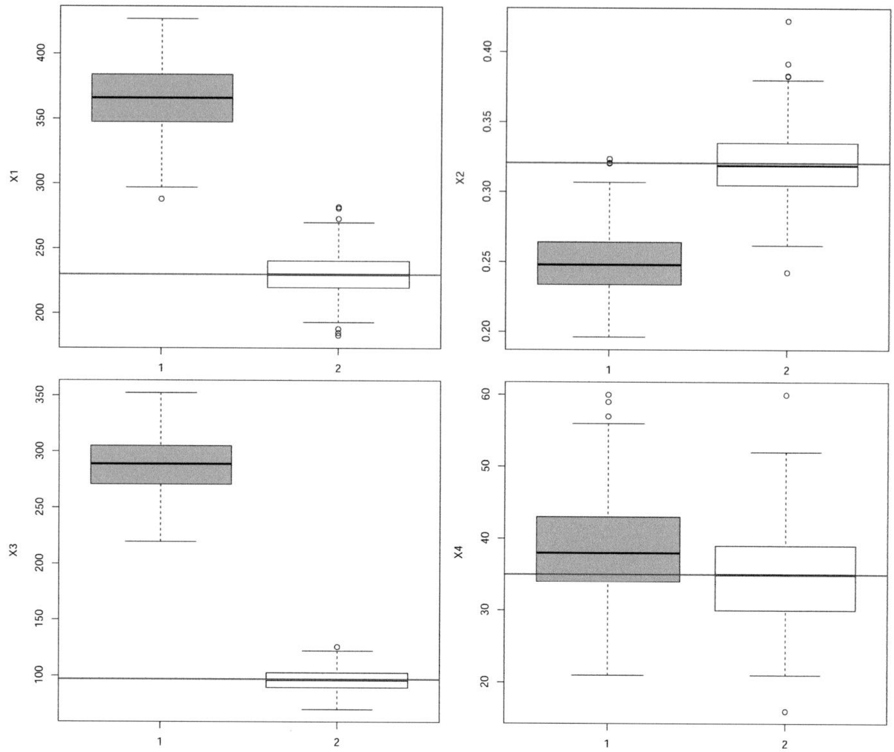

FIGURE 7.32
Comparison of the MPL and MCML estimates for the Selommes landscapes through their sufficient statistics distribution.

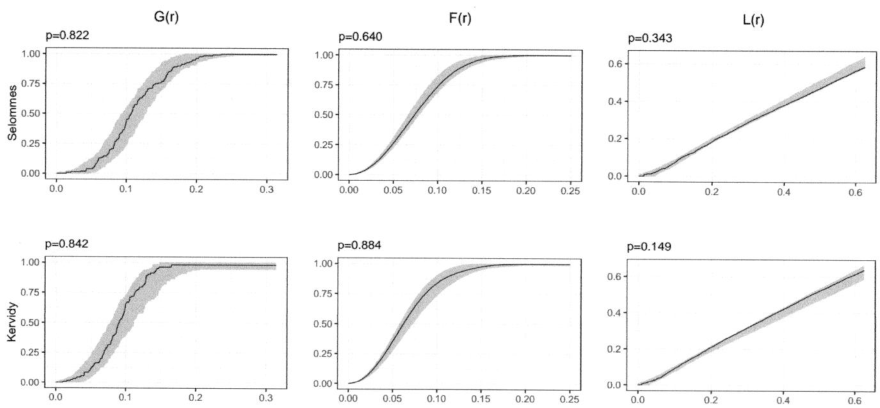

FIGURE 7.33
Estimates of summary statistics (black lines) calculated for cell centroids of the two landscapes.

geometric characteristics of a cell, such as area, perimeter, width and length of the smallest rectangle containing it, and isoperimetric quotient, were taken into account to construct empirical distribution functions to be integrated into global envelope tests. For the two landscapes considered, with the exception of the isoperimetric quotient, all the tests constructed using the aforementioned statistics did not allow the null hypothesis to be rejected, thus validating the fitted model. These tests also rejected the model for a third landscape pattern with high anisotropies. In this context, Gibbs T–tesselation models appear to be a modelling tool capable of capturing and reproducing important features of agricultural landscapes. Further related research directions should include the improvement of the convergence of their simulation dynamics and the construction of model validation procedures.

A

Appendices

A.1 DRlib C++ library

DRlib is an object oriented **C++** library for point pattern analysis. It provides programming code for building point process models, together with simulation algorithms and statistical inference procedures. Among the existing software libraries for point pattern analysis, the main reference is perhaps the **R** library **spatstat** [18, 12].

The **DRlib** library project is complementary to the existing tools in the sense that it provides the user with accessible, reliable and efficient **C++** code that allows intensive simulation-based Bayesian inference for marked Gibbs point processes with multiple interactions. The construction of a related **Python** library is also under development.

The origin of the **DRlib** project has its roots in the **MPPLIB C++** library developed by [193, 194, 195, 281]. The **MPPLIB** library is mainly dedicated to the implementation of exact simulation algorithms for marked point processes with interactions. The main contributors were: A. Steenbeck, M. N. M. van Lieshout, R. S. Stoica, P. Gregori, K. Berthelsen.

The current version of the **DRlib** library is freely available at:

`https://gitlab.univ-lorraine.fr/labos/iecl/drlib`

The **DRlib** is an undergoing project. The present contributors are: D. Gemmerlé, R. S. Stoica, D. Astaburuaga, N. Gillot and C. Reype.

DOI: 10.1201/9781003380382-A

A.2 MHDeath procedure

Here, the pseudo-code and the proof of the algorithm given by [151, 152, 182] are presented. This algorithm allows to compute only the part of permutations necessary for the simulation within the Exact Metropolis-Hastings algorithm in Section 5.2.2.

Algorithm A.1.
$\eta_{\boldsymbol{y}} = MHDeath(\boldsymbol{x}, \boldsymbol{y}, \eta_x, \eta_z)$

Let $\boldsymbol{x} \subseteq \boldsymbol{y} \subseteq \boldsymbol{z}$ *be three objects configurations, and let* $\{\eta_x\} \subseteq \boldsymbol{x}$, $\{\eta_y\} \subseteq \boldsymbol{y}$ *and* $\{\eta_z\} \subseteq \boldsymbol{z}$ *the candidates to be deleted from their respective configurations. If* $\eta_w\} = \emptyset$, *this means that there is no candidate for death retained in the "population"* w *(with* $w = \boldsymbol{x}, \boldsymbol{y}$ *or* $\boldsymbol{z}$*). Suppose that*

η_z *is chosen uniformly in* $\boldsymbol{z}$,
$\eta_z \in \boldsymbol{y} \Rightarrow \eta_y = \eta_z$ *if* η_z *exists,*
$\eta_y \in \boldsymbol{x} \Rightarrow \eta_x = \eta_y$ *is* η_y *exists,*
a death is proposed from $\boldsymbol{x}$, *then* η_x *is chosen uniformly in* $\boldsymbol{x}$.

Conditionally to the values $(\boldsymbol{x}, \boldsymbol{y}, \eta_x, \eta_z)$ *and to the fact that a death is proposed for* $\boldsymbol{y}$, *the candidate* η_y *is chosen uniformly in* $\boldsymbol{y}$ *by the following mechanism:*

$$\eta_y = \begin{cases} \eta_z & \textit{if } \eta_z \in \boldsymbol{y} \\ \eta_x & \textit{if } \eta_z \notin \boldsymbol{y} \textit{ and } \omega \leq n(\boldsymbol{x})/n(\boldsymbol{y}) \\ \eta'_y & \textit{otherwise} \end{cases}$$

with ω *uniform in* $[0, 1]$, η'_y *chosen uniformly in* $\boldsymbol{y} \setminus \boldsymbol{x}$ *and by considering* $n(\boldsymbol{x})/n(\boldsymbol{y}) = 0$ *if* $n(\boldsymbol{y}) = 0$.

Proof. We show that under the previous assumptions, Algorithm A.1 chooses η_y uniformly over $\boldsymbol{y}$ [182]. To do this let

$$\begin{aligned} &\mathbb{P}(\eta_y = y_0) = \\ &= \mathbb{P}(\eta_y = y_0 | \eta_z \in \boldsymbol{y})\mathbb{P}(\eta_z \in \boldsymbol{y}) + \mathbb{P}(\eta_y = y_0 | \eta_z \notin \boldsymbol{y})\mathbb{P}(\eta_z \notin \boldsymbol{y}) \end{aligned}$$

For the first term in the right side of the previous equation, we have

$$\mathbb{P}(\eta_y = y_0 | \eta_z \in \boldsymbol{y})\mathbb{P}(\eta_z \in \boldsymbol{y}) = \mathbb{P}(\eta_z = y_0) = \frac{1}{n(\boldsymbol{z})},$$

while for the second, we obtain

$$
\begin{aligned}
&\mathbb{P}(\eta_y = y_0 | \eta_z \notin \boldsymbol{y})\mathbb{P}(\eta_z \notin \boldsymbol{y}) = \\
&= \mathbb{P}(\eta_y = y_0 | \eta_z \notin \boldsymbol{y}, \omega \leq n(\boldsymbol{x})/n(\boldsymbol{y}))\mathbb{P}(\omega \leq n(\boldsymbol{x})/n(\boldsymbol{y}) | \eta_z \notin \boldsymbol{y})\mathbb{P}(\eta_z \notin \boldsymbol{y}) + \\
&+ \mathbb{P}(\eta_y = y_0 | \eta_z \notin \boldsymbol{y}, \omega > n(\boldsymbol{x})/n(\boldsymbol{y}))\mathbb{P}(\omega > n(\boldsymbol{x})/n(\boldsymbol{y}) | \eta_z \notin \boldsymbol{y})\mathbb{P}(\eta_z \notin \boldsymbol{y}) \\
&= \mathbb{P}(\eta_y = y_0 | \eta_z \notin \boldsymbol{y}, \omega \leq n(\boldsymbol{x})/n(\boldsymbol{y}))\frac{n(\boldsymbol{x})}{n(\boldsymbol{y})}\frac{n(\boldsymbol{z} \setminus \boldsymbol{y})}{n(\boldsymbol{z})} + \\
&+ \mathbb{P}(\eta_y = y_0 | \eta_z \notin \boldsymbol{y}, \omega > n(\boldsymbol{x})/n(\boldsymbol{y}))\frac{n\boldsymbol{y} \setminus (\boldsymbol{x})}{n(\boldsymbol{y})}\frac{n(\boldsymbol{z} \setminus \boldsymbol{y})}{n(\boldsymbol{z})}.
\end{aligned}
$$

Now, if $y_0 \in \boldsymbol{x}$ we have

$$\mathbb{P}(\eta_y = y_0 | \eta_z \notin \boldsymbol{y}, \omega > n(\boldsymbol{x})/n(\boldsymbol{y})) = 0,$$

hence

$$
\begin{aligned}
&\mathbb{P}(\eta_y = y_0 | \eta_z \notin \boldsymbol{y}, \omega \leq n(\boldsymbol{x})/n(\boldsymbol{y}))\frac{n(\boldsymbol{x})}{n(\boldsymbol{y})}\frac{n(\boldsymbol{z} \setminus \boldsymbol{y})}{n(\boldsymbol{z})} \\
&= \frac{1}{n(\boldsymbol{x})}\frac{n(\boldsymbol{x})}{n(\boldsymbol{y})}\frac{n(\boldsymbol{z} \setminus \boldsymbol{y})}{n(\boldsymbol{z})} \\
&= \frac{n(\boldsymbol{z} \setminus \boldsymbol{y})}{n(\boldsymbol{z})n(\boldsymbol{y})}.
\end{aligned}
$$

If $y_0 \notin \boldsymbol{x}$

$$\mathbb{P}(\eta_y = y_0 | \eta_z \notin \boldsymbol{y}, \omega \leq n(\boldsymbol{x})/n(\boldsymbol{y})) = 0,$$

and this gives

$$
\begin{aligned}
&\mathbb{P}(\eta_y = y_0 | \eta_z \notin \boldsymbol{y}, \omega > n(\boldsymbol{x})/n(\boldsymbol{y}))\frac{n(\boldsymbol{x})}{n(\boldsymbol{y})}\frac{n(\boldsymbol{z} \setminus \boldsymbol{y})}{n(\boldsymbol{z})} = \\
&= \frac{1}{n(\boldsymbol{y} \setminus \boldsymbol{x})}\frac{n(\boldsymbol{y} \setminus \boldsymbol{x})}{n(\boldsymbol{y})}\frac{n(\boldsymbol{z} \setminus \boldsymbol{y})}{n(\boldsymbol{z})} \\
&= \frac{n(\boldsymbol{z} \setminus \boldsymbol{y})}{n(\boldsymbol{z})n(\boldsymbol{y})}.
\end{aligned}
$$

By taking the sum of all the terms, the result is obtained:

$$\mathbb{P}(\eta_y = y_0) = \frac{1}{n(\boldsymbol{z})} + \frac{n(\boldsymbol{z} \setminus \boldsymbol{y})}{n(\boldsymbol{z})n(\boldsymbol{y})} = \frac{1}{n(\boldsymbol{y})}.$$

□

Bibliography

[1] K. Adamczyk-Chauvat, M. Kassa, J. Papaïx Kiên Kiêu, and R. S. Stoica. Statistical inference for random T-tessellations models. Application to agricultural landscape modeling. *Annals of the Institute of Statistical Mathematics*, 76(3):447–479, 2024.

[2] T. Arak, P. Clifford, and D. Surgailis. Point-based polygonal models for random graphs. *Advances in Applied Probability*, 25(2):348–372, 1993.

[3] D. Astaburuaga, R. S. Stoica, D. Gemmerlé, and F. Cuevas Pacheco. Inhomogeneous interacting marked point processes for studying morphostructures in paleobiological data. In *Proceedings RING*. Preprint HAL, hal-04653458, 2024.

[4] Y. F. Atchadé, N. Lartillot, and C. P. Robert. Bayesian computation for statistical models with intractable normalizing constants. *Brazilian Journal of Probability and Statistics*, 27(4):416–436, 2013.

[5] K. B. Athreya, H. Doss, and J. Sethuraman. On the convergence of the Markov chain simulation method. *The Annals of Statistics*, 24:69–100, 1996.

[6] G. J. Babu and E. D. Feigelson. Spatial point processes in astronomy. *Journal of Statistical Planning and Inference*, 50:311–326, 1996.

[7] F. Baccelli and B. Blaszczyszyn. *Stochastic Geometry and Wireless Networks: Volume I Theory*. Now Publishers, 2010.

[8] F. Baccelli and B. Blaszczyszyn. *Stochastic Geometry and Wireless Networks: Volume II Applications*. Now Publishers, 2010.

[9] F. Baccelli and P. Bremaud. *Elements of Queueing Theory. Palm Martingale Calculus and Stochastic Recurrences. Second Edition*. Springer, 2003.

[10] A. Baddeley, I. Bárány, R. Schneider, and W. Weil. *Stochastic Geometry*. Lecture Notes in Mathematics, Springer, 2007.

[11] A. Baddeley, P. Gregori, J. Mateu, R. Stoica, and D. Stoyan (eds.). *Case Studies in Spatial Point Process Modeling*. Lecture Notes in Statistics, Springer, 2006.

[12] A. Baddeley, E. Rubak, and R. Turner. *Spatial Point Patterns. Methodology and Applications with R.* Chapman and Hall/CRC, 2016.

[13] A. J. Baddeley. Spatial sampling and censoring. In O. Barndorff-Nielsen, W.S. Kendall, and M.N.M. van Lieshout, editors, *Stochastic Geometry, Likelihood and Computation.* CRC Press/Chapman and Hall, Boca Raton, 1999.

[14] A. J. Baddeley and I. S. Molchanov. Averaging of random sets based on their distance functions. *Journal of Mathematical Imaging and Vision*, 8:79–92, 1998.

[15] A. J. Baddeley and J. Møller. Nearest-neighbour Markov point processes and random sets. *International Statistical Review*, 57(2):89–121, 1989.

[16] A. J. Baddeley, J. Møller, and A. G. Pakes. Properties of residual for spatial point processes. *Annals of the Institute of Statistical Mathematics*, 60:627–649, 2008.

[17] A. J. Baddeley and B. W. Silverman. A cautionary example on the use of second-order methods for analyzing point patterns. *Biometrics*, 40:1089–1093, 1984.

[18] A. J. Baddeley and R. Turner. **spatstat** : an **r** package for analyzing spatial point patterns. *Journal of Statistical Software*, 12(6):1–42, 2005.

[19] A. J. Baddeley, R. Turner, J. Møller, and M. Hazelton. Residual analysis for spatial point processes. *Journal of the Royal Statistical Society, Series B*, 67(5):617–666, 2005.

[20] A. J. Baddeley and T. R. Turner. Practical maximum pseudolikelihood for spatial point patterns (with discussion). *Australian and New Zealand Journal of Statistics*, 42(3):283–322, 2000.

[21] A. J. Baddeley and M. N. M. van Lieshout. Area-interaction point processes. *Annals of the Institute of Statistical Mathematics*, 47:601–619, 1995.

[22] R. Bardenet, F. Lavancier, X. Mary, and A. Vasseur. On a few statistical applications of determinantal point processes. *EASAIM: Proceedings and Surveys*, 60:180–202, 2017.

[23] A. A. Barker. Monte Carlo calculations of the radial distribution function for a proton electron plasma. *Australian Journal of Physics*, 18:119–133, 1965.

[24] O. E. Barndorff-Nielsen, W. S. Kendall, and M. N. M. van Lieshout (eds.). *Stochastic Geometry Likelihood and Computation.* Chapman and Hall/CRC, Boca Raton, 1999.

[25] N. Bartoli and P. Del Moral. *Simulation et algorithmes stochastiques. Une introduction avec applications.* Cépaduès-Editions, 2001.

[26] M. A. Beaumont, J.-M. Cornuet, C. P. Robert, and J.-M. Marin. Adaptative approximate bayesian computation. *Biometrika*, 96(4):983–990, 2009.

[27] R. Beck, L. Dobos, T. Budavári, A. S. Szalay, and I. Csabai. Photometric redshifts for the SDSS Data Release 12. *Monthly Notices of the Royal Astronomical Society*, 460(2):1371–1381, 2016.

[28] T. Bedford and J. Van Den Berg. A remark on the Van Lieshout and Baddeley J-function for point processes. *Advances in Applied Probability*, 29(1):19–25, 1997.

[29] M. Berman and T. R. Turner. Approximating point process likelihoods with glim. *Journal of the Royal Statistical Society. Series C (Applied Statistics)*, 41:31–38, 1992.

[30] K. K. Berthelsen and J. Møller. A primer on perfect simulation for spatial point processes. *Bulletin of the Brazilian Mathematical Society*, 33:351–367, 2002.

[31] K. K. Berthelsen and J. Møller. Likelihood and non-parametric Bayesian MCMC inference for spatial point processes based on perfect simulation and path sampling. *Scandinavian Journal of Statistics*, 30:549–564, 2003.

[32] J. Besag. Spatial interaction and the statistical analysis of lattice systems. *Journal of the Royal Statistical Society, Series B*, 36(2):192–236, 1974.

[33] J. Besag. Statistical analysis of non-lattice data. *The Statistician*, 24(3):179–195, 1975.

[34] J. Besag. Some methods of statistical analysis for spatial data. *Bulletin International Statistical Institute*, 47:77–92, 1977.

[35] J. Besag, R. Milne, and S. Zachary. Point process limits of lattice processes. *Journal of Applied Probability*, 19(1):210–216, 1982.

[36] G. Biau, F. Cérou, and A. Guyader. New insights into approximate bayesian computation. *Annales de l'Institut Henri Poincaré, Probabilités et Statistiques*, 51(1):376–403, 2015.

[37] P. Billingsley. *Probability and Measure.* John Wiley and Sons, 1995.

[38] J.-M. Billiot, J.-F. Coeurjolly, and R. Drouilhet. Maximum pseudolikelihood estimator for exponential family models of marked gibbs point processes. *Electronic Journal of Statistics*, 2:234–264, 2008.

[39] C. A. N. Biscio and F. Lavancier. Quantifying repulsiveness of determinantal point processes. *Bernoulli*, 22(4):2001–2028, 2016.

[40] C. A. N. Biscio and F. Lavancier. Contrast estimation for parametric stationary determinantal point processes. *Scandinavian Journal of Statistics*, 44:204–229, 2017.

[41] M. G. B. Blum. Approximate bayesian computation: A nonparametric perspective. *Journal of the American Statistical Association*, 105(491):1178–1187, 2010.

[42] François Bonneau and Dietrich Stoyan. Directional pair-correlation analysis of fracture networks. *Journal of Geophysical Research: Solid Earth*, 127(9):1–31, 2022.

[43] A. Brix and J. Chadoeuf. Spatio-temporal modelling of weeds by shot-noise g cox processes. *Biometrical Journal*, 44(1):83–99, 2001.

[44] A. Brix and J. Møller. Space-time multi type log gaussian cox processes with a view to modelling weeds. *Scandinavian Journal of Statistics*, 28(3):471–488, 2001.

[45] R. F. Brøndum and E. Rubak. *Stochastic channel modelling: a Bayesian approach using reversible jump Markov chain Monte Carlo methods.* Master thesis - University of Aalborg, 2006.

[46] P. Brémaud. *Point Processes and Queues. Martingale Dynamics.* Springer, 1981.

[47] P. Brémaud. *Point Process Calculus in Time and Space. An Introduction with Applications.* Springer, 2020.

[48] A. Caimo and N. Friel. Bayesian inference for exponential random graph models. *Social Networks*, 33:41–55, 2011.

[49] Y. Capowiez, A. Pierret, O. Daniel, P. Monestiez, and A. Kretzschmar. 3D skeleton reconstructions of natural earthworm burrow systems using CAT scan images of soil cores. *Biology and Fertility of Soils*, 27:51–59, 1998.

[50] P. Cattiaux, L. Colombani, and M. Costa. Limit theorems for Hawkes processes including inhibition. *Stochastic Processes and their Applications*, 149:404–426, 2022.

[51] J. Chadoeuf, A. Kretzschmar, M. Goulard, and K. R. J. Smettem. Analyse de la liaison entre un processus de galeries et la fissuration d'un sol. *Biometrics*, 49(4):1237–1244, 1993.

[52] J. Chadoeuf, A. Kretzschmar, M. Goulard, and K. R. J. Smettem. Description of the spatial interaction between earthworm burrows and cracks at the soil surface. In A. J. Ringrose-Voase and G. S. Humphreys, editors, *Soil Micromorphology: Studies in Management and Genesis. Proceedings IX International Working Meeting on Soil Micromorphology*, volume 22 of *Developments in Soil Science*, pages 521–530, Amsterdam, 1994. Elsevier.

[53] Y. C. Chin and A. J. Baddeley. On connected component Markov point processes. *Advances in Applied Probability (SGSA)*, 31:279–282, 1999.

[54] S. N. Chiu, D. Stoyan, W. S. Kendall, and J. Mecke. *Stochastic Geometry and its Applications. Third Edition.* John Wiley and Sons, 2013.

[55] P. G. Ciarlet. *Introduction à l'analyse numérique matricielle et à l'optimisation.* Masson, Paris, 1994.

[56] J.-F. Coeurjolly, D. Dereudre, R. Drouilhet, and F. Lavancier. Takacs–Fiksel method for stationary marked Gibbs point processes. *Scandinavian Journal of Statistics*, 39(3):416–443, 2012.

[57] J.-F. Coeurjolly and R. Drouilhet. Asymptotic properties of the maximum pseudo-likelihood estimator for stationary Gibbs point processes including the Lennard-Jones model. *Electronic Journal of Statistics*, 4:677–706, 2010.

[58] J.-F. Coeurjolly and F. Lavancier. Residuals and goodness-of-fit tests for stationary marked gibbs point processes. *Journal of the Royal Statistical Society, Series B*, 75(2):247–276, 2013.

[59] P. Coles and B. Jones. A lognormal model for the cosmological mass distribution. *Monthly Notices of the Royal Astronomical Society*, 248(1):1–13, 1991.

[60] D. Coupier and D. Dereudre. Continuum percolation for quermass interaction model. *Electronic Journal of Probability*, 19:1–19, 2014.

[61] O. Cronie and M. N. M. van Lieshout. A J-function for inhomogeneous spatio-temporal point processes. *Scandinavian Journal of Statistics*, 42(2):562–579, 2015.

[62] O. Cronie and M. N. M. van Lieshout. Summary statistics for inhomogeneous marked point processes. *Annals of the Institute of Statistical Mathematics*, 68(4):905–928, 2016.

[63] Ottmar Cronie, Julia Jansson, and Konstantinos Konstantinou. Discussion of the paper "Marked spatial point processes: current state and extensions to point processes on linear networks". *Journal of Agricultural, Biological and Environmental Statistics*, 29:379–388, 2024.

[64] A. Cuevas and R. Fraiman. A plug-in approach to support estimation. *The Annals of Statistics*, 25:2300–2312, 1997.

[65] A. Cuevas, W. Gonzalez-Manteiga, and A. Rodriguez-Casal. Plug-in estimation of general level sets. *Australian & New Zealand Journal of Statistics*, 48:7–19, 2006.

[66] D. J. Daley and G. Last. Descending chains, the lilypond model, and mutual-nearest-neighbour matching. *Advances in Applied Probability (SGSA)*, 37(3):604–628, 2005.

[67] D. J. Daley, H. Stoyan, and D. Stoyan. Fraction of a poisson germ model with maximally non-overlapping spherical grains. *Advances in Applied Probability (SGSA)*, 31(3):610–624, 1999.

[68] D. J. Daley and D. Vere-Jones. *An Introduction to the Theory of Point Processes, Volume 1 : Elementary Theory and Methods, Second Edition.* Springer, 2003.

[69] D. J. Daley and D. Vere-Jones. *An Introduction to the Theory of Point Processes, Volume 2 : General Theory and Structure, Second Edition.* Springer, 2008.

[70] H. A. David. *Order Statistics.* John Wiley and Sons, 1981.

[71] Yu. Davydov and A. Illig. Ergodic properties of crystallization processes. *Journal of Mathematical Sciences*, 163(4):375–381, 2009.

[72] Yu. Davydov and V. Paulauskas. On the estimation of the parameters of multivariate stable distributions. *Acta Applicandae Mathematicae*, 58:107–124, 1999.

[73] Yu. Davydov and V. Paulauskas. Recent results on p-stable convex compact sets with applications. *Fields Institute Communications*, 44:127–139, 2004.

[74] B. Delyon, M. Lavielle, and E. Moulines. Convergence of a stochastic approximation version of the EM algorithm. *The Annals of Statistics*, 27(1):94–128, 1999.

[75] A. P. Dempster, N. M. Laird, and D. B. Rubin. Maximum likelihood from incomplete data via the EM algorithm. *Journal of the Royal Statistical Society, Series B*, 39(1):1–38, 1977.

[76] D. Dereudre. The existence of quermass-interaction processes for nonlocally stable interaction and nonbounded convex grains. *Advances in Applied Probability*, 41(3):664–681, 2009.

[77] D. Dereudre. Variational principle for gibbs point processes with finite range interaction. *Electronic Communications in Probability*, 21(10):1–11, 2016.

[78] D. Dereudre. Introduction to the theory of Gibbs point processes. In D. Coupier, editor, *Stochastic Geometry Modern Research Frontiers*. Springer, 2019.

[79] D. Dereudre, F. Lavancier, and K. S. Helisová. Estimation of the intensity parameter of the germ-grain quermass-interaction model when the number of germs is not observed. *Scandinavian Journal of Statistics*, 41:809–829, 2014.

[80] X. Descombes, R. D. Morris, J. Zerubia, and M. Berthod. Estimation of Markov random field prior parameters using Markov chain Monte Carlo maximum likelihood. *IEEE Transactions on Image Processing*, 8:954–963, 1999.

[81] X. Descombes, R. S. Stoica, L. Garcin, and J. Zerubia. A RJMCMC algorithm for object processes in image processing. *Monte Carlo Methods and Applications*, 7(1-2):149–156, 2001.

[82] X. Descombes, R. S. Stoica, and J. Zerubia. Two Markov point processes for simulating line networks. In *International Conference on Image Processing Proceedings*, Kobe, 1999. IEEE.

[83] X. Descombes, M. N. M. van Lieshout, R. S. Stoica, and J. Zerubia. Parameter estimation by a Markov chain Monte Carlo technique for the candy model. In *Statistical Signal Processing Proceedings*, Singapore, 2001. IEEE.

[84] P. J. Diggle. *Statistical Analysis of Spatial and Spatio-Temporal Point Patterns, 3rd Edition*. Chapman and Hall, 2014.

[85] P. J. Diggle, T. Fiksel, P. Grabarnik, Y. Ogata, D. Stoyan, and M. Tanemura. On parameter estimation for pairwise interaction point processes. *International Statistical Review*, 62(1), 1994.

[86] M. Eckardt and M. Moradi. Marked spatial point processes: current state and extensions to point processes on linear networks. *Journal of Agricultural, Biological and Environmental Statistics*, 29:346–378, 2024.

[87] M. Eckardt and M. Moradi. Rejoinder on "Marked spatial point processes: current state and extensions to point processes on linear networks". *Journal of Agricultural, Biological and Environmental Statistics*, 29:405–416, 2024.

[88] D. Coupier (ed.). *Stochastic Geometry Modern Research Frontiers*. Springer, 2019.

[89] X. Descombes (ed.). *Stochastic Geometry for Image Analysis (Digital Signal and Image Processing series)*. John Wiley and Sons, 2012.

[90] M. Einasto, L.J. Liivamägi, E. Tempel, E. Saar, J. Vennik, P. Nurmi, M. Gramann, J. Einasto, E. Tago, P. Heinämäki, A. Ahvensalmi, and V.J. Martinez. Multimodality of rich clusters from the sdss dr8 within the supercluster-void network. *Astronomy and Astrophysics*, 542, 2012.

[91] R. J. Erskine. Mastitis control in dairy herds. In O.M. Radostis, editor, *Herd Health : Food Animal Production Medecine*. W.B. Saunders Company, Philadelphia, 2001.

[92] E. Everhart. An efficient integrator that uses Gauss-Radau spacings. In A. Carusi and G.-B. Valsecchi, editors, *Dynamics of Comets: Their Origin and Evolution*, pages 185–202, 1985.

[93] E. Feigelson and G. Babu. *Modern Statistical Methods for Astronomy With R Applications*. Cambridge University Press, 2012.

[94] P. A. Ferrari, R. Fernandez, and N. L. Garcia. Perfect simulation for interacting point processes, loss networks and Ising models. *Stochastic Processes and their Applications*, 102:63–88, 2002.

[95] T. Fiksel. Estimation of interaction potentials of Gibbsian point processes. *Statistics*, 19(1):77–86, 1988.

[96] D. Flipo. *Chaînes de Markov. Agrégation de Mathématiques*. Cours polycopié, Université Lille 1, 2007.

[97] G. Fort, E. Moulines, and P. Priouret. Convergence of adaptive and interacting Markov chain Monte Carlo algorithms. *The Annals of Statistics*, 39(6):3262–3289, 2011.

[98] R. Foxall and A. J. Baddeley. Nonparametric measures of association between a spatial point process and a random set, with geological applications. *Journal of the Royal Statistical Society. Series C (Applied Statistics)*, 51(2):165–182, 2002.

[99] O. Frank and D. Strauss. Markov graphs. *Journal of the American Statistical Association*, 81(395):832–842, 1986.

[100] E. Gabriel and P. J. Diggle. Second-order analysis of inhomogeneous spatio-temporal point process data. *Statistica Neerlandica*, 63(1):34–51, 2009.

[101] C. Gaetan and X. Guyon. *Spatial Statistics and Modeling*. Springer Series in Statistics, Springer, 2010.

[102] M. A. Gallego, M. V. Ibá nez, and A. Simó. Inhomogeneous k-function for germ-grain models. *Spatial Statistics*, 18:489–504, 2016.

[103] A. E. Gelfand and B. P. Carlin. Maximum-likelihood estimation for constrained- or missing- data models. *Canadian Journal of Statistics*, 21:303–311, 1993.

[104] A. E. Gelfand, P. J. Diggle, M. Fuentes, and P. Guttorp (eds.). *Handbook of Spatial Statistics.* Chapman and Hall/CRC, Boca Raton, 2010.

[105] S. Geman and D. Geman. Stochastic relaxation, gibbs distributions and the bayesian restoration of images. *IEEE Transactions on Pattern Analysis and Machine Intelligence*, 6(6):721–741, 1984.

[106] H.-O. Georgii. *Gibbs Measures and Phase Transitions. Second Edition.* De Gruyter, 2011.

[107] C. J. Geyer. On the convergence of Monte Carlo maximum likelihood calculations. *Journal of Royal Statistical Society, Series B*, 54(1):261–274, 1994.

[108] C. J. Geyer. Likelihood inference for spatial point processes. In O. Barndorff-Nielsen, W.S. Kendall, and M.N.M. van Lieshout, editors, *Stochastic Geometry, Likelihood and Computation.* CRC Press/Chapman and Hall, Boca Raton, 1999.

[109] C. J. Geyer and J. Møller. Simulation procedures and likelihood inference for spatial point processes. *Scandinavian Journal of Statistics*, 21(4):359–373, 1994.

[110] C. J. Geyer and E. A. Thompson. Constrained Monte Carlo maximum likelihood for dependent data. *Journal of Royal Statistical Society, Series B*, 54(3):657–699, 1992.

[111] N. Gillot, R. S. Stoica, and D. Gemmerlé. Study the galaxy distribution characterisation via bayesian statistical learning of spatial marked point processes. In *Proceedings RING.* Preprint HAL, hal-04163649, 2023.

[112] N. Gillot, R. S. Stoica, A. Särkkä, and D. Gemmerlé. Spatial point process modelling and bayesian inference for large data sets. In *Proceedings RING.* Preprint HAL, hal-04645186, 2024.

[113] G. L. Gimel'farb. *Image Textures and Gibbs Random Fields.* Springer, 1999.

[114] M. Goulard, A. Särkkä, and P. Grabarnik. Parameter estimation for marked gibbs point processes through the maximum pseudo-likelihood method. *Scandinavian Journal of Statistics*, 23(3):365–379, 1996.

[115] P. Grabarnik and A. Pommerening. Discussion of the paper "Marked spatial point processes: current state and extensions to point processes on linear networks". *Journal of Agricultural, Biological and Environmental Statistics*, 29:389–391, 2024.

[116] P. J. Green. Reversible jump Markov chain Monte Carlo computation and Bayesian model determination. *Biometrika*, 82(4):711–732, 1995.

[117] P. Gregori, J. Mateu, and R. S. Stoica. A marked point process for modelling three dimensional patterns. In A. Baddeley, P. Gregori, J. Mateu, R. S. Stoica, and D. Stoyan, editors, *Spatial Point Process Modelling and its Applications*, Castellon, Spain, 2004. Publicacions de la Universitat Jaume I.

[118] A. Grelaud, C. P. Robert, J.-M. Marin, F. Rodolphe, and J.-F. Taly. ABC likelihood-free methods for model choice in Gibbs random fields. *Bayesian Analysis*, 4(2):317–336, 2009.

[119] U. Grenander and M. I. Miller. Representations of knowledge in complex systems. *Journal of the Royal Statistical Society, Series B*, 56(4):549–603, 1994.

[120] M. G. Gu and S. Li. A stochastic approximation algorithm for maximum-likelihood estimation with incomplete data. *The Canadian Journal of Statistics*, 26(4):567–582, 1998.

[121] X. Guyon. *Random Fields on a Network Modeling, Statistics and Applications*. Springer, 1995.

[122] H. Haario and E. Saksman. Simulated annealing process in general state space. *Advances in Applied Probability*, 23(4):866–893, 1991.

[123] H. Haario and E. Saksman. Weak convergence of the simulated annealing process in general state space. *Annales Academiae Scientiarum Fennicae*, 17:39–50, 1992.

[124] H. Haario, E. Saksman, and J. Tamminen. An adaptive Metropolis algorithm. *Bernoulli*, 7(2):223–242, 2001.

[125] O. Häggström and K. Nelander. Exact sampling for from anti-monotone systems. *Statistica Neerlandica*, 52(3):360–380, 1998.

[126] O. Häggström and K. Nelander. On exact simulation of Markov random fields using coupling from the past. *Scandinavian Journal of Statistics*, 26(3):395–411, 1999.

[127] O. Häggström, M. N. M. van Lieshout, and J. Møller. Characterization results and Markov chain Monte Carlo algorithms including exact simulation for some spatial point processes. *Bernoulli*, 5(4):641–658, 1999.

[128] B. Hajek. Cooling schedules for optimal annealing. *Mathematics of Operations Research*, 13(2):311–329, 1988.

[129] W. K. Hastings. Monte Carlo sampling methods using Markov chains and their applications. *Biometrika*, 57(1):97–109, 1970.

[130] A. G. Hawkes. Point spectra of some mutually exciting point processes. *Journal of the Royal Statistical Society. Series B (Methodological)*, 33(3):438–443, 1971.

[131] A. G. Hawkes. Spectra of some self-exciting and mutually exciting point processes. *Biometrika*, 58(1):83–90, 1971.

[132] A. G. Hawkes and D. Oakes. A cluster process representation of a self-exciting process. *Journal of Applied Probability*, 11(3):493–503, 1974.

[133] D. Hearn and M. P. Baker. *Computer Graphics. C Version. Second Edition.* Prentice-Hall, 1997.

[134] P. Heinrich, R. S. Stoica, and V. C. Tran. Level sets estimation and vorob'ev expectation of random compact sets. *Spatial Statistics*, 2:47–61, 2012.

[135] G. Hellmund, M. Prokešová, and E. B. Vedel Jensen. Lévy-based cox point processes. *Advances in Applied Probability (SGSA)*, 40(3):603–629, 2008.

[136] J.-L. Henry. Inférence pour les processus ponctuels: ABC shadow. Master's thesis, M2R Maths en action, 2019. Advisors: R. S. Stoica and M. Deaconu (Université de Lorraine, Inria Nancy).

[137] J. Ben Hough, M. Krishnapur, Y. Peres, and B. Virág. Determinantal processes and independence. *Probability Surveys*, 3:206–229, 2006.

[138] D. T. Hristopulos. *Radom Fields for Spatial Data Modeling.* Springer, 2020.

[139] M. L. Huber. *Perfect Simulation.* CRC Press, 2016.

[140] L. Hurtado-Gil, R. S. Stoica, V. J. Martínez, and P. Arnalte-Mur. Morpho-statistical characterisation of the spatial galaxy distribution through gibbs point processes. *Monthly Notices of the Royal Astronomical Society*, 507(2):1710–1722, 2021.

[141] A. Iftimi, O. Cronie, and F. Montes. Second-order analysis of marked inhomogeneous spatiotemporal point processes: Applications to earthquake data. *Scandinavian Journal of Statistics*, 46(3):661–685, 2019.

[142] J. Illian, A. Penttinen, H. Stoyan, and D. Stoyan. *Statistical Analysis and Modelling of Spatial Point Patterns.* John Wiley and Sons, 2008.

[143] J. Jensen and H. R. Kunsch. On asymptotic normality of pseudo likelihood estimates for pairwise interaction processes. *Annals of the Institute of Statistical Mathematics*, 46:475–486, 1994.

[144] J. Jensen and J. Møller. Pseudolikelihood for exponential family models of spatial point processes. *The Annals of Applied Probability*, 1(3):445–461, 1991.

[145] D. Jeulin. *Morphological Models of Random Structures.* Springer, 2021.

[146] B. J. Jones. *Precision Cosmology. The First Half Million Yeas.* Cambridge University Press, 2017.

[147] J. Kahn. Existence of Gibbs measure for a model of T-tessellation. *arXiv:1012.2182v1*, 2010.

[148] O. Kallenberg. An informal guide to the theory of conditioning in point processes. *International Statistical Review*, 52(2):151–164, 1984.

[149] F. P. Kelly and B. D. Ripley. A note on strauss's model for clustering. *Biometrika*, 63:357–360, 1976.

[150] W. S. Kendall. Perfect simulation for the area-interaction point process. In L. Accardi and C. Heyde, editors, *Probability towards 2000.* Springer-Verlag, 1998.

[151] W. S. Kendall and J. Møller. Perfect metropolis-hastings simulation of locally stable point processes. *Research Report R-99-2001, Department of Mathematical Sciences, Aalborg University*, 1999.

[152] W. S. Kendall and J. Møller. Perfect simulation using dominating processes on ordered spaces, with application to locally stable point processes. *Advances in Applied Probability (SGSA)*, 32:844–865, 2000.

[153] W. S. Kendall and G. Montana. Small sets and Markov transition densities. *Stochastic Processes and their Applications*, 99(2):177–194, 2002.

[154] W. S. Kendall, M. N. M. van Lieshout, and A. J. Baddeley. Quermass-interaction processes: conditions for stability. *Advances in Applied Probability*, 31(2):315–342, 1999.

[155] J. F. C. Kingman. *Poisson Processes.* Clarendon Press, Oxford, 1993.

[156] J. F. C. Kingman. Poisson processes revisited. *Probability and Mathematical Statistics*, 26(1):77–95, 2006.

[157] S. Kirkpatrick, C. D. Gelatt Jr., and M. P. Vecchi. Optimization by simulated annealing. *Science*, 220(4598):671–680, 1983.

[158] K. Kiêu and K. Adamczyk-Chauvat. A completely random T-tessellation model and Gibbsian extensions. *Preprint HAL*, hal-02793013, 2015.

[159] K. Kiêu, K. Adamczyk-Chauvat, H. Monod, and R. S. Stoica. A completely random T-tessellation model and Gibbsian extensions. *Spatial Statistics*, 6:118–138, 2013.

[160] F. Kleinschroth, J. R. Healey, S. Gourlet-Fleury, F. Mortier, and R. S. Stoica. Effects of logging on roadless space in intact forest landscapes of the congo basin. *Conservation Biology*, 31(2):469–480, 2017.

[161] S. A. Kovyazin. On the limit behavior of a class of empirical means of a random set. *Theory of Probability and its Applications*, 30(4):814–820, 1986.

[162] A. Kretzschmar and F. Aries. An analysis of the structure of the burrow system of the giant gippsland earthworm megascolides australis McCOY, 1878 using 3D-images. *Soil Biology and Biochemistry*, 24(12):1583–1586, 1992.

[163] M. Kruuse, E. Tempel, R. Kipper, and R. S. Stoica. Photometric redshift galaxies as tracers of the filamentary network. *Astronomy and Astrophysics*, 625(A130):1–13, 2019.

[164] R. Kugel and R. van de Weygaert. Cosmic web dynamics: Forces and strains. *Preprint ArXiv, arXiv:2407.16489*, 2024.

[165] C. Lantuéjoul. *Geostatistical Simulation Models and Algorithms*. Springer-Verlag, Berlin, 2002.

[166] Q. Laporte. *Étude morpho-statistique des réseaux sociaux. Application aux collaborations inter-organisationnelles.* PhD thesis - Université de Lorraine, 2021.

[167] Q. Laporte, R. S. Stoica, M. Clausel, F. Charoy, and G. Oster. Morpho-statistical description of networks through graph modelling and bayesian inference. *IEEE Transactions on Network Science and Engineering*, 9(4):2123–2138, 2022.

[168] G. Last and M. Penrose. *Lectures on the Poisson Process*. Cambridge University Press, 2018.

[169] F. Lavancier and J. Møller. Modelling aggregation on the large scale and regularity on the small scale in spatial point pattern datasets. *Scandinavian Journal of Statistics*, 43(2):587–609, 2016.

[170] F. Lavancier, J. Møller, and E. Rubak. Determinantal point process models and statistical inference. *Journal of Royal Statistical Society: Series B (Statistical Methodology)*, 77(4):853–877, 2015.

[171] F. Lavancier, A. Poinas, and R. Waagepetersen. Adaptive estimating function inference for nonstationary determinantal point processes. *Scandinavian Journal of Statistics*, 48(1):87–107, 2021.

[172] F. Lavancier and E. Rubak. On simulation of continuous determinantal point processes. *Pre-print Arxiv, arXiv:2301.11081*, 2023.

[173] A. B. Lawson and D. G. T. Denison. Spatial cluster modelling: an overview. In A. B. Lawson and D. G. T. Denison, editors, *Spatial Cluster Modelling*. Chapman and Hall/CRC, 2002.

[174] A. B. Lawson and M. Kulldorff. A review of cluster detection methods. In A. Lawson, A. Biggeri, D. Böhning, E. Lesaffre, J-F. Viel, and R. Bertollini, editors, *Disease Mapping and Risk Assessment for Public Health*. John Wiley and Sons, 1999.

[175] D. Li and P. Barmby. Gibbs point process model for young star clusters in M33. *Monthly Notices of the Royal Astronomical Society*, 501(3):3472–3489, 2021.

[176] S. Z. Li. *Markov Random Field Modeling*. Springer-Verlag, London, 2009.

[177] F. Liang. A double Metropolis–Hastings sampler for spatial models with intractable normalizing constants. *Journal of Statistical Computation and Simulation*, 80:1007–1022, 2010.

[178] F. Liang, I. H. Jin, Q. Song, and J. S. Liu. Adaptive exchange algorithm for sampling from Distributions with intractable normalizing constants. *Journal of the American Statistica Association*, 111:377–393, 2016.

[179] N. I. Libeskind, R. van de Weygaert, M. Cautun, B. Falck, E. Tempel, T. Abel, M. Alpaslan, M. A. Aragón-Calvo, J. E. Forero-Romero, R. Gonzalez, S.Gottlöber, O. Hahn, W. A. Hellwing, Y. Hoffman, B. J. T. Jones, F. Kitaura, A. Knebe, S. Manti, M. Neyrinck, S. E. Nuza, N. Padilla, E. Platen, N. Ramachandra, A. Robotham, E. Saar, S. Shandarin, M. Steinmetz, R. S. Stoica, T. Sousbie, and G. Yepes. Tracing the cosmic web. *Monthly Notices of the Royal Astronomical Society*, 473(1):1195–1217, 2018.

[180] M. N. M. van Lieshout. Stochastic annealing for nearest neighbour point processes with application to object recognition. *Advances in Applied Probability*, 26:281–300, 1994.

[181] M. N. M. van Lieshout. *Markov Point Processes and their Applications*. Imperial College Press, London, 2000.

[182] M. N. M. van Lieshout. A note on exact simulation of marked point processes via the Metropolis-Hastings algorithm. *Private Communication, CWI Amsterdam*, 2002.

[183] M. N. M. van Lieshout. A J-function for marked point patterns. *Annals of the Institute of Statistical Mathematics*, 58(4):235–259, 2006.

[184] M. N. M. van Lieshout. A J-function for inhomogeneous point processes. *Statistica Neerlandica*, 65(2):183–201, 2011.

[185] M. N. M. van Lieshout. Nonparametric indices of dependence between components for inhomogeneous multivariate random measures and marked sets. *Scandinavian Journal of Statistics*, 45(4):985–1015, 2018.

[186] M. N. M. van Lieshout. *Spatial Statistics A Concise Introduction.* Chapman and Hall, 2019.

[187] M. N. M. van Lieshout. Discussion of the paper "Marked spatial point processes: current state and extensions to point processes on linear networks". *Journal of Agricultural, Biological and Environmental Statistics*, 29:395–398, 2024.

[188] M. N. M. van Lieshout. Non-parametric adaptive bandwidth selection for kernel estimators of spatial intensity functions. *Annals of the Institute of Statistical Mathematics*, 76:313–331, 2024.

[189] M. N. M. van Lieshout and A. J. Baddeley. A nonparametric measure of spatial interaction in point patterns. *Statistica Neerlandica*, 50(3):344–361, 1996.

[190] M. N. M. van Lieshout and A. J. Baddeley. Indices of dependence between types in multivariate point patterns. *Scandinavian Journal of Statistics*, 26(4):511–532, 1999.

[191] M. N. M. van Lieshout and A. J. Baddeley. Extrapolating and interpolating spatial patterns. In A. B. Lawson and D. G. T. Denison, editors, *Spatial Cluster Modelling.* CRC Press/Chapman and Hall, Boca Raton, 2002.

[192] M. N. M. van Lieshout and R. S. Stoica. The Candy model revisited: properties and inference. *Statistica Neerlandica*, 57:1–30, 2003.

[193] M. N. M. van Lieshout and R. S. Stoica. Perfect simulation for marked point processes. *CWI Report,PNA-R0306*, 2003.

[194] M. N. M. van Lieshout and R. S. Stoica. Exact Metropolis-Hastings sampling for marked point processes using a C++ library. *CWI Report,PNA-R0403*, 2004.

[195] M. N. M. van Lieshout and R. S. Stoica. Perfect simulation for marked point processes. *Computational Statistics and Data Analysis*, 51:679–698, 2006.

[196] M. N. M. van Lieshout and E. W. van Zwet. Exact sampling from conditional Boolean models with applications to maximum likelihood inference. *Advances in Applied Probability*, 33:339–353, 2001.

[197] S. Liu. *Lois stables et processus ponctuels : liens et estimation des paramètres.* PhD thesis - Université Lille 1, 2010.

[198] L. Ljung, G. Pflug, and H. Walk. *Stochastic Approximation and Optimization of Random Systems.* Springer Basel AG, 1992.

[199] O. Macchi. The coincidence approach to stochastic point processes. *Advances in Applied Probability*, 7(1):83–122, 1975.

[200] J.-M. Marin, P. Pudlo, C. P. Robert, and R. J. Ryder. Approximate bayesian computational methods. *Statistics and Computing*, 22(6):1167–1180, 2012.

[201] V. J. Martinez and E. Saar. *Statistics of the Galaxy Distribution.* Chapman and Hall, 2002.

[202] V. J. Martinez, E. Saar, E. Martinez-Gonzalez, and M.-J. Pons-Borderia. *Data Analysis in Cosmology.* Springer, 2009.

[203] S. Mase. Marked gibbs processes and asymptotic normality of maximum pseudo-likelihood estimators. *Mathematische Nachrichten*, 209(1):151–169, 2000.

[204] D. Mason and W. Polonik. Asymptotic normality of plug-in level set estimates. *Annals of Applied Probability*, 19(3):1108–1142, 2009.

[205] J. Mateu and F. Montes. Likelihood inference for Gibbs processes in the analysis of spatial point patterns. *International Statistical Review*, 69(1):81–104, 2001.

[206] J. Mateu and F. Montes. Pseudo-likelihood inference for Gibbs processes with exponential families through generalized linear models. *Statistical Inference for Stochastic Processes*, 4:125–154, 2001.

[207] Georges Matheron. *Random Sets and Integral Geometry.* John Wiley and Sons, 1975.

[208] K. Matthes, J. Kerstan, and J. Mecke. *Infinitely Divisible Point Processes.* John Wiley and Sons, 1978.

[209] P. McCullagh and J. Møller. The permanental process. *Advances in Applied Probability (SGSA)*, 38(4):873–888, 2006.

[210] K. L. Mengersen and R. L. Tweedie. Rates of convergence of the Hastings and Metropolis algorithms. *The Annals of Statistics*, 24(1):101–121, 1996.

[211] N. Metropolis, A. W. Rosenbluth, M. N. Rosenbluth, A. H. Teller, and E. Teller. Equation of state calculations by fast computing machines. *The Journal of Chemical Physics*, 21(6):1087–1092, 1953.

[212] S. Meyn and R. L. Tweedie. *Markov Chains and Stochastics Stability.* Cambridge University Press, 2009.

[213] R. E. Miles and M. S. Mackisack. A large class of random tessellations with the classic Poisson polygon distributions. *Forma*, 17:1–17, 2002.

[214] A. Yu. Mitrophanov. Sensitivity and convergence of uniformly ergodic Markov chains. *Journal of Applied Probability*, 42(4):1003–1014, 2005.

[215] I. Molchanov. Uniform laws of large numbers for empirical associated functionals of random closed sets. *Theory of Probability and its Applications*, 32:556–560, 1987.

[216] I. Molchanov. Empirical estimation of distribution quantiles of random closed sets. *Theory of Probability and its Applications*, 35:594–600, 1990.

[217] I. Molchanov. *Statistics of the Boolean Model for Practitioners and Mathematicians.* John Wiley and Sons, 1997.

[218] I. Molchanov. A limit theorem for solutions of inequalities. *Scandinavian Journal of Statistics*, 25:235–242, 1998.

[219] I. Molchanov. Random closed sets: results and problems. In O. Barndorff-Nielsen, W. S. Kendall, and M. N. M. van Lieshout, editors, *Stochastic Geometry, Likelihood and Computation.* CRC Press/Chapman and Hall, Boca Raton, 1999.

[220] I. Molchanov. *Theory of Random Sets. Second Edition.* Springer, 2017.

[221] J. Møller. Markov chain Monte Carlo and spatial point processes. In O. Barndorff-Nielsen, W. S. Kendall, and M. N. M. van Lieshout, editors, *Stochastic Geometry, Likelihood and Computation.* CRC Press/Chapman and Hall, Boca Raton, 1999.

[222] J. Møller and K. S. Helisová. Power diagrams and interaction processes for unions of discs. *Advances in Applied Probability*, 40(2):321–347, 2008.

[223] J. Møller and K. S. Helisová. Likelihood inference for unions of interacting discs. *Scandinavian Journal of Statistics*, 37:365–381, 2010.

[224] J. Møller, A. N. Pettitt, R. W. Reeves, and K. K. Berthelsen. An efficient Markov chain Monte Carlo method for distributions with intractable normalizing constants. *Biometrika*, 93:451–458, 2006.

[225] J. Møller and J. G. Rasmussen. Perfect simulation of Hawkes processes. *Advances in Applied Probability*, 37(3):629–646, 2005.

[226] J. Møller and J. G. Rasmussen. Approximate simulation of hawkes processes. *Methodology and Computing in Applied Probability*, 8:53–64, 2006.

[227] J. Møller, A. R. Syversveen, and R. P. Waagepetersen. Log Gaussian Cox processes. *Scandinavian Journal of Statistics*, 25(3):451–482, 1998.

[228] J. Møller and G.-L. Torrisi. The pair correlation function of spatial hawkes processes. *Statistics and Probability Letters*, 77(10):995–1003, 2007.

[229] J. Møller and R. P. Waagepetersen. *Statistical Inference and Simulation for Spatial Point Processes*. Chapman and Hall/CRC, Boca Raton, 2004.

[230] A. Monfort. *Cours de statistique mathématique*. Economica, 1997.

[231] R. A. Moyeed and A. J. Baddeley. Stochastic approximation of the MLE for a spatial point pattern. *Scandinavian Journal of Statistics*, 18:39–50, 1991.

[232] D. J. Murdoch and P. J. Green. Exact sampling from continuous state space. *Scandinavian Journal of Statistics*, 25(3):483–502, 1998.

[233] I. Murray, Z. Ghahramani, and D. J. C. MacKay. MCMC for doubly-intractable distributions. In *Proceedings of the 22nd Annual Conference on Uncertainty in Artificial Intelligence*, page 359–366, 2006.

[234] M. Myllymäki, P. Grabarnik, H. Seijo, and D. Stoyan. Deviation test construction and power comparison for marked spatial point patterns. *Spatial Statistics*, pages 19–34, 2015.

[235] M. Myllymäki, T. Mrkvička, P. Grabarnik, H. Seijo, and U. Hahn. Global envelope tests for spatial processes. *Journal of the Royal Statistical Society, Series B*, 79(2):381–404, 2017.

[236] Mari Myllymäki, Tuomas Rajala, and Aila Särkkä. Discussion of the paper "Marked spatial point processes: current state and extensions to point processes on linear networks". *Journal of Agricultural, Biological and Environmental Statistics*, 29:399–404, 2024.

[237] J. Neyman and E. L. Scott. Statistical approach to problems of cosmology. *Journal of the Royal Statistical Society, Series B*, 20(1):1–43, 1958.

[238] E. Nummelin. *General Irreducible Markov Chains and Non-negative Operators*. Cambridge University Press, 1984.

[239] Y. Ogata. The asymptotic behaviour of maximum likelihood estimators for stationary point processes. *Annals of the Institute of Statistical Mathematics*, 30:243–261, 1978.

[240] Y. Ogata. Statistical models for earthquake occurrences and residual analysis for point processes. *Journal of the American Statistical Association*, 83(401):9–27, 1988.

[241] Y. Ogata. Space-time point processes for earthquake occurrences. *Annals of the Institute of Statistical Mathematics*, 50:379–402, 1998.

[242] Y. Ogata. Statistics of earthquake activity: models and methods. *Annual Review of Earth and Planetary Sciences*, 45:497–527, 2017.

[243] Y. Ogata and M. Tanemura. Estimation for interaction potentials of spatial point processes through the maximum likelihood procedure. *Annals of the Institute of Statistical Mathematics*, 33:315–338, 1981.

[244] Y. Ogata and M. Tanemura. Likelihood analysis of spatial point patterns. *Journal of Royal Statistical Society, Series B*, 46:496–518, 1984.

[245] Y. Ogata and M. Tanemura. Estimation of interaction potentials of marked spatial point patterns through the maximum likelihood method. *Biometrics*, 41:421–433, 1985.

[246] Y. Ogata and M. Tanemura. Likelihood estimation of soft-core interaction potentials for gibbsian point patterns. *Annals of the Institute of Statistical Mathematics*, 41:583–600, 1989.

[247] P. J. E. Peebles. *The Large-Scale Structure of the Universe.* Princeton University Press, 1980.

[248] A. Penttinen. Modelling interactions in spatial point patterns: parameter estimation by the maximum likelihood method. *Jyväskylä Studies in Computer Science, Economics and Statistics*, 7:1–105, 1984.

[249] A. Poinas and F. Lavancier. Asymptotic approximation of the likelihood of stationary determinantal point processes. *Scandinavian Journal of Statistics*, 50(2):842–874, 2023.

[250] W. H. Press, B. P. Flannery, and S. A. Teukolsky. *Numerical Recipes in C : The Art of Scientific Computing.* Cambridge University Press, 1988.

[251] C. J. Preston. Spatial birth-and-death processes. *Bulletin of the International Statistical Institute*, 46:371–391, 1977.

[252] J.G. Propp and D.B. Wilson. Exact sampling with coupled Markov chains and applications to statistical mechanics. *Random Structures and Algorithms*, 9:223–252, 1996.

[253] A. Reinhart. A review of self-exciting spatio-temporal point processes and their applications. *Statistical Science*, 33(3):299–318, 2018.

[254] R.-D. Reiss. *A Course on Point Processes.* Springer-Verlag, New-York, 1993.

[255] P. Reynaud-Bouret, V. Rivoirard, and C. Tuleau-Malot. Inference of functional connectivity in Neurosciences via Hawkes processes. In *2013 IEEE Global Conference on Signal and Information*, Austin, Texas, USA, 2013. IEEE.

[256] P. Rigollet and R. Vert. Optimal rates for plug-in estimators of density level sets. *Bernoulli*, 15:1154–1178, 2009.

[257] B. D. Ripley. The second-order analysis of stationary point processes. *Journal of Applied Probability*, 13:255–266, 1976.

[258] B. D. Ripley. Modelling spatial patterns. *Journal of the Royal Statistical Society, Series B*, 39:172–212, 1977.

[259] B. D. Ripley. *Statistical Inference for Spatial Processes.* Cambridge University Press, 1988.

[260] B. D. Ripley. *Pattern Recognition and Neural Networks.* Cambridge University Press, 1996.

[261] B. D. Ripley and F. P. Kelly. Markov point processes. *Journal of the London Mathematical Society*, 15:188–192, 1977.

[262] C. P. Robert and G. Casella. *Monte Carlo Statistical Methods - Second Edition.* Springer, 2004.

[263] G. O. Roberts and A. F. M. Smith. Simple conditions for the convergence of the Gibbssampler and Metropolis-Hastings algorithms. *Stochastic Processes and their Applications*, 49:207–216, 1994.

[264] G. O. Roberts and R. L. Tweedie. Geometric convergence and central limit theorems for multidimensional Hastings and Metropolis algorithms. *Biometrika*, 83(1):95–110, 1996.

[265] L. Roques and F. Hamel. Mathematical analysis of the optimal habitat configurations for species persistence. *Mathematical Biosciences*, 210:34–59, 2007.

[266] L. Roques and R. S. Stoica. Species persistence decreases with habitat fragmentation : an analysis in periodic stochastic environments. *Journal of Mathematical Biology*, 55(2):189–205, 2007.

[267] S. M. Ross. *Simulation - Fifth Edition.* Academic Press, Elsevier, 2013.

[268] D. Ruelle. *Statistical Mechanics: Rigorous Results. Second Edition.* W. A. Benjamin, 1974.

[269] D. Ruelle. *Statistical Mechanics : Rigorous Results.* Imperial College Press, World Scientific Publishing, 1999.

[270] L. A. Santaló. *Integral Geometry and Geometric Probability.* Addison - Wesley Publishing Company, 1976.

[271] A. Särkkä. Pseudo-likelihood approach for Gibbs point processes in connection with field observations. *Statistics*, 26(1):89–97, 1995.

[272] A. Särkkä and E. Tomppo. Modelling interactions between trees by means of field observations. *Forest Ecology and Management*, 108(1):57–62, 1998.

[273] D. A. Saunders, R. J. Hobbs, and C. R. Margules. Biological consequences of ecosystem fragmentation : a review. *Conservation Biology*, 5:18–32, 1991.

[274] R. Schneider and W. Weil. *Stochastic and Integral Geometry.* Springer, 2008.

[275] F. P. Schoenberg. Comment on "A review of self-exciting spatio-temporal point processes and their applications" by A. Reinhart. *Statistical Science*, 3(3):325–326, 2018.

[276] T. Shirai and Y. Takahashi. Random point fields associated with certain fredholm determinants i: fermion, poisson and boson point processes. *Journal of Functional Analysis*, 205:414–463, 2003.

[277] S. Shirota and A. E. Gelfand. Approximate Bayesian computation and model assessment for repulsive spatial point processes. *Journal of Computational and Graphical Statistics*, pages 1–12, 2017.

[278] A. Singh, C. Scott, and R. Nowak. Adaptive Hausdorff estimation of density level sets. *Annals of Statistics*, 37:2760–2782, 2009.

[279] A.V. Skorokhod. Asymptotic formulas for stable distribution laws. In *Selected Translations in Mathematical Statistics and Probability 1*, pages 157–161, 1961.

[280] J. Sorce, R. S. Stocia, and E. Tempel. Statistically bias-minimized peculiar velocity catalogs from gibbs point processes and bayesian inference. *Astronomy and Astrophysics*, 679 (A1):1–25.

[281] A. G. Steenbeek, M. N. M. van Lieshout, R. S. Stoica, P. Gregori, and K. K. Berthelsen. Mpplib, a c++ library for marked point processes. *CWI*, 2003.

[282] R. S. Stoica. *Processus Ponctuels pour l'Extraction de Réseaux Linéïques.* PhD thesis - Université de Nice Sophia-Antipolis, 2001.

[283] R. S. Stoica. Marked point processes for statistical and morphological analysis of astronomical data. *The European Physical Journal Special Topics*, 186:123–165, 2010.

[284] R. S. Stoica. *Modélisation probabiliste et inférence statistique pour l'analyse des données spatialisées.* Habilitation à Diriger des Recherches thesis - Université de Lille, 2014.

[285] R. S. Stoica, F. Chatelain, and M. Sigelle. Inférence paramétriques pour les processus ponctuels marqués en analyse d'images. In X. Descombes, editor, *Applications de la géométrie stochastique à l'analyse d'images (Série Signal et Image IC2).* Hermès Lavoisier, 2011.

[286] R. S. Stoica, F. Chatelain, and M. Sigelle. Parametric inference for marked point processes in image analysis. In X. Descombes, editor, *Stochastic Geometry for Image Analysis (Digital Signal and Image Processing Series).* John Wiley and Sons, 2012.

[287] R. S. Stoica, M. Deaconu, A. Philippe, and L. Hurtado-Gil. Shadow simulated annealing: a new algorithm for approximate bayesian inference of gibbs point processes. *Spatial Statistics*, 43, 2021.

[288] R. S. Stoica, X Descombes, M. N. M. van Lieshout, and J. Zerubia. An application of marked point processes to the extraction of linear networks from images. In J. Mateu and F. Montes, editors, *Spatial Statistics through Applications.* WIT Press, Southampton, UK, 2002.

[289] R. S. Stoica, X. Descombes, and J. Zerubia. Road extraction in remote sensed images using a stochastic geometry framework. In *Bayesian Inference and Maximum Entropy Methods in Science and Engineering : 20th International Workshop*, volume 560, pages 531–542. AIP Conference Proceeding, 2001.

[290] R. S. Stoica, X. Descombes, and J. Zerubia. A Gibbs point process for road extraction in remotely sensed images. *International Journal of Computer Vision*, 57:121–136, 2004.

[291] R. S. Stoica, E. Gay, and A. Kretzschmar. Cluster detection in spatial data based on Monte Carlo inference. *Biometrical Journal*, 49(2):1–15, 2007.

[292] R. S. Stoica, P. Gregori, and J. Mateu. Simulated annealing and object point processes: tools for analysis of spatial patterns. *Stochastic Processes and their Applications*, 115:1860–1882, 2005.

[293] R. S. Stoica, S. Liu, Yu. Davydov, M. Fouchard, A. Vienne, and G. B. Valsecchi. Order statistics and heavy-tailed distributions for planetary perturbations on Oort cloud comets. *Astronomy and Astrophysics*, 513(A14):1–9, 2010.

[294] R. S. Stoica, V. J. Martinez, J. Mateu, and E. Saar. Detection of cosmic filaments using the Candy model. *Astronomy and Astrophysics*, 434:423–432, 2005.

[295] R. S. Stoica, V. J. Martinez, and E. Saar. A three dimensional object point process for detection of cosmic filaments. *Journal of the Royal Statistical Society. Series C (Applied Statistics)*, 55:189–205, 2007.

[296] R. S. Stoica, V. J. Martinez, and E. Saar. Filaments in observed and mock galaxy catalogues. *Astronomy and Astrophysics*, 510(A38):1–12, 2010.

[297] R. S. Stoica, A. Philippe, P. Gregori, and J. Mateu. ABC shadow algorithm: a tool for statistical analysis of spatial patterns. *Statistics and Computing*, 27:1225–1238, 2017.

[298] R. S. Stoica, E. Tempel, L. J. Liivamägi, G. Castellan, and E. Saar. Spatial patterns analysis in cosmology based on marked point processes. In D. Fraix-Burnet and D. Valls-Gabaud, editors, *Statistics for Astrophysics. Methods and Applications of the Regression*. European Astronomical Society Publication Series EDP Sciences, 2015.

[299] D. Stoyan. Discussion of the paper "Marked spatial point processes: current state and extensions to point processes on linear networks". *Journal of Agricultural, Biological and Environmental Statistics*, 29:392–394, 2024.

[300] D. Stoyan, W.S. Kendall, and J. Mecke. *Stochastic Geometry and its Applications*. John Wiley and Sons, 1995.

[301] D. Stoyan and I. S. Molchanov. Set-valued means of random particles. *Journal of Mathematical Imaging and Vision*, 7:111–121, 1997.

[302] D. J. Strauss. A model for clustering. *Biometrika*, 62:467–475, 1975.

[303] R. Takacs. Estimator for the pair potential of a Gibbsian point process. *Statistics*, 17(3):429–433, 1986.

[304] R. Takacs and T. Fiksel. Interaction pair-potentials for a system of ants' nests. *Biometrical Journal*, 28:1007–1013, 1986.

[305] F. Taty Moukati, R. S. Stoica, F. Bonneau, X. Wu, and G. Caumon. From fault likelihood to fault networks: stochastic seismic interpretation through a marked point process with interactions. *Preprint HAL, hal-04513328. To appear in Mathematical Geosciences*, 2024.

[306] E. Tempel, R. Kipper, E. Saar, M. Bussov, A. Hektor, and J. Pelt. Galaxy filaments as pearl necklaces? *Astronomy and Astrophysics*, 572(A8):1–8, 2014.

[307] E. Tempel, M. Kruuse, R. Kipper, T. Tuvikene, J. G. Sorce, and R. S. Stoica. Bayesian group finder based on marked point processes. method and application to the 2mrs data set. *Astronomy and Astrophysics*, 618(A61):1–18, 2018.

[308] E. Tempel, R. S. Stoica, R. Kipper, and E. Saar. Bisous model - detecting filamentary pattern in point processes. *Astronomy and Computing*, 16:17–25, 2016.

[309] E. Tempel, R. S. Stoica, and E. Saar. Evidence for spin alignment of spiral and elliptical galaxies in filaments. *Monthly Notices of the Royal Astronomical Society*, 428:1827–1836, 2013.

[310] E. Tempel, R. S. Stoica, E. Saar, V. J. Martinez, L. J. Liivamägi, and G. Castellan. Detecting filamentary pattern in the cosmic web: a catalogue of filaments for the SDSS. *Monthly Notices of the Royal Astronomical Society*, 438(4):3465–3482, 2014.

[311] E. Thönnes. A primer on perfect simulation. In K. R. Mecke and D. Stoyan, editors, *Statistical Physics and Spatial Statistics*. Springer-Verlag, Berlin, 2000.

[312] L. Tierney. Markov chains for exploring posterior distributions. *The Annals of Statistics*, 22(4):1701–1728, 1994.

[313] S. Torquato. *Random Heterogeneous Materials Microstructure and Macroscopic Properties*. Springer, 2002.

[314] A. Tsybakov. On nonparametric estimation of density level sets. *The Annals of Statistics*, 25:948–969, 1997.

[315] N. Vafaei, M. Ghorbani, M. Ganji, and M. Myllymäki. Spatio-temporal determinantal point processes. *Pre-print ArXiv*, arXiv:2301.02353.

[316] N. Vihrs, J. Møller, and A. E. Gelfand. Approximate Bayesian inference for a spatial point process model exhibiting regularity and random aggregation. *Scandinavian Journal of Statistics*, pages 1–26, 2021.

[317] B. Widom and J. S. Rowlinson. New model for the study of liquid-vapor phase transition. *The Journal of Chemical Physics*, 52:1670–1684, 1970.

[318] G. Winkler. *Image Analysis, Random Fields and Markov Chain Monte Carlo Methods (Second Edition)*. Springer, 2003.

[319] X. Wu, Y. Shi, S. Fomel, L. Liang, Q. Zhang, and A. Z. Yusifov. FaultNet3D: predicting fault probabilities, strikes, and dips with a single convolutional neural network. *IEEE Transactions on Geoscience and Remote Sensing*, 57(11):9138–9155, 2019.

[320] Donald G. York, J. Adelman, Jr. John E. Anderson, Scott F. Anderson, James Annis, Neta A. Bahcall, J. A. Bakken, Robert Barkhouser, Steven Bastian, Eileen Berman, William N. Boroski, Steve Bracker, Charlie Briegel, John W. Briggs, J. Brinkmann, Robert Brunner, Scott Burles, Larry Carey, Michael A. Carr, Francisco J. Castander, Bing Chen, Patrick L. Colestock, A. J. Connolly, J. H. Crocker,

István Csabai, Paul C. Czarapata, John Eric Davis, Mamoru Doi, Tom Dombeck, Daniel Eisenstein, Nancy Ellman, Brian R. Elms, Michael L. Evans, Xiaohui Fan, Glenn R. Federwitz, Larry Fiscelli, Scott Friedman, Joshua A. Frieman, Masataka Fukugita, Bruce Gillespie, James E. Gunn, Vijay K. Gurbani, Ernst de Haas, Merle Haldeman, Frederick H. Harris, J. Hayes, Timothy M. Heckman, G. S. Hennessy, Robert B. Hindsley, Scott Holm, Donald J. Holmgren, Chi hao Huang, Charles Hull, Don Husby, Shin-Ichi Ichikawa, Takashi Ichikawa, Željko Ivezić, Stephen Kent, Rita S. J. Kim, E. Kinney, Mark Klaene, A. N. Kleinman, S. Kleinman, G. R. Knapp, John Korienek, Richard G. Kron, Peter Z. Kunszt, D. Q. Lamb, B. Lee, R. French Leger, Siriluk Limmongkol, Carl Lindenmeyer, Daniel C. Long, Craig Loomis, Jon Loveday, Rich Lucinio, Robert H. Lupton, Bryan MacKinnon, Edward J. Mannery, P. M. Mantsch, Bruce Margon, Peregrine McGehee, Timothy A. McKay, Avery Meiksin, Aronne Merelli, David G. Monet, Jeffrey A. Munn, Vijay K. Narayanan, Thomas Nash, Eric Neilsen, Rich Neswold, Heidi Jo Newberg, R. C. Nichol, Tom Nicinski, Mario Nonino, Norio Okada, Sadanori Okamura, Jeremiah P. Ostriker, Russell Owen, A. George Pauls, John Peoples, R. L. Peterson, Donald Petravick, Jeffrey R. Pier, Adrian Pope, Ruth Pordes, Angela Prosapio, Ron Rechenmacher, Thomas R. Quinn, Gordon T. Richards, Michael W. Richmond, Claudio H. Rivetta, Constance M. Rockosi, Kurt Ruthmansdorfer, Dale Sandford, David J. Schlegel, Donald P. Schneider, Maki Sekiguchi, Gary Sergey, Kazuhiro Shimasaku, Walter A. Siegmund, Stephen Smee, J. Allyn Smith, S. Snedden, R. Stone, Chris Stoughton, Michael A. Strauss, Christopher Stubbs, Mark SubbaRao, Alexander S. Szalay, Istvan Szapudi, Gyula P. Szokoly, Anirudda R. Thakar, Christy Tremonti, Douglas L. Tucker, Alan Uomoto, Dan Vanden Berk, Michael S. Vogeley, Patrick Waddell, Shu i Wang, Masaru Watanabe, David H. Weinberg, Brian Yanny, and Naoki Yasuda. The Sloan Digital Sky Survey: technical summary. *The Astronomical Journal*, 120(3):1579–1587, 2000.

[321] L. Younes. Estimation and annealing for Gibbsian fields. *Annales de l'Institut Henri Poincaré, Probabilités et Statistiques*, 24:269–294, 1988.

[322] L. Younes. Parametric inference for imperfectly observed Gibbsian fields. *Probability Theory and Related Fields*, 82:625–645, 1989.

[323] L. Younes. Maximum likelihood estimation for Gibbsian fields. *Lecture Notes-Monograph Series, Spatial Statistics and Imaging*, 20:403–426, 1991.

[324] J. Zhuang. Second-order residual analysis of spatio-temporal point processes and applications in model evaluation. *Journal of the Royal Statistical Society, Series B*, 6(4):635–653, 2006.

Index

.ct Safety Concerns and Information please contact our
.sentative GPSR@taylorandfrancis.com Taylor & Francis
.ig GmbH, Kaufingerstraße 24, 80331 München, Germany